Dynamical Systems and Evolution Equations

Theory and Applications

MATHEMATICAL CONCEPTS AND METHODS IN SCIENCE AND ENGINEERING

Series Editor: **Angelo Miele**
Mechanical Engineering and Mathematical Sciences, Rice University

1 **INTRODUCTION TO VECTORS AND TENSORS** • Volume 1: Linear and Multilinear Algebra • *Ray M. Bowen and C.-C. Wang*

2 **INTRODUCTION TO VECTORS AND TENSORS** • Volume 2: Vector and Tensor Analysis • *Ray M. Bowen and C.-C. Wang*

3 **MULTICRITERIA DECISION MAKING AND DIFFERENTIAL GAMES** • *Edited by George Leitmann*

4 **ANALYTICAL DYNAMICS OF DISCRETE SYSTEMS** • *Reinhardt M. Rosenberg*

5 **TOPOLOGY AND MAPS** • *Taqdir Husain*

6 **REAL AND FUNCTIONAL ANALYSIS** • *A. Mukherjea and K. Pothoven*

7 **PRINCIPLES OF OPTIMAL CONTROL THEORY** • *R. V. Gamkrelidze*

8 **INTRODUCTION TO THE LAPLACE TRANSFORM** • *Peter K. F. Kuhfittig*

9 **MATHEMATICAL LOGIC:** An Introduction to Model Theory • *A. H. Lightstone*

11 **INTEGRAL TRANSFORMS IN SCIENCE AND ENGINEERING** • *Kurt Bernardo Wolf*

12 **APPLIED MATHEMATICS:** An Intellectual Orientation • *Francis J. Murray*

14 **PRINCIPLES AND PROCEDURES OF NUMERICAL ANALYSIS** • *Ferenc Szidarovszky and Sidney Yakowitz*

16 **MATHEMATICAL PRINCIPLES IN MECHANICS AND ELECTROMAGNETISM,** Part A: Analytical and Continuum Mechanics • *C.-C. Wang*

17 **MATHEMATICAL PRINCIPLES IN MECHANICS AND ELECTROMAGNETISM,** Part B: Electromagnetism and Gravitation • *C.-C. Wang*

18 **SOLUTION METHODS FOR INTEGRAL EQUATIONS:** Theory and Applications • *Edited by Michael A. Golberg*

19 **DYNAMIC OPTIMIZATION AND MATHEMATICAL ECONOMICS** • *Edited by Pan-Tai Liu*

20 **DYNAMICAL SYSTEMS AND EVOLUTION EQUATIONS:** Theory and Applications • *J. A. Walker*

21 **ADVANCES IN GEOMETRIC PROGRAMMING** • *Edited by Mordecai Avriel*

Dynamical Systems and Evolution Equations

Theory and Applications

J. A. Walker
Northwestern University
Evanston, Illinois

PLENUM PRESS · NEW YORK AND LONDON

Library of Congress Cataloging in Publication Data

Walker, John Andrew, 1939-
Dynamical systems and evolution equations.

(Mathematical concepts and methods in science and engineering; v. 20)
Includes bibliographical references and index.
1. Differentiable dynamical systems. 2. Differential topology. 3. Liapunov functions. 4. Topological dynamics. I. Title.
QA614.8.W34 514'.7 79-18064
ISBN 978-1-4684-1038-9 **ISBN 978-1-4684-1036-5** (eBook)
DOI 10.1007/978-1-4684-1036-5

Softcover reprint of the hardcover 1st edition 1980

A Division of Plenum Publishing Corporation
227 West 17th Street, New York, N.Y. 10011

Preface

This book grew out of a nine-month course first given during 1976–77 in the Division of Engineering Mechanics, University of Texas (Austin), and repeated during 1977–78 in the Department of Engineering Sciences and Applied Mathematics, Northwestern University. Most of the students were in their second year of graduate study, and all were familiar with Fourier series, Lebesgue integration, Hilbert space, and ordinary differential equations in finite-dimensional space. This book is primarily an exposition of certain methods of topological dynamics that have been found to be very useful in the analysis of physical systems but appear to be well known only to specialists. The purpose of the book is twofold: to present the material in such a way that the applications-oriented reader will be encouraged to apply these methods in the study of those physical systems of personal interest, and to make the coverage sufficient to render the current research literature intelligible, preparing the more mathematically inclined reader for research in this particular area of applied mathematics.

We present only that portion of the theory which seems most useful in applications to physical systems. Adopting the view that the world is deterministic, we consider our basic problem to be predicting the future for a given physical system. This prediction is to be based on a known equation of evolution, describing the forward-time behavior of the system, but it is to be made without explicitly solving the equation. This objective leads us to employ some of the ideas of topological dynamics, usually in conjunction with certain functions called Liapunov functions. Our general ideas are described in Chapter I, which provides some early motivation and demonstration while working only with the finite-dimensional problem. However, the ideas discussed are not intrinsically restricted by dimensionality.

Chapters II–IV extend the introductory treatment of Chapter I to a general metric space framework. Chapter II summarizes much of the mathematics needed for this extension, Chapter III discusses abstract evolution equations in Banach spaces, and Chapter IV describes several of the more useful ideas of topological dynamics. The material in Chapter IV forms the heart of our subject, and our basic tools are Liapunov functions. Because of a lack of space, as well as the state of the art, attention is focused throughout on the autonomous case (dynamical systems), and only brief comments are directed to the nonautonomous case (processes) after Chapter I. Numerous examples and exercises are provided, and considerable emphasis is placed on means of overcoming various technical difficulties that arise in applications.

With increasing frequency, these methods and ideas are being applied to the study of concrete physical systems. A few of the many recent applications to physical problems are discussed in Chapter V. This chapter also mentions discretization and approximation.

I wish to thank the U.S. National Science Foundation for supporting much of the work that led to this book. Special thanks are due to Lyle Clark of the University of Texas; without his strong encouragement this book would not have been written. I am indebted to Judy Piehl of Northwestern University for her skill, patience, and constant good cheer during the preparation of the manuscript.

J. A. Walker
Evanston, Illinois

Contents

Chapter I. Evolution Equations on $\mathfrak{R}^n$ 1

1. The Space $\mathfrak{R}^n$ 2
2. Evolution Equations on $\mathfrak{R}^n$. 3
3. Finite Escape Times. 9
4. Processes on $\mathfrak{R}^n \times \mathfrak{R}$ 12
5. Linear Dynamical Systems 15
6. The Simplest Class of Processes 17
7. Stability of a Particular Motion: Processes 19
8. Stability of Equilibrium: Dynamical Systems. 23
9. Stability of Equilibrium: Linear Dynamical Systems 30
10. Quadratic Liapunov Functions and Linearization 35
11. The Invariance Principle and Asymptotic Behavior 38
12. Comments and Extensions 42

References 47

Chapter II. Preliminaries for Abstract Evolution Equations 49

1. Abstract Spaces 51
2. Functions. 59
3. Linear Functions. 70
4. Differentiation of Functions 75
5. Abstract Evolution Equations 81

References 83

Chapter III. Abstract Dynamical Systems and Evolution Equations 85

1. Dynamical Systems and C_0-Semigroups. 86
2. Linear Dynamical Systems 87
3. Generation of Linear Dynamical Systems. 91
4. Choosing the State Space in Applications. 103
5. Generation of Nonlinear Dynamical Systems. 116
6. Comments and Extensions 129

References 135

Chapter IV. Some Topological Dynamics 137
1. Liapunov Functions and Positive Invariance 138
2. Computation of $\dot{V}$ 146
3. Stability and Liapunov's Direct Method 154
4. Positive Limit Sets and the Invariance Principle 164
5. Orbital Precompactness and Use of the Invariance Principle 175
6. Comments and Extensions 193
References 196

Chapter V. Applications and Special Topics 199
1. A Feedback Control Problem 200
2. The Thermoelastic Stability Problem 203
3. The Viscoelastic Stability Problem 207
4. A Fission Reactor Stability Problem 210
5. A Supersonic Panel-Flutter Problem 216
6. Discrete Dynamical Systems 222
7. Finite-Dimensional Approximation 225
References 229

INDEX 233

I

Evolution Equations on $\mathcal{R}^n$

Under the assumption that the future time behavior of a physical system can be accurately described by a differential "equation of evolution," we wish to be able to use this equation to predict the future for that physical system, given initial data about the "initial state" of the system, as well as data about the "future" of all external agents that affect the system. We believe that any physical system *has* a future, which is necessarily of infinite extent unless the system can "explode." We further believe that, at least on a macroscopic level, sufficient data of the aforementioned type determine a *unique* future for the system.

Given some evolution equation, thought to model a particular physical system, we believe that we must have existence and uniqueness of forward-time solutions for all physically possible data; otherwise, the equation cannot be an accurate model of the physical system. Moreover, if such solutions do not continue forward in time forever, we may conclude either that the model is wrong or that the physical system can explode, in the sense that the variables being used to describe the system become meaningless in finite time.

Given an equation of evolution that passes inspection on these grounds and may therefore actually describe our physical system, we then wish to use the forward-time solution (for given data) to predict the future of the physical system. As the given data are seldom exactly right, and as it is seldom practical to explicitly obtain all solutions corresponding to all possible data, we wish to have alternative methods of "predicting the future" (qualitatively, at least) that permit some uncertainty in the data and do not require explicit solution of the equation of evolution.

In this chapter, we shall consider such questions under the assumption that the "state" of the system at any given time is completely defined by a finite number n of real numbers; in particular, we are interested in ordinary differential equations on $\mathcal{R}^n$.

1. The Space $\mathcal{R}^n$

We denote by $\mathcal{R}$ (respectively, $\hat{\mathcal{C}}$) the set of real numbers (respectively, complex numbers) equipped with the usual rules for addition and multiplication, and $|\alpha|$ denotes the absolute value (respectively, modulus) of $\alpha \in \mathcal{R}$ (respectively, $\alpha \in \hat{\mathcal{C}}$). $\mathcal{R}^+$ will denote the set $[0, \infty) \subset \mathcal{R}$.

$\mathcal{R}^n$ will represent the finite-dimensional linear space of ordered n-tuples of real numbers:

$$(x_1, x_2, \ldots, x_n) = x \in \mathcal{R}^n = \mathcal{R} \times \mathcal{R} \times \cdots \times \mathcal{R}$$

(n times, with n finite).

Upon equipping $\mathcal{R}^n$ with the Euclidean norm

$$\|x\| = \left(x_1^2 + x_2^2 + \cdots + x_n^2\right)^{1/2},$$

$\mathcal{R}^n$ is called a Euclidean space; it becomes a Hilbert space upon defining the inner product

$$\langle x, y\rangle = x_1 y_1 + x_2 y_2 + \cdots + x_n y_n, \quad x, y \in \mathcal{R}^n.$$

There are other norms that could be defined on $\mathcal{R}^n$, the general requirements being that $\|\cdot\|: \mathcal{R}^n \to \mathcal{R}^+$, $\|x\| = 0$ if and only if $x = 0$, $\|x + y\| \leqslant \|x\| + \|y\|$ for all $x, y \in \mathcal{R}^n$, and $\|\alpha x\| = |\alpha|\,\|x\|$ for every $x \in \mathcal{R}^n$, $\alpha \in \mathcal{R}$. For example, we could define, for $x = (x_1, \ldots, x_n) \in \mathcal{R}^n$,

$$\|x\| \equiv |x_1| + |x_2| + \cdots + |x_n|,$$

or, instead,

$$\|x\| \equiv \max_{m=1,2,\ldots,n} |x_m|.$$

There are many possible choices, but all norms on $\mathcal{R}^n$ are equivalent, i.e., if $\|\cdot\|$ and $|\cdot|$ are both possible norms, then there exist real numbers $c_2 > c_1 > 0$ such that

$$c_1\|x\| \leqslant |x| \leqslant c_2\|x\| \qquad \text{for every} \quad x \in \mathcal{R}^n.$$

Only certain norms on $\mathcal{R}^n$ are compatible with an inner product, that is, are such that $\|x\|^2 = \langle x, x\rangle$. The general requirements for an inner product $\langle\cdot, \cdot\rangle: \mathcal{R}^n \times \mathcal{R}^n \to \mathcal{R}$ are that for all $x, y \in \mathcal{R}^n$ and all $\alpha \in \mathcal{R}$,

$$\langle \alpha x, y\rangle = \alpha\langle x, y\rangle, \quad \langle x, y\rangle = \langle y, x\rangle,$$

$$\langle x + y, z\rangle = \langle x, z\rangle + \langle y, z\rangle, \qquad \langle x, x\rangle > 0 \quad \text{if} \quad x \neq 0.$$

If a linear operator $B: \mathcal{R}^n \to \mathcal{R}^n$ (a real $n \times n$ matrix) is such that in terms of the Euclidean inner product,

$$\langle x, By \rangle = \langle Bx, y \rangle \qquad \text{for all} \quad x, y \in \mathcal{R}^n,$$

$$\langle x, Bx \rangle > 0 \qquad \text{for every nonzero} \quad x \in \mathcal{R}^n,$$

then another inner product could be defined as

$$\langle\langle x, y \rangle\rangle \equiv \langle x, By \rangle \qquad \text{for all} \quad x, y \in \mathcal{R}^n,$$

and the compatible norm would be defined as $|x| \equiv (\langle\langle x, x \rangle\rangle)^{1/2}$.

In this chapter, $\|\cdot\|$ and $\langle \cdot, \cdot \rangle$ can be thought of as representing the Euclidean norm and inner product, respectively, although this need not be assumed unless it is specifically stated.

Exercise 1.1. Let the norm $\|\cdot\|: \mathcal{R}^n \to \mathcal{R}^+$ be compatible with the inner product $\langle \cdot, \cdot \rangle: \mathcal{R}^n \times \mathcal{R}^n \to \mathcal{R}$. Show that, for all $x, y \in \mathcal{R}^n$,

$$|\langle x, y \rangle| \leqslant \|x\| \, \|y\|.$$

This is known as the *Schwarz inequality*. *Hint*: Note that $\|x\| + \|y\| \geqslant \|x \pm y\|$ and $\|x \pm y\|^2 = \langle x \pm y, x \pm y \rangle$.

Exercise 1.2. Given a norm $\|\cdot\|: \mathcal{R}^n \to \mathcal{R}^+$, show that there exists a compatible inner product if and only if the given norm satisfies the *parallelogram law*

$$\|x + y\|^2 + \|x - y\|^2 = 2\|x\|^2 + 2\|y\|^2$$

for all $x, y \in \mathcal{R}^n$. *Hint*: To show sufficiency, define

$$\langle x, y \rangle \equiv \tfrac{1}{4}\|x + y\|^2 - \tfrac{1}{4}\|x - y\|^2$$

for all $x, y \in \mathcal{R}^n$.

2. Evolution Equations on $\mathcal{R}^n$

We are interested in ordinary differential equations of the form

$$\dot{x}(t) = f(x(t), t), \qquad t \in \mathcal{R},$$
$$x(t_0) = x_0 \in \mathcal{R}^n, \qquad t_0 \in \mathcal{R}, \tag{1}$$

where $f: \mathcal{R}^n \times \mathcal{R} \to \mathcal{R}$, $x(\cdot): \mathcal{R} \to \mathcal{R}^n$, and $\dot{x}(t)$ is defined as

$$\dot{x}(t) \equiv \lim_{h \to 0} \frac{1}{h}\left[x(t + h) - x(t)\right], \qquad t \in \mathcal{R}.$$

We shall call $x(t)$ the *state* at *time* t, $x(t_0)$ is the *initial* state, and t_0 is the initial time; x_0, t_0, and the function f are presumed to be given *a priori*.

The form (1) is quite general; for example, consider

$$\ddot{y}(t) + g(y(t), \dot{y}(t), t) = 0, \qquad t \in \mathcal{R},$$

$$y(t_0) = y_0 \in \mathcal{R}, \quad \dot{y}(t_0) = v_0 \in \mathcal{R}, \qquad t_0 \in \mathcal{R},$$

and define $x = (y, v) \in \mathcal{R}^2$. Then we have

$$\dot{x}(t) = \begin{bmatrix} \dot{y}(t) \\ \dot{v}(t) \end{bmatrix} = \begin{bmatrix} v(t) \\ -g(y(t), v(t), t) \end{bmatrix} = f(x(t), t), \qquad t \in \mathcal{R},$$

$$x(t_0) = \begin{bmatrix} y_0 \\ v_0 \end{bmatrix} \in \mathcal{R}^2, \qquad t_0 \in \mathcal{R}.$$

Definition 2.1. A function $x(\cdot)$: $(\mathcal{I} \subset \mathcal{R}) \to \mathcal{R}^n$ is a *solution* of (1) *on* a (connected) *interval* $\mathcal{I}$, $t_0 \in \mathcal{I}$, if $x(t_0) = x_0$, $x(\cdot)$ is differentiable on $\mathcal{I}$, and $\dot{x}(t) = f(x(t), t)$ for every $t \in \mathcal{I}$.

Example 2.1. Let f: $\mathcal{R}^1 \times \mathcal{R} \to \mathcal{R}^1$ be given by $f(x, t) = x^2$, $x \in \mathcal{R}$, $t \in \mathcal{R}$. Let $t_0 \in \mathcal{R}$ and note that (1) becomes

$$\dot{x}(t) = (x(t))^2, \qquad t \in \mathcal{R},$$

$$x(t_0) = x_0 \in \mathcal{R}, \qquad t_0 \in \mathcal{R}.$$

Note that $x(t) \equiv 0$ is a solution on the interval $\mathcal{I} = (-\infty, \infty) = \mathcal{R}$ for $x_0 = 0$. If $x_0 \neq 0$, a solution is given by

$$x(t) = x_0/[1 - (t - t_0)x_0]$$

on the (maximal) interval

$$\mathcal{I} = \begin{cases} (-\infty, t_0 + 1/x_0) & \text{for } x_0 > 0, \\ (t_0 + 1/x_0, \infty) & \text{for } x_0 < 0. \end{cases}$$

A few solutions are sketched in Figure 1. For $x_0 > 0$, we say that the solution has *forward finite escape time* $\tau_e^+ = t_0 + 1/x_0$; for $x_0 < 0$, the solution has *backward finite escape time* $\tau_e^- = t_0 + 1/x_0$. We have demonstrated the existence (but not uniqueness) of a solution for each $(x_0, t_0) \in \mathcal{R}^1 \times \mathcal{R}$, and we have found its *maximal* interval of definition.

Example 2.2. Let f: $\mathcal{R}^1 \times \mathcal{R} \to \mathcal{R}^1$ be given by

$$f(x, t) = \begin{cases} +x^{1/2} & \text{for } x \geqslant 0, \\ 0 & \text{for } x < 0, \end{cases} \quad t \in \mathcal{R}.$$

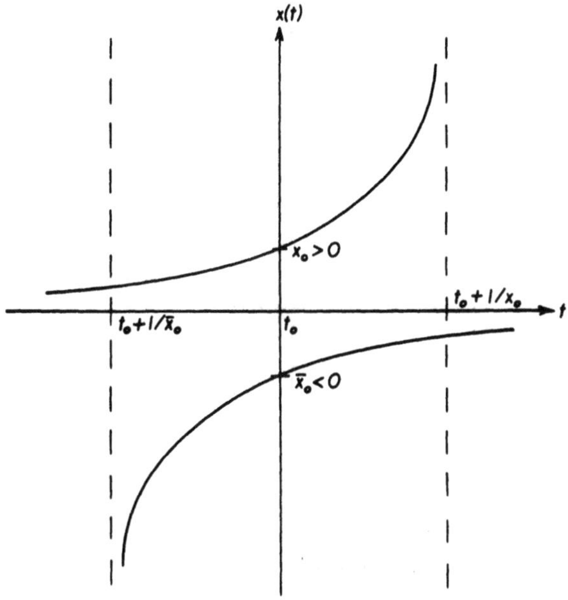

Fig. 1. Solutions for Example 2.1.

If $x_0 < 0$, $x(t) \equiv x_0$ is a solution on $\mathcal{I} = (-\infty, \infty) = \mathcal{R}$ for any $t_0 \in \mathcal{R}$. If $x_0 = 0$, additional solutions are provided by

$$x(t) = \begin{cases} 0 & \text{for} \quad t \leqslant t_0 + c, \\ (t - t_0 - c)^2/4 & \text{for} \quad t > t_0 + c, \end{cases}$$

for any $c \in \mathcal{R}^+$. If $x_0 > 0$, a solution

$$x(t) = \begin{cases} 0 & \text{for} \quad t \leqslant t_0 - 2x_0^{1/2} \\ (t - t_0 + 2x_0^{1/2})^2/4 & \text{for} \quad t > t_0 - 2x_0^{1/2} \end{cases}$$

exists on $\mathcal{I} = (-\infty, \infty) = \mathcal{R}$. A few solutions are sketched in Figure 2. We have found at least one solution for every (x_0, t_0) and have shown that its maximal interval of definition is $\mathcal{I} = \mathcal{R}$. We have not shown uniqueness; in fact, we have found that for $x_0 = 0$ there are many solutions.

Exercise 2.1. Given some bounded $g: \mathcal{R} \to \mathcal{R}^n$ that has an isolated discontinuity at $t = 1$, with $g(1^-) \neq g(1^+)$, prove that the problem

$$\dot{x}(t) = g(t), \qquad t \in \mathcal{I} \subset \mathcal{R},$$

$$x(0) = x_0 \in \mathcal{R}^n,$$

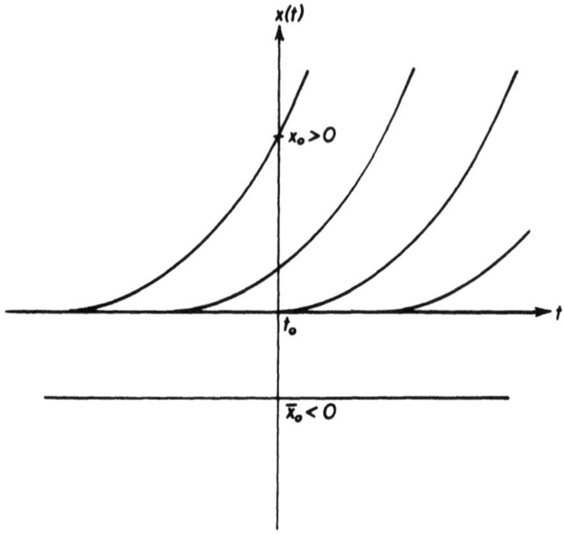

Fig. 2. Solutions for Example 2.2.

does not have a solution (as we have defined a solution) on the interval $\mathcal{I} = [0, 2)$ (or on any other $\mathcal{I} = [0, \tau)$ for $\tau > 1$). (Do you think that our definition of *solution* may be more restrictive than is desirable for some problems?)

In Examples 2.1 and 2.2 we have demonstrated existence by displaying a solution. As this can rarely be done, we need a more general approach to the question of existence of solutions. This is provided by Peano's existence theorem.

Theorem 2.1. If $f: \mathcal{R}^n \times \mathcal{R} \to \mathcal{R}^n$ is continuous at (x_0, t_0), then there *exists* at least one solution of (1), defined on an interval $\mathcal{I} \subset \mathcal{R}$ having t_0 as an interior point.

In Ref. 1 this result is proved by using polygonal approximations; a more abstract proof, using the Schauder fixed-point theorem, is given in Ref. 2. Theorem 2.1 gives no indication of the length of the maximal interval of definition; we know only that it contains some neighborhood of t_0. Somewhat more information is provided by the following continuation theorem.

Theorem 2.2. Let $f: \mathcal{R}^n \times \mathcal{R} \to \mathcal{R}^n$ be continuous on an open set $\mathcal{D}$ in $\mathcal{R}^n \times \mathcal{R}$. Given $(x_0, t_0) \in \mathcal{D}$, each solution $x(\cdot)$ of (1) is defined on an open interval $\mathcal{I} = (a, b)$ such that $(x(t), t)$ approaches the boundary of $\mathcal{D}$ as $t \nearrow b$ and as $t \searrow a$, where $-\infty \leqslant a < t_0 < b \leqslant +\infty$.

This theorem requires some explanation. The "extended real numbers" a and b depend on the particular solution $x(\cdot)$ and are not necessarily finite unless $\mathfrak{D}$ is a bounded set. By $t \nearrow b$ we mean that t approaches b through values less than b if $b < \infty$, and t approaches ∞ if $b = \infty$. Also, by "$(x(t), t)$ approaches the boundary of $\mathfrak{D}$ as $t \nearrow b$," we mean that either $b = \infty$ or else $b < \infty$ and, given any closed and bounded subset $\mathfrak{D}_0$ of $\mathfrak{D}$, $(x(t), t) \notin \mathfrak{D}_0$ for some $t < b$ arbitrarily near b. If $\mathfrak{D} = \mathfrak{R}^n \times \mathfrak{R}$, it follows that $\mathfrak{I}$ is *maximal* for $x(\cdot)$, in the sense that $x(\cdot)$ cannot be extended (as a solution) to a properly larger interval of definition. Proofs of Theorem 2.2 can be found in Refs. 1–4.

We note that in Examples 2.1 and 2.2 we may choose the set $\mathfrak{D}$ to be all of $\mathfrak{R}^1 \times \mathfrak{R}$, since f is everywhere continuous.

In the attempt to obtain information about the maximal interval of definition of a solution of (1), it is helpful to use a "cylindrical box" $\mathfrak{D}$ in Theorem 2.2; that is, for fixed (x_0, t_0), consider the set

$$\mathfrak{D} = \{x \mid \|x - x_0\| < \beta\} \times (\alpha_1, \alpha_2) \subset \mathfrak{R}^n \times \mathfrak{R}$$

for some fixed numbers $\beta > 0$, $\alpha_1 < t_0 < \alpha_2$. If $f: \mathfrak{R}^n \times \mathfrak{R} \to \mathfrak{R}^n$ is continuous on this set $\mathfrak{D}$, then we see by Theorem 2.2 that a solution of (1) exists on an interval $\mathfrak{I} \subset \mathfrak{R}$ such that either $[t_0, \alpha_2) \subset \mathfrak{I}$ or $\lim_{p\to\infty}\|x(t_p)\| = \beta$ for some sequence $\{t_p\} \subset (t_0, \alpha_2)$. Similarly, either $(\alpha_1, t_0] \subset \mathfrak{I}$ or $\lim_{m\to\infty}\|x(t_m)\| = \beta$ for some sequence $\{t_m\} \subset (\alpha_1, t_0)$. Hence, if by some means one can show that $x(t)$ does not approach the "wall" defined by $\|x - x_0\| = \beta$, as t both increases and decreases from t_0, then $\mathfrak{I}$ contains the interval (α_1, α_2). Later we shall take up this type of question again.

We now consider the question of uniqueness of solutions.

Definition 2.2. A function $f: \mathfrak{R}^n \times \mathfrak{R} \to \mathfrak{R}$ is said to be *locally Lipschitz continuous with respect to x at* $(x, t) \in \mathfrak{R}^n \times \mathfrak{R}$ if there exist numbers $\varepsilon(x, t) > 0$ and $\kappa(x, t) < \infty$ such that

$$\|f(y, t) - f(x, t)\| \leqslant \kappa\|y - x\|$$

for every $y \in \mathfrak{R}^n$ such that $\|y - x\| \leqslant \varepsilon$.

If $f(x, t)$ has continuous first partial derivatives with respect to x in a neighborhood of (x_0, t_0), then f is locally Lipschitz continuous with respect to x at (x_0, t_0).

Theorem 2.3. Let $\mathfrak{D}$ be an open set in $\mathfrak{R}^n \times \mathfrak{R}$. If $f: \mathfrak{R}^n \times \mathfrak{R} \to \mathfrak{R}^n$ is continuous on $\mathfrak{D}$ and locally Lipschitz continuous with respect to x at every $(x, t) \in \mathfrak{D}$, then for each $(x_0, t_0) \in \mathfrak{D}$ there exists a *unique* solution

$x(\cdot) = \varphi(\cdot, x_0, t_0)$ of (1) on an interval $\mathfrak{I} = (a, b)$ such that $(x(t), t)$ approaches the boundary of $\mathfrak{D}$ as $t \nearrow b$ and as $t \searrow a$; moreover, φ is (jointly) continuous with respect to (t, x_0, t_0).

Theorem 2.3 is known as the Picard–Lindelöf theorem. It provides the local existence result of Theorem 2.1 as well as the continuation result of Theorem 2.2; moreover, through use of the additional Lipschitz continuity condition, Theorem 2.3 provides uniqueness and continuous dependence of the solution $x(t) = \varphi(t, x_0, t_0)$ on the triple (t, x_0, t_0). Continuous dependence on the first argument t is obviously required by our definition of solution (Definition 2.1), but the result here is much stronger; in particular, under the assumption of Theorem 2.3, we are assured that solutions depend continuously on the initial state x_0 and initial time t_0, as well as on the present time t.

Proofs of Theorem 2.3 may be found in many texts on finite-dimensional ordinary differential equations (see Refs. 1–4). The following exercise shows that some of the additional conclusions of Theorem 2.3 can be proved in a rather simple way.

Exercise 2.2. Let $\hat{\mathfrak{D}} \subset \tilde{\mathfrak{D}} \subset \mathfrak{D} \subset \mathfrak{R}^n \times \mathfrak{R}$, with $\mathfrak{D}$ and $\hat{\mathfrak{D}}$ open, $\tilde{\mathfrak{D}}$ compact (closed and bounded). Let f: $\mathfrak{R}^n \times \mathfrak{R} \to \mathfrak{R}$ be continuous on $\mathfrak{D}$ and locally Lipschitz with respect to x on $\mathfrak{D}$. By a simple result from analysis, compactness of $\tilde{\mathfrak{D}} \subset \mathfrak{D}$ implies that f is uniformly Lipschitz with respect to x on $\tilde{\mathfrak{D}}$, i.e., there exists a number $\kappa < \infty$ such that $\|f(x, t) - f(y, t)\| \leqslant \kappa\|x - y\|$ for all $(x, t) \in \tilde{\mathfrak{D}}$ and $(y, t) \in \tilde{\mathfrak{D}}$. Use this fact to show that if $(x_0, t_0) \in \hat{\mathfrak{D}}$, $(y_0, t_0) \in \hat{\mathfrak{D}}$, and corresponding solutions $x(t)$, $y(t)$ for (1) do exist on some (common) connected interval $\mathfrak{I}$ containing t_0, then

$$\langle x(t) - y(t), x(t) - y(t)\rangle \leqslant \exp\{2\kappa|t - t_0|\}\langle x_0 - y_0, x_0 - y_0\rangle$$

for all $t \in \mathfrak{I}$. Show that this implies that for every $(x_0, t_0) \in \mathfrak{D}$ such that a solution $x(t)$ exists, $x(t) = u(t - t_0, x_0, t_0)$ is unique and depends continuously on x_0. *Hint*: If $g(\cdot) : (\mathfrak{I} \subset \mathfrak{R}) \to \mathfrak{R}^n$ is differentiable, then

$$2\langle g(t), \dot{g}(t)\rangle = \frac{d}{dt}\langle g(t), g(t)\rangle = \frac{d}{dt}\|g(t)\|^2$$

for all $t \in \mathfrak{I}$, using the norm that is compatible with the inner product; moreover, $\langle g(t), \dot{g}(t)\rangle \leqslant \|g(t)\|\,\|\dot{g}(t)\|$.

Reconsidering Example 2.1, we see that $\partial f/\partial x = 2x$ is continuous, and thus Theorem 2.3 assures uniqueness; hence, the solutions we found were all that there were. In Example 2.2, f is locally Lipschitz with respect to x for all $x \neq 0$, but not for $x = 0$. Hence, Theorem 2.3 does not assure uniqueness for $(x_0, t_0) = (0, t_0)$, and we found that in fact such an initial condition does not lead to a unique solution in Example 2.2.

Theorem 2.3 guarantees that for each $(x_0, t_0) \in \mathfrak{D}$ there exists a solution defined on an open interval $\mathcal{I} = (a, b)$, $a < t_0 < b$, where $(x(t), t) \in \mathfrak{D}$ for all $t \in (a, b)$; furthermore, this solution is unique. As this also implies that there exists one and only one solution on the interval $[t_0, b)$, we have *forward* existence and *forward* uniqueness; similarly, since there is one and only one solution on the interval $(a, t_0]$, we have *backward* existence and *backward* uniqueness. It should be kept in mind that Theorem 2.3 guarantees both forward and backward existence and uniqueness, as well as continuous dependence of solutions on the initial $(x_0, t_0) \in \mathfrak{D}$. In physical problems, assurances of backward existence and uniqueness are unnecessary, and sometimes untrue. It follows that the conclusions, as well as the assumptions, of Theorems 2.1–2.3 may be stronger than is desirable for some physical problems. We shall return to these considerations on several occasions.

3. Finite Escape Times

Suppose for the moment that $f: \mathfrak{R}^n \times \mathfrak{R} \to \mathfrak{R}^n$ happens to be both continuous on all of $\mathfrak{R}^n \times \mathfrak{R}$ and locally Lipschitz continuous with respect to x on all of $\mathfrak{R}^n \times \mathfrak{R}$. Can we conclude that for given (x_0, t_0) the unique solution guaranteed by Theorem 2.3 has maximal interval of definition $\mathcal{I} = \mathfrak{R}$? The answer is no, since Example 2.1 contradicts such a conclusion. However, by considering a family of bounded sets $\mathfrak{D} \subset \mathfrak{R}^n \times \mathfrak{R}$ in the form of the cylindrical boxes described earlier, we can conclude the following: The maximal interval of definition $\mathcal{I} = (a, b)$, $a < t_0 < b$, where either $b = \infty$ or $\|x(t_p)\| \to \infty$ as $p \to \infty$ for some sequence $\{t_p\}$ with $t_0 < t_p < b$, $t_p \to b$ as $p \to \infty$; in the latter case, $x(\cdot)$ is said to have the *forward finite escape time b*. Similarly, either $a = -\infty$ or $\|x(t_m)\| \to \infty$ as $m \to \infty$ for some sequence $\{t_m\}$ with $a < t_m < t_0$, $t_m \to a$ as $m \to \infty$; in the latter case, $x(\cdot)$ has the *backward finite escape time a*.

Exercise 3.1. Consider the two-dimensional problem

$$\ddot{y}(t) + y(t) - y^3(t) = 0, \qquad t \in \mathfrak{R},$$

$$y(0) = y_0, \qquad \dot{y}(0) = v_0, \qquad (y_0, v_0) \in \mathfrak{R}^2.$$

(a) For what values of (y_0, v_0) are we sure of the local existence and uniqueness of solutions?

(b) For what values of (y_0, v_0) are we sure that solutions continue forward in time indefinitely (in particular, no forward finite escape times)?

(c) For what values of (y_0, v_0) are the solutions such that

(i) $y(t) \equiv$ constant?

(ii) $|y(t)| \to \infty$ as t increases?

(iii) $y(t) \equiv y(t+\tau)$ for some τ (periodic function)?

(iv) $\lim_{t\to\infty} y(t)$ exists?

Hint: The function $V(y, v) = \frac{1}{2}(v^2 + y^2 - y^4/2)$ is a *first integral* of the equation, i.e., along any solution,

$$\frac{d}{dt} V(y(t), \dot{y}(t)) \equiv 0 \qquad \text{(verify this).}$$

It will be helpful to sketch the curves $V = c$ in the (y, v)-plane for various values of the constant c.

For a given *physical* system, the existence or nonexistence of solutions with *forward* finite escape times depends almost entirely on our original choice of the physical variables making up the state $x(t)$. That is, unless the state is defined so that $x(t)$ might become physically meaningless in finite time, no solution of an accurate mathematical model should have a forward finite escape time. Results of the following type are useful for investigating this important question.

Theorem 3.1. Let there exist real numbers α, β and an inner product $\langle \cdot, \cdot \rangle: \mathcal{R}^n \times \mathcal{R}^n \to \mathcal{R}$, with compatible norm $\|\cdot\|$, such that $\langle x, f(x, t)\rangle \leqslant \alpha\langle x, x\rangle$ for all $t > t_0$ and all $x \in \mathcal{R}^n$ with $\|x\| > \beta$. Then, for this value of t_0, no solution of (1) has forward finite escape time.

Proof. For $x(\cdot)$ a solution of (1) having maximal interval of definition (α_1, α_2) with $\alpha_1 < t_0 < \alpha_2$, let $\hat{\beta} \equiv \max\{\beta, \|x_0\|\}$, and note that if $\|x(t)\| \leqslant \hat{\beta}$ for all $t \in [t_0, \alpha_2)$, then α_2 is not a finite escape time; hence, by continuity of $x(\cdot)$ on (α_1, α_2), the alternative is that there exist t_1, t_2 such that $t_0 \leqslant t_1 < t_2 \leqslant \alpha_2$, $\|x(t_1)\| = \hat{\beta}$, and $\|x(t)\| > \hat{\beta}$ for all $t \in (t_1, t_2)$. In this case, for $t \in (t_1, t_2)$, our assumption implies that

$$\begin{aligned} 2\alpha\|x(t)\|^2 &\geqslant 2\langle x(t), f(x(t), t)\rangle \\ &= \langle x(t), \dot{x}(t)\rangle + \langle \dot{x}(t), x(t)\rangle \\ &= \frac{d}{dt}\langle x(t), x(t)\rangle = \frac{d}{dt}\|x(t)\|^2. \end{aligned}$$

Employing Gronwall's lemma (Ref. 2), we find that $\|x(t)\| \leqslant \|x(t_1)\|e^{\alpha t}$ for all $t \in [t_1, t_2)$, and either the estimate holds on (t_1, α_2), implying that α_2 is not a finite escape time, or there exists $t_3 \in [t_2, \alpha_2)$ such that $\|x(t_3)\| = \hat{\beta}$. In the latter case the argument can be repeated; therefore, $x(\cdot)$ does not have a forward finite escape time. □

Reconsidering Example 2.1, with $f(x, t) = x^2$, $x \in \mathcal{R}^1$, do there exist $\alpha < \infty$, $\beta < \infty$, such that $\langle x, f(x, t)\rangle = x^3 \leqslant \alpha\langle x, x\rangle = \alpha x^2$ for all x such that $|x| > \beta$? Clearly the answer is negative, and there we found that there do exist solutions having forward finite escape times. Reconsidering Example 2.2, we find that

$$\langle x, f(x,t)\rangle = \begin{cases} x^{3/2} & \text{for } x \geqslant 0, \\ 0 & \text{for } x < 0, \end{cases}$$

$$\leqslant x^2 \qquad \text{for all } x \text{ such that } |x| > 1.$$

Hence, the conditions of Theorem 3.1 are satisfied with $\alpha = \beta = 1$, and Example 2.2 admits no solutions having forward finite escape times, as we have previously established. We emphasize that Theorem 3.1 provides a sufficient condition for nonexistence of forward finite escape times; it is not a necessary condition, even with the assumptions weakened as in the following problem.

Exercise 3.2. Prove the following extension of Theorem 3.1: Let there exist continuous functions $\alpha: \mathfrak{R} \to \mathfrak{R}$, $\beta: \mathfrak{R} \to \mathfrak{R}^+$ and an inner product $\langle\ ,\ \rangle: \mathfrak{R}^n \times \mathfrak{R}^n \to \mathfrak{R}$, with compatible norm $\|\cdot\|$, such that $\langle x, f(x,t)\rangle \leqslant \alpha(t)\langle x, x\rangle$ for all $t > t_0$ and all $x \in \mathfrak{R}^n$ with $\|x\| > \beta(t)$. Then, for this value of t_0, no solution of (1) can have a forward finite escape time.

In Theorem 3.1, the choice of the inner product $\langle\cdot,\cdot\rangle$ is sometimes critical.

Exercise 3.3. Use Theorem 3.1 to show that no solution of

$$\begin{bmatrix} \dot z(t) \\ \dot y(t) \end{bmatrix} = \begin{bmatrix} -y(t)\left[(z(t))^2 + (y(t))^2\right] \\ z(t)\left[(z(t))^2 + (y(t))^2\right] \end{bmatrix}, \qquad t \in \mathfrak{R},$$

$$\begin{bmatrix} z(t_0) \\ y(t_0) \end{bmatrix} = \begin{bmatrix} z_0 \\ y_0 \end{bmatrix} \in \mathfrak{R}^2, \qquad t_0 \in \mathfrak{R},$$

has forward finite escape time. How free is your choice of the inner product used here?

Exercise 3.4. For $\varepsilon \geqslant 0$, show that no solution of van der Pol's equation,

$$\ddot y(t) + \varepsilon\left[(y(t))^2 - 1\right]\dot y(t) + y(t) = 0, \qquad t \in \mathfrak{R},$$

$$\begin{bmatrix} y(t_0) \\ \dot y(t_0) \end{bmatrix} = \begin{bmatrix} y_0 \\ v_0 \end{bmatrix} \in \mathfrak{R}^2, \qquad t_0 \in \mathfrak{R},$$

has forward finite escape time. *Remark*: For $\varepsilon < 0$, some solutions *do* have forward finite escape times.

The sufficient condition of Theorem 3.1 is based on a special type of "differential inequality" (see Section I.6 of Ref. 2); at the expense of using more complicated differential inequalities, one can derive somewhat more general sufficient conditions for the nonexistence of solutions having forward finite escape times. We shall obtain more general results in Section 12.

4. Processes on $\mathcal{R}^n \times \mathcal{R}$

Suppose that by some means (e.g., Theorems 2.3 and 3.1) we can show that equation (1) possesses a forward-unique solution $x(\cdot) = \varphi(\cdot, x_0, t_0)$, defined on the interval $\mathcal{I} = [t_0, \infty)$, for *every* $(x_0, t_0) \in \mathcal{R}^n \times \mathcal{R}$. Then defining $u(\tau, x_0, t_0) \equiv \varphi(\tau + t_0, x_0, t_0)$, $\tau \in \mathcal{R}^+$, we see that the mapping $u: \mathcal{R}^+ \times \mathcal{R}^n \times \mathcal{R} \to \mathcal{R}^n$ satisfies

$$u(0, x_0, t_0) = x_0 \quad \text{and} \quad u(\tau + s, x_0, t_0) = u(\tau, u(s, x_0, t_0), s + t_0)$$

for all $\tau, s \in \mathcal{R}^+$, $x_0 \in \mathcal{R}^n$, $t_0 \in \mathcal{R}$. Suppose further that by some means (e.g., Theorem 2.3) we can also show that u is continuous on $\mathcal{R}^+ \times \mathcal{R}^n \times \mathcal{R}$. Then (1) is said to *generate* a process u.

Definition 4.1. A *process* on $\mathcal{R}^n \times \mathcal{R}$ is a mapping $u: \mathcal{R}^+ \times \mathcal{R}^n \times \mathcal{R} \to \mathcal{R}^n$ (defined on all of $\mathcal{R}^+ \times \mathcal{R}^n \times \mathcal{R}$) such that

(i) $u(\cdot, x_0, t_0,): \mathcal{R}^+ \to \mathcal{R}^n$ is continuous (right-continuous at $\tau = 0$),
(ii) $u(\tau, \cdot, \cdot): \mathcal{R}^n \times \mathcal{R} \to \mathcal{R}^n$ is continuous,
(iii) $u(0, x_0, t_0) = x_0$,
(iv) $u(\tau + s, x_0, t_0) = u(\tau, u(s, x_0, t_0), s + t_0)$,

for all $\tau, s \in \mathcal{R}^+$, $x_0 \in \mathcal{R}^n$, $t_0 \in \mathcal{R}$.

We did not generate a process in Example 2.1 or 2.2; nevertheless, if (1) describes the behavior of a physical system that is not "explosive," and if all values of $(x_0, t_0) \in \mathcal{R}^n \times \mathcal{R}$ are physically possible initial conditions, then the properties (i)–(iv) of Definition 4.1 are essentially those that we physically expect of the solutions, with $x(t) = u(t - t_0, x_0, t_0)$, $t \geqslant t_0$. Therefore, if (1) is a model of a physical system, but the solutions do not have all these properties (in particular, if forward uniqueness is violated, or if the interval of definition does not contain all $t \geqslant t_0$), then it is a reasonable conclusion that either the model is inaccurate or the state $x(t)$ may become physically meaningless in finite time.

Exercise 4.1. Let f: $\mathcal{R}^n \times \mathcal{R} \to \mathcal{R}^n$ be continuous on all of $\mathcal{R}^n \times \mathcal{R}$ and uniformly Lipschitz continuous with respect to x on all of $\mathcal{R}^n \times \mathcal{R}$. Prove that (1) generates a process u: $\mathcal{R}^+ \times \mathcal{R}^n \times \mathcal{R} \to \mathcal{R}^n$.

Whether or not we have a process on $\mathcal{R}^n \times \mathcal{R}$, let us henceforth concentrate on evolution equations describing only forward evolution in time; that is, rather than (1), our real interest lies in the equation

$$\dot{x}(t) = f(x(t), t), \qquad t \geqslant t_0 \in \mathcal{R},$$

$$x(t_0) = x_0 \in \mathcal{R}^n, \tag{2}$$

and a solution is now defined only on the interval $[t_0, b)$ for some $b > t_0$ (possibly, $b = \infty$) with $\dot{x}(t_0)$ denoting the right derivative at t_0.

To simplify matters somewhat, consider (2) and suppose that f does not depend explicitly on t; that is, let f: $\mathcal{R}^n \to \mathcal{R}^n$ and consider the evolution equation

$$\dot{x}(t) = f(x(t)), \qquad t \geqslant t_0 \in \mathcal{R},$$

$$x(t_0) = x_0 \in \mathcal{R}^n. \tag{2*}$$

This type of evolution equation is said to be *autonomous*, because the initial time t_0 is immaterial and only $t - t_0$ need be considered; to see this, define $\tau = t - t_0$, $y(\tau) = x(\tau + t_0)$, and note that (2*) implies that

$$\dot{y}(\tau) = f(y(\tau)), \qquad \tau \in \mathcal{R}^+,$$

$$y(0) = x_0 \in \mathcal{R}^n,$$

and the solutions $y(\cdot)$ do not depend on the value of t_0. Consequently, if (2*) generates a process u: $\mathcal{R}^+ \times \mathcal{R}^n \times \mathcal{R} \to \mathcal{R}^n$, then $u(t - t_0, x_0, t_0)$ is independent of the third argument! This very special type of process is often called a *semiflow* or *semidynamical system* (Ref. 5); here we will simply call it a *dynamical system*.

Definition 4.2. A *dynamical system* on $\mathcal{R}^n$ is a mapping u: $\mathcal{R}^+ \times \mathcal{R}^n \to \mathcal{R}^n$ (defined on all of $\mathcal{R}^+ \times \mathcal{R}^n$) such that

(i) $u(\cdot, x)$: $\mathcal{R}^+ \to \mathcal{R}^n$ is continuous (right-continuous at $t = 0$),
(ii) $u(t, \cdot)$: $\mathcal{R}^n \to \mathcal{R}^n$ is continuous,
(iii) $u(0, x) = x$,
(iv) $u(t + s, x) = u(t, u(s, x))$,

for all $t, s \in \mathcal{R}^+$, $x \in \mathcal{R}^n$.

As a special case, we note that if f: $\mathcal{R}^n \to \mathcal{R}^n$ is everywhere locally Lipschitz continuous, and if no solution of (3) has forward finite escape time, then Theorem 2.3 states that the autonomous evolution equation

$$\dot{x}(t) = f(x(t)), \qquad t \in \mathcal{R}^+,$$
$$x(0) = x_0 \in \mathcal{R}^n, \tag{3}$$

generates a dynamical system on $\mathcal{R}^n$.

As a dynamical system seems less complicated (algebraically) than a process, one is led to ask whether a process on $\mathcal{R}^n$ might be viewed instead as a dynamical system, on a possibly more complicated state space. The answer is affirmative, and there are several ways of accomplishing this objective; the least complicated approach is suggested in the following exercise. Unfortunately, such "simplification" is more apparent than real; the approach of the following exercise turns out to be of little help in applications requiring the study of the behavior of solutions for large values of time. Consequently, we will preserve the distinction between a process and a dynamical system.

Exercise 4.2. Let u: $\mathcal{R}^+ \times \mathcal{R}^n \times \mathcal{R} \to \mathcal{R}^n$ be a process, and define a mapping U: $\mathcal{R}^+ \times \mathcal{R}^{n+1} \to \mathcal{R}^{n+1}$ by its components,

$$U_i(\tau, z) \equiv u_i(\tau, x, t_0), \qquad i = 1, 2, \ldots, n,$$
$$U_{n+1}(\tau, z) \equiv z_{n+1} + \tau,$$

for all $\tau \in \mathcal{R}^+$, $x \in \mathcal{R}^n$, $t_0 \in \mathcal{R}$, where

$$(z_1, z_2, \ldots, z_n, z_{n+1}) \equiv (x_1, x_2, \ldots, x_n, t_0).$$

(i) Show that U is a dynamical system on $\mathcal{R}^{n+1}$.
(ii) If u is generated by (2), what evolution equation generates U?
(iii) Show that every *motion* $U(\cdot, z)$: $\mathcal{R}^+ \to \mathcal{R}^{n+1}$ of U is unbounded.

We note that properties (i)–(iv) of Definition 4.2 have been shown (Ref. 6) to imply (joint) continuity of a dynamical system u: $\mathcal{R}^+ \times \mathcal{R}^n \to \mathcal{R}$. From the foregoing exercise, it then follows that properties (i)–(iv) of Definition 4.1 imply (joint) continuity of a process u: $\mathcal{R}^+ \times \mathcal{R}^n \times \mathcal{R} \to \mathcal{R}^n$.

We also remark that it is sometimes desirable to consider a process u on $\mathcal{R}^n \times \mathcal{R}_T$, rather than on $\mathcal{R}^n \times \mathcal{R}$, with $\mathcal{R}_T \equiv \{t_0 \in \mathcal{R} | t_0 > T\}$ for some given $T \in \mathcal{R}$. Then we consider a mapping u: $\mathcal{R}^+ \times \mathcal{R}^n \times \mathcal{R}_T \to \mathcal{R}^n$ having properties (i)–(iv) of Definition 4.1 only for all τ, $s \in \mathcal{R}^+$, $x_0 \in \mathcal{R}^n$,

$t_0 \in \mathcal{R}_T$. For physical systems this modification seems necessary, since every physical system (even the entire universe) has been in existence for only a finite length of time.

5. Linear Dynamical Systems

We have seen that the autonomous equation (3) may or may not generate a dynamical system u on $\mathcal{R}^n$; in the case of Examples 2.1 and 2.2, it does not. For $A: \mathcal{R}^n \to \mathcal{R}^n$ a given linear operator (constant real $n \times n$ matrix), we now consider the linear autonomous equation

$$\dot{x}(t) = Ax(t), \qquad t \in \mathcal{R}^+,$$
$$x(0) = x_0 \in \mathcal{R}^n. \tag{4}$$

First we shall show that (4) does generate a dynamical system u on $\mathcal{R}^n$.

For each $t \in \mathcal{R}^+$, define a linear operator $e^{tA}: \mathcal{R}^n \to \mathcal{R}^n$ by

$$e^{tA} \equiv \sum_{k=0}^{\infty} \frac{t^k}{k!} A^k,$$

where $A^0 \equiv I$, $A^1 = A$, $A^2 = AA$, etc. To see that this definition makes sense, we employ the linear operator norm (see Chapter II) and note that for $0 \leqslant m < p$, and each $t \in \mathcal{R}^+$,

$$\left\| \sum_{k=m}^{k=p} \frac{t^k}{t!} A^k \right\| \leqslant \sum_{k=m}^{k=p} \frac{t^k}{k!} \|A\|^k.$$

Therefore, $\sum_{k=0}^{m}(t^k/k!)A^k$ converges in the uniform operator topology (see Chapter II) as $m \to \infty$ because $\sum_{k=0}^{m}(t^k/k!)\|A\|^k$ converges (to $e^{t\|A\|} \in \mathcal{R}^+$); moreover, $\|e^{tA}\| \leqslant e^{t\|A\|}$ for every $t \in \mathcal{R}^+$. We also note that $Ae^{tA} = e^{tA}A$, $e^{0A} = I$, and

$$\frac{d}{dt} e^{tA} x_0 = \sum_{k=1}^{\infty} \frac{kt^{k-1}}{k!} A^k x_0 = \sum_{m=0}^{\infty} \frac{t^m}{m!} A^{m+1} x_0$$
$$= e^{tA} A x_0 = A e^{tA} x_0$$

for every $x_0 \in \mathcal{R}^n$; hence, $e^{tA}x_0$ is a solution of (4) for each $x_0 \in \mathcal{R}^n$, and, by Theorem 2.3, it is unique.

Defining $u: \mathcal{R}^+ \times \mathcal{R}^n \to \mathcal{R}^n$ by

$$u(t,x) \equiv e^{tA}x \qquad \text{for all} \quad t \in \mathcal{R}^+, x \in \mathcal{R}^n,$$

we see by Theorem 2.3 that u has all the properties (i)–(iv) required by Definition 4.2; alternatively, we can show (ii) and (iii) directly by noting that $u(0,x) = e^{0A}x = x$ and

$$\begin{aligned}\|u(t,x) - u(t,y)\| &= \|e^{tA}x - e^{tA}y\| \\ &= \|e^{tA}(x-y)\| \leqslant \|e^{tA}\|\,\|x-y\| \\ &\leqslant e^{t\|A\|}\|x-y\|\end{aligned}$$

for all $x, y \in \mathcal{R}^n$, $t \in \mathcal{R}^+$. Moreover, since

$$\begin{aligned}e^{tA}e^{\tau A} &= \left(\sum_{k=0}^{\infty} \frac{t^k}{k!}A^k\right)\left(\sum_{m=0}^{\infty} \frac{\tau^m}{m!}A^m\right) \\ &= \sum_{l=0}^{\infty}\left(\sum_{k=0}^{l} \frac{t^k\tau^{l-k}}{k!\,(l-k)!}A^l\right) \qquad (l \equiv k+m) \\ &= \sum_{l=0}^{\infty} \frac{(t+\tau)^l}{l!}A^l = e^{(t+\tau)A},\end{aligned}$$

we see that $u(t+\tau, x) = u(t, u(\tau, x))$ for all $t, \tau \in \mathcal{R}^+$, $x \in \mathcal{R}^n$, thereby directly verifying property (iv) in Definition 4.2. Finally, noting that for fixed $t \in \mathcal{R}^+$, $x \in \mathcal{R}^n$,

$$\begin{aligned}\|u(t+\tau, x) - u(t,x)\| &= \|e^{(t+\tau)A}x - e^{tA}x\| \\ &= \|e^{tA}(e^{\tau A}x - x)\| \leqslant \|e^{tA}\|\,\|e^{\tau A}x - x\| \\ &\leqslant e^{t\|A\|}\|e^{\tau A} - I\|\,\|x\| \\ &\leqslant e^{t\|A\|}(e^{\tau\|A\|} - 1)\|x\| \to 0 \text{ as } \tau \to 0,\end{aligned}$$

we directly verify property (i); hence, u is a dynamical system on $\mathcal{R}^n$. We also see that u is a linear dynamical system.

Definition 5.1. A dynamical system u on $\mathcal{R}^n$ is *linear* if

$$u(t, \alpha x + \beta y) = \alpha u(t,x) + \beta u(t,y)$$

for all $\alpha, \beta \in \mathcal{R}$, $t \in \mathcal{R}^+$, and $x, y \in \mathcal{R}^n$.

We have shown that every linear autonomous evolution equation (4) generates a linear dynamical system on $\mathcal{R}^n$; in fact, the converse is also true. A result in Chapter III, based on the theory of C_0-semigroups (see Ref. 7), implies that every linear dynamical system is generated by some linear autonomous equation of the form (4) (see Theorem III.2.1); in particular, A is the linear operator defined by

$$Ax = \lim_{t \searrow 0} \frac{1}{t}\left[u(t, x) - x\right],$$

and the limit exists for every $x \in \mathcal{R}^n$. This is a remarkable result, for it shows that the properties of Definition 4.2, combined with the linearity condition of Definition 5.1, are sufficient to insure that $u(\cdot, x)$ is differentiable everywhere on $\mathcal{R}^+$! Similarly nice results do not hold in general for nonlinear dynamical systems or for general processes.

There are several ways of "explicitly" representing e^{tA}, other than by the power series definition. In terms of the Laplace transform L, it is easily shown by using equation (4) that $L[e^{tA}] = (sI - A)^{-1}$ for all s sufficiently far to the right in the complex plane $\hat{\mathcal{C}}$; therefore,

$$e^{tA} = L^{-1}\left[(sI - A)^{-1}\right], \qquad t \in \mathcal{R}^+.$$

For $n > 1$ there are very few evolution equations for which a complete general solution can be "explicitly" determined; (4) is an exception to this rule. Another significant exception is discussed in the following section.

Exercise 5.1. Let A be represented by a symmetric real matrix, $A = A^T$. Then all eigenvalues λ_p of A are real, and the corresponding eigenvectors g_p are mutually orthogonal (under the Euclidean inner product $\langle \cdot, \cdot \rangle$) and span $\mathcal{R}^n$; we may assume that $\| g_p \| = 1$. Show that, for all $t \in \mathcal{R}^+$, $x \in \mathcal{R}^n$,

$$e^{tA}x = \sum_{p=1}^{n} \langle g_p, x \rangle \exp\{\lambda_p t\}\, g_p.$$

6. The Simplest Class of Processes

We now consider the simplest type of process on $\mathcal{R}^n \times \mathcal{R}$ that is not a dynamical system. Basically, such a process corresponds to a "forced linear autonomous" evolution equation; however, in view of the point raised in Exercise 2.1, let us postpone making a connection with a differential equation. Instead consider a given linear operator $A\colon \mathcal{R}^n \to \mathcal{R}^n$ (constant real $n \times n$ matrix) and a given function $g\colon \mathcal{R} \to \mathcal{R}^n$, Lebesgue integrable on bounded sets in $\mathcal{R}$ (see Ref. 8), and define

$$u(\tau, x_0, t_0) \equiv e^{\tau A}x_0 + \int_{t_0}^{t_0+\tau} \exp\{(t_0 + \tau - s)A\}\, g(s)\, ds$$

for each $\tau \in \mathcal{R}^+$, $x_0 \in \mathcal{R}^n$, $t_0 \in \mathcal{R}$.

We note that u: $\mathcal{R}^+ \times \mathcal{R}^n \times \mathcal{R} \to \mathcal{R}^n$ is continuous in the first argument, affine in the second argument, and $u(0, x_0, t_0) = x_0$ for every $x_0 \in \mathcal{R}^n$, $t_0 \in \mathcal{R}$. It is also easily verified that u is jointly continuous in the last pair of arguments and

$$u(\tau, s, x_0, t_0) = u(\tau, u(s, x_0, t_0), s + t_0)$$

$$\text{for all } \tau, s \in \mathcal{R}^+, x_0 \in \mathcal{R}^n, t_0 \in \mathcal{R};$$

hence, by Definition 4.1, u is a process on $\mathcal{R}^n \times \mathcal{R}$.

Exercise 6.1. Show that u meets conditions (ii) and (iv) of Definition 4.1.

Defining $x(t) = u(t - t_0, x_0, t_0)$ for $t \geq t_0$, some $x_0 \in \mathcal{R}^n$, $t_0 \in \mathcal{R}$, it is not difficult to verify that $x(\cdot)$ satisfies the integral equation

$$x(t) = x_0 + \int_{t_0}^{t} [Ax(s) + g(s)]\, ds \tag{5}$$

for all $t \geq t_0$. Note that $x(\cdot)$ may not be differentiable at every $t \geq t_0$, because we have assumed only that g is integrable on bounded sets. If we instead impose the stronger condition that g: $\mathcal{R} \to \mathcal{R}^n$ is everywhere continuous, the integral equation (5) implies that $\dot{x}(t)$ exists for all $t \geq t_0$ and, in addition,

$$\dot{x}(t) = Ax(t) + g(t) \qquad \text{for every} \quad t \geq t_0 \in \mathcal{R},$$

$$x(t_0) = x_0 \in \mathcal{R}^n. \tag{6}$$

Hence, if g is merely integrable on bounded sets, the process u is generated by the integral equation (5); if g is everywhere continuous, the process u is generated by the (stronger) differential equation (6).

It is apparent that any solution of (5) (necessarily continuous) need not be everywhere differentiable and, even if it is, need not satisfy (6) if g is not everywhere continuous. Exercise 2.1 should now be transparent. It should also be apparent that if a given process is not a linear dynamical system, it may not be generated by any differential equation; i.e., its motions may not be solutions of any differential equation, in the sense of Definition 2.1.

7. Stability of a Particular Motion: Processes

Given a process u: $\mathfrak{R}^+ \times \mathfrak{R}^n \times \mathfrak{R} \to \mathfrak{R}^n$ and fixed $x_0 \in \mathfrak{R}^n$, $t_0 \in \mathfrak{R}$, the mapping $u(\cdot, x_0, t_0)$: $\mathfrak{R}^+ \to \mathfrak{R}^n$ is the *motion* that originates at (x_0, t_0). If $u(\tau, x_e, t_0) \equiv x_e$ for all $\tau \in \mathfrak{R}^+$, $t_0 \in \mathfrak{R}$, and some $x_e \in \mathfrak{R}^n$, then x_e is an *equilibrium*. There are many questions that can be asked about each particular motion of u; one of the more interesting is the question of stability (Refs. 1–3, 5, 9–17).

Definition 7.1. The motion $u(\cdot, x_0, t_0)$ is *stable* relative to t_0 if, for each $\varepsilon > 0$, there exists $\delta(\varepsilon) > 0$ such that $\|x - x_0\| < \delta$ implies $\|u(\tau, x, t_0) - u(\tau, x_0, t_0)\| < \varepsilon$ for every $\tau \in \mathfrak{R}^+$. The motion u is *unstable* relative to t_0 if it is not stable relative to t_0.

Basically, the idea of stability is that all motions that "start" sufficiently close (to the given x_0 at the same initial t_0) will stay as close as desired to the given motion over all future times. Note that this is a much stronger condition than the (related) continuity condition (ii) of Definition 4.1. Also note that some motions may be stable while others may be unstable, relative to the same t_0; furthermore, the motion $u(\cdot, x_0, t_0)$ may be stable relative to t_0, but $u(\cdot, x_0, t_1)$ might be unstable relative to $t_1 \neq t_0$. The last remark is obviated if u is a dynamical system rather than a general process; then all mention of t_0 can be removed from the definition of stability of a particular motion $u(\cdot, x_0)$: $\mathfrak{R}^+ \to \mathfrak{R}^n$.

If all motions are not explicitly known, verification of the condition in Definition 7.1 presents a problem; obviously, if we know nothing, we can conclude nothing. Usually, the minimum information required is explicit knowledge of the particular motion $u(\cdot, x_0, t_0)$ and an "equation of perturbation" that describes the time behavior of $u(\tau, x, t_0) - u(\tau, x_0, t_0)$ for all $\tau \in \mathfrak{R}^+$ and all sufficiently small $\|x - x_0\|$.

Exercise 7.1. Given $x_0 \in \mathfrak{R}^n$, $t_0 \in \mathfrak{R}$, and a process u: $\mathfrak{R}^+ \times \mathfrak{R}^n \times \mathfrak{R} \to \mathfrak{R}^n$, show that another process v: $\mathfrak{R}^+ \times \mathfrak{R}^n \times \mathfrak{R}_{t_0} \to \mathfrak{R}$ is defined by

$$v(\tau, y_0, s_0) \equiv u(\tau, y_0 + u(s_0 - t_0, x_0, t_0), s_0) - u(\tau + s_0 - t_0, x_0, t_0),$$

$$\tau \in \mathfrak{R}^+, y_0 \in \mathfrak{R}^n, s_0 > t_0, \tag{7}$$

for the given values of x_0, t_0. Furthermore, show that the motion $u(\cdot, x_0, t_0)$ is stable relative to t_0 if and only if the motion $v(\cdot, 0, t_0) \equiv 0$ is stable relative to t_0. Recall the remarks of Section 4 and note that $\mathfrak{R}_{t_0} \equiv [t_0, \infty)$.

On the basis of the results of this exercise, we see that we wish to investigate the stability of the *equilibrium* $y = 0$ of the process v. To do this, suppose that u is generated by equation (2) and the motion $u(\cdot, x_0, t_0)$ is explicitly known; then define the function h: $\mathcal{R}^n \times \mathcal{R}^+ \to \mathcal{R}^n$ by

$$h(y, \tau) \equiv f(u(\tau, x_0, t_0) + y, \tau + t_0) - f(u(\tau, x_0, t_0), \tau + t_0),$$
$$y \in \mathcal{R}^n, \tau \in \mathcal{R}^+, \tag{8}$$

and consider the *equation of perturbation*

$$\dot{y}(\tau) = h(y(\tau), \tau), \qquad \tau \in \mathcal{R}^+,$$
$$y(0) = y_0. \tag{9}$$

Exercise 7.2. Assuming that (2) generates u, show that $y(\cdot) \equiv v(\cdot, x_0, t_0)$ is the unique solution of (9) on $\mathcal{R}^+$, where v is defined by (7).

We now see that the stability question

$$(\|x - x_0\| < \delta) \overset{?}{\Rightarrow} (\|u(\tau, x, t_0) - u(\tau, x_0, t)\| < \varepsilon \quad \text{for all} \quad \tau \in \mathcal{R}^+)$$

is equivalent to the question

$$(\|y_0\| < \delta) \overset{?}{\Rightarrow} (\|v(\tau, y_0, t_0)\| < \varepsilon \quad \text{for all} \quad \tau \in \mathcal{R}^+),$$

with v defined by (7), and this is in turn equivalent to the question

$$(\|y_0\| < \delta) \overset{?}{\Rightarrow} (\|y(\tau)\| < \varepsilon \quad \text{for all} \quad \tau \in \mathcal{R}^+),$$

where $y(\cdot)$ satisfies (9). This seems to simplify the problem of investigating the stability properties of a particular motion $u(\cdot, x_0, t_0)$ of a dynamical system u that is generated by a differential equation. The objective is to change the stability problem for the given motion $u(\cdot, x_0, t_0)$, of the process u, into a stability problem for the motion $v(\cdot, 0, t_0)$ of the related process v; the advantage is that the motion $v(\cdot, 0, t_0)$ is an equilibrium of v, since $v(\tau, 0, s_0) \equiv 0$ for all $\tau \in \mathcal{R}^+$, $s_0 \geqslant t_0$. Further, if u is generated by (2), this question can be answered by studying the perturbation equation (9). The primary difficulty is that, in order to define the function h of equation (8), it is usually necessary to know the complete motion $u(\cdot, x_0, t_0)$: $\mathcal{R}^+ \to \mathcal{R}^n$; knowledge of only x_0 and t_0 is usually not enough. In certain special cases, this difficulty does not arise.

Example 7.1. Suppose that we wish to investigate the stability properties of a particular solution of the problem

$$\ddot{z}(t) + \gamma z(t) + z(t) = \sin 2t, \qquad t \geqslant t_0,$$
$$z(0) = z_0 \in \mathcal{R}, \qquad \dot{z}(0) = v_0 \in \mathcal{R},$$

where γ is a given real number. Suppose that the particular solution of interest is

$$z_p(t) = (2\gamma \cos 2t - 3 \sin 2t)/(9 + 4\gamma^2), \qquad t \geqslant 0,$$

corresponding to the initial conditions

$$z_p(0) = 2\gamma/(9 + 4\gamma^2), \qquad \dot{z}_p(0) = -6/(9 + 4\gamma^2), \qquad t_0 = 0.$$

Placing the problem in the form (2), we have

$$\dot{x}(t) = Ax(t) + g(t), \qquad t \geqslant t_0,$$

$$x(t_0) = x_0 \in \mathfrak{R}^2,$$

where

$$x_0 = \begin{bmatrix} z_0 \\ v_0 \end{bmatrix}, \quad A \equiv \begin{bmatrix} 0 & 1 \\ -1 & -\gamma \end{bmatrix}, \quad g(t) \equiv \begin{bmatrix} 0 \\ \sin 2t \end{bmatrix}, \qquad t \in \mathfrak{R}.$$

We are interested in the stability (with respect to $t_0 = 0$) of the particular solution

$$x_p(t) = \frac{1}{9 + 4\gamma^2} \begin{bmatrix} 2\gamma \cos 2t - 3 \sin 2t \\ -4\gamma \sin 2t - 6 \cos 2t \end{bmatrix}, \qquad t \geqslant 0,$$

corresponding to

$$x_0 = \frac{1}{9 + 4\gamma^2} \begin{bmatrix} 2\gamma \\ -6 \end{bmatrix}, \qquad t_0 = 0.$$

By the results of Section 6, we see that the evolution equation for $x(\cdot)$ generates a process u on $\mathfrak{R}^2 \times \mathfrak{R}$, given by

$$u(\tau, x_0, t_0) = e^{\tau A} x_0 + \int_{t_0}^{t_0 + \tau} \exp\{(t_0 + \tau - s)A\}\, g(s)\, ds.$$

Defining another process v: $\mathfrak{R}^+ \times \mathfrak{R}^2 \times \mathfrak{R}^+ \to \mathfrak{R}^2$ by (7), we find that

$$v(\tau, y_0, s_0) = e^{\tau A} y_0$$

for all $\tau \in \mathfrak{R}^+$, $y_0 \in \mathfrak{R}^2$, $s_0 \in \mathfrak{R}^+$. Using (8) we see that

$$h(y, \tau) = Ay$$

for all $y \in \mathfrak{R}^2$, $\tau \in \mathfrak{R}^+$. Hence, both the process v and the equation of perturbation

$$\dot{y}(\tau) = Ay(\tau), \qquad \tau \in \mathfrak{R}^+,$$

$$y(0) = y_0,$$

turn out to be independent of the particular motion $u(\cdot, x_0, t_0) = x_p(\cdot)$ that we have chosen to study!

Furthermore, v is actually a dynamical system on $\mathcal{R}^2$, and the question of stability of $x_p(\cdot)$ (or of any other particular motion of u) is equivalent to the stability question for the motion $v(\cdot, 0) \equiv 0$ of v: $\mathcal{R}^+ \times \mathcal{R}^2 \to \mathcal{R}^2$. That is, given $\varepsilon > 0$, does there exist $\delta > 0$ such that $\|y_0\| < \delta$ implies $\|y(\tau)\| < \varepsilon$ for all $\tau \geqslant 0$? One way to decide this question is to find $e^{\tau A}$, since $y(\tau) = e^{\tau A} y_0$ is the unique solution of the perturbation equation (9).

If $|\gamma| \neq 2$ it is easily verified that $\exp\{\tau A\} x_0 = \exp\{\lambda_1 \tau\} x_1 + \exp\{\lambda_2 \tau\} x_2$ for some $x_1, x_2 \in \mathcal{C}^2$, satisfying $x_1 + x_2 = x_0$ and $\det|\lambda_p I - A| = 0$, $Ax_p = \lambda_p x_p$, $p = 1, 2$. From $\det|\lambda_p I - A| = 0$ we find that $\lambda_p = [-\gamma \pm (\gamma^2 - 4)^{1/2}]/2$; therefore, if $\gamma \geqslant 0$ and $\gamma \neq 2$, the trivial motion $v(\cdot, 0) \equiv 0$ of the perturbation equation is stable and, therefore, so is the motion $u(\cdot, x_0, t_0) = x_p(\cdot)$ of u (and, for that matter, so is every other motion of u). On the other hand, if $\gamma < 0$ and $\gamma \neq -2$, we conclude instability.

If $|\gamma| = 2$, it is easily verified that $e^{\tau A} x_0 = e^{\lambda \tau}(x_0 - \tau x_1)$ for some $x_1 \in \mathcal{C}^2$, $\lambda \in \mathcal{C}$, satisfying

$$\det|\lambda I - A| = 0, \qquad x_1 = (\lambda I - A) x_0.$$

Solving the first of these we find that $\lambda = -\gamma/2$, implying stability for $\gamma = 2$ and instability for $\gamma = -2$.

Unfortunately, it is usually necessary to know (explicitly) what a particular motion *is* before its stability properties can be investigated by using the perturbation equation (9); this follows from the definition (8) of h: $\mathcal{R}^n \times \mathcal{R}^+ \to \mathcal{R}^n$. In fact, differing motions often have differing stability properties and, therefore, differing perturbation equations. Example 7.1 provides an exception to these complications; the most general known exception is essentially covered in Exercise 7.3.

Definition 7.2. A function F: $\mathcal{R}^n \to \mathcal{R}^m$ is *affine* if $F(x) = Ax + F(0)$ for all $x \in \mathcal{R}^n$, where A: $\mathcal{R}^n \to \mathcal{R}^m$ is some linear operator.

Exercise 7.3. Let f: $\mathcal{R}^n \times \mathcal{R} \to \mathcal{R}^n$ be continuous and such that $f(\cdot, t)$: $\mathcal{R}^n \to \mathcal{R}^n$ is affine at each $t \in \mathcal{R}$, i.e., $f(x, t) = A(t)x + f(0, t)$, with $f(0, \cdot)$: $\mathcal{R} \to \mathcal{R}^n$ continuous and $A(t)$ a real $n \times n$ matrix with continuous time-varying elements.

(a) Show that (2) defines a process u: $\mathcal{R}^+ \times \mathcal{R}^n \times \mathcal{R} \to \mathcal{R}^n$ and the mapping $u(\tau, \cdot, t_0)$: $\mathcal{R}^n \to \mathcal{R}^n$ is affine for each $\tau \in \mathcal{R}^+$, $t_0 \in \mathcal{R}$.

(b) Fixing $x_0 \in \mathcal{R}^n$, $t_0 \in \mathcal{R}$, and defining the process v: $\mathcal{R}^+ \times \mathcal{R}^n \times \mathcal{R}_{t_0}$ by (7), show that the mapping $v(\tau, \cdot, s_0)$: $\mathcal{R}^n \to \mathcal{R}^n$ is linear for each $\tau \in \mathcal{R}^+$, $s_0 \in \mathcal{R}_{t_0}$.

(c) For the same x_0, t_0, find the function h of equation (8), and write the perturbation equation (9).

(d) Show that if any motion $u(\cdot, x_0, t_0)$ of u is stable (respectively, unstable) relative to t_0, then every motion $u(\cdot, x, t_0)$ of u is stable (respectively, unstable) relative to the same t_0.

Having transformed the question of stability of a particular motion $u(\cdot, x_0, t_0)$ of a process u into the question of stability of the equilibrium motion $v(\cdot, 0, t_0) \equiv 0$ of the related process v of (7), we require a means of investigating stability of an equilibrium. It is obvious that explicit solution of the perturbation equation (9) for all y_0, as in Example 7.1, is hardly a practical method of stability analysis. The basic method of investigating stability of an equilibrium is Liapunov's Direct Method (Refs. 2, 5, 11–17) and its various extensions (Refs. 5, 12). Although applicable to the investigation of the stability of an equilibrium of a general process (Ref. 5), this approach is much easier to explain, understand, and apply when we restrict our attention to dynamical systems. In the following sections, that is what we shall do.

Exercise 7.4. Given that the process $u: \mathcal{R}^+ \times \mathcal{R}^n \times \mathcal{R} \to \mathcal{R}^n$ is generated by (2), and that the process $v: \mathcal{R}^+ \times \mathcal{R}^n \times \mathcal{R}_{t_0} \to \mathcal{R}^n$ defined by (7) is actually a dynamical system, show that the perturbation equation (9) must be autonomous and must generate $v: \mathcal{R}^+ \times \mathcal{R}^n \to \mathcal{R}^n$.

8. Stability of Equilibrium: Dynamical Systems

Whether a process having an equilibrium arises from an equation of perturbation, as in Section 7, or directly from the physical problem, it is often true that the only motions whose stability properties are of significant interest happen to be equilibria. A given process u can have one, none, or many equilibria. If a process $u: \mathcal{R}^+ \times \mathcal{R}^n \times \mathcal{R} \to \mathcal{R}^n$ is generated by (2), then all equilibria are easily found as solutions x_e of the equation $f(x_e, t) \equiv 0$ for all $t \in \mathcal{R}$. We assume here that there exists at least one equilibrium x_e and, rather restrictively, we consider only dynamical systems. We can now specialize Definition 7.1 to the case of an equilibrium x_e of a dynamical system $u: \mathcal{R}^+ \times \mathcal{R}^n \to \mathcal{R}^n$.

Definition 8.1. The equilibrium x_e is *stable* if, for each $\varepsilon > 0$, there exists $\delta(\varepsilon) > 0$ such that $\|x - x_e\| < \delta$ implies $\|u(t, x) - x_e\| < \varepsilon$ for all $t \in \mathcal{R}^+$. The equilibrium x_e is *unstable* if it is not stable.

Exercise 8.1. Let some eigenvalue λ_p of the linear operator $A: \mathcal{R}^n \to \mathcal{R}^n$ have positive real part. Show that the equilibrium at $x = 0$ of the dynamical system u generated by (4) is unstable.

If x_e is stable, somewhat more detailed properties are of interest.

Definition 8.2. The equilibrium x_e is *asymptotically stable* if it is stable and there exists $\gamma > 0$ such that $\|x - x_e\| < \gamma$ implies $\|u(t, x) - x_e\| \to 0$ as $t \to \infty$. This property is said to be *global* if γ can be chosen as $\gamma = \infty$.

Definition 8.3. The equilibrium x_e is *exponentially stable* if there exist $\gamma > 0$, $\alpha > 0$, $M < \infty$, such that $\|x - x_e\| < \gamma$ implies

$$\|u(t, x) - x_e\| \leqslant Me^{-\alpha t}\|x - x_e\|$$

for all $t \in \mathcal{R}^+$. This property is *global* if, for every $\gamma > 0$, there exist suitable $M(\gamma) < \infty$, $\alpha(\gamma) > 0$.

We note that exponential stability is stronger than asymptotic stability, which is stronger than stability. In the case of exponential stability, we are often interested in the value of α; sometimes we are also interested in the values of M and γ; but, unlike α, the permissible values of M and γ depend on the choice of the norm $\|\cdot\|$.

The key to investigating the stability properties of a given equilibrium x_e, of a given dynamical system u: $\mathcal{R}^+ \times \mathcal{R}^n \to \mathcal{R}^n$, is the determination of a suitable Liapunov function.

Definition 8.4. Let u be a dynamical system on $\mathcal{R}^n$, let V: $\mathcal{R}^n \to \mathcal{R}$ be continuous, and define

$$\dot{V}(x) \equiv \liminf_{t \searrow 0} \frac{1}{t}\left[V(u(t, x)) - V(x)\right]$$

for all $x \in \mathcal{R}^n$. Let $\mathcal{G}$ be a subset of $\mathcal{R}^n$ such that $\dot{V}(x) \leqslant 0$ for all $x \in \mathcal{G}$. Then V is a continuous *Liapunov function* for u on $\mathcal{G}$.

The useful property of a Liapunov function is that, as we shall prove in Chapter IV, if $x_0 \in \mathcal{G}$ and $u(t, x_0) \in \mathcal{G}$ for all $t \in [0, T)$, then $V(u(\cdot, x))$: $\mathcal{R}^+ \to \mathcal{R}$ is nonincreasing on $[0, T)$. Consequently, as we shall show in Chapter IV, if $\mathcal{G}$ happens to be *defined* to be a disjoint component of the open set $\{x \in \mathcal{R}^n | V(x) < \beta\}$ *or* of the closed set $\{x \in \mathcal{R}^n | V(x) \leqslant \beta\}$, for some $\beta \leqslant \infty$, we can conclude that $u(t, x_0) \in \mathcal{G}$ for all $t \in \mathcal{R}^+$ if $x_0 \in \mathcal{G}$! We shall use these facts here.

Regarding the computation of $\dot{V}(x)$, we note that it is everywhere well defined provided that we admit $\pm\infty$ as "values." If V has continuous partial derivatives with respect to the components $x_1, x_2, \ldots, x_n$ of x, at all $x \in \mathcal{R}^n$, and if u is generated by (3), then $V(u(\cdot, x))$: $\mathcal{R}^+ \to \mathcal{R}$ has a right derivative at $t = 0$, equal to $\dot{V}(x)$, and

$$\dot{V}(x) = \sum_{i=1}^{n} \frac{\partial V(x)}{\partial x_i} f_i(x) = \langle \nabla V(x), f(x) \rangle,$$

where ∇V denotes the gradient and $\langle \cdot, \cdot \rangle$ is the Euclidean inner product.

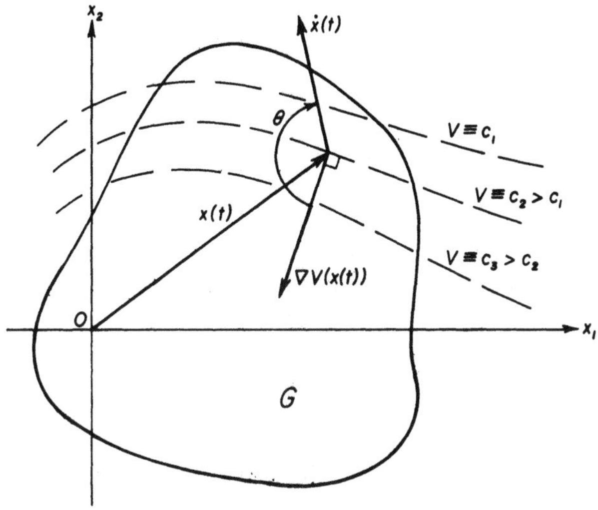

Fig. 3. Geometric interpretation.

Computation of $\dot{V}(x)$ under less restrictive assumptions is discussed in Chapter IV.

Geometrically, the idea of a Liapunov function for u is that, on $\mathcal{G}$, the motions cannot cross the level surfaces of V in an "uphill" direction. See Figure 3 for the case $\mathfrak{R}^n = \mathfrak{R}^2$, and note that

$$\dot{V}(x(t)) = \langle \nabla V(x(t)), \dot{x}(t) \rangle \leqslant 0$$

if and only if $\pi/2 \leqslant \theta \leqslant 3\pi/2$.

The following two theorems embody what is known as Liapunov's Direct Method (Refs. 2, 5, 11–17). Proofs are not provided here, because they follow easily from the proofs of comparable theorems in Chapter IV, using the local compactness of $\mathfrak{R}^n$.

Theorem 8.1. Let V be a continuous Liapunov function for u on an open set $\mathcal{G}$ containing the equilibrium x_e. If $V(x) > V(x_e)$ for all $x \in \mathcal{G} \setminus \{x_e\}$, then x_e is stable. If, in addition, $\dot{V}(x) < 0$ for all $x \in \mathcal{G} \setminus \{x_e\}$, then x_e is asymptotically stable. If, in addition,

$$\left.\begin{aligned} \dot{V}(x) &\leqslant -\beta(V(x) - V(x_e)) \\ c_1\|x - x_e\|^k \leqslant V(x) - V(x_e) &\leqslant c_2\|x - x_e\|^k \end{aligned}\right\} \text{ for all } x \in \mathcal{G}$$

and some numbers $c_2 \geq c_1 > 0, \beta > 0, k > 0$, then x_e is exponentially stable with

$$\|u(t, x) - x_e\| \leq (c_2/c_1)^{1/k} e^{-\beta t/k} \|x - x_e\|$$

for all $t \geq 0$ and all $x \in \mathcal{G}_\alpha$, where $\mathcal{G}_\alpha$ is a disjoint component of the set $\{x \in \mathcal{R}^n | V(x) < \alpha\}$, some α such that $x_e \in \mathcal{G}_\alpha \subset \mathcal{G}$.

Theorem 8.2. Let V be a continuous Liapunov function for u on an open set $\mathcal{G}$ containing the equilibrium x_e. If $\dot{V}(x) < 0$ for all $x \in \mathcal{G} \setminus \{x_e\}$, and if each neighborhood of x_e contains some x such that $V(x) < V(x_e)$, then x_e is unstable.

Example 8.1. (i) For nonzero $\alpha \in \mathcal{R}$, consider

$$\dot{x}(t) = \alpha x(t), \qquad t \in \mathcal{R}^+,$$
$$x(0) = x_0 \in \mathcal{R},$$

and define $V(x) = -\alpha x^2$ for $x \in \mathcal{R}$. Noting that

$$\dot{V}(x) = \frac{\partial V(x)}{\partial x} \cdot \alpha x = -2\alpha^2 x^2,$$

we see that $V: \mathcal{R} \to \mathcal{R}$ is a Liapunov function on all of $\mathcal{R}$ for the dynamical system u generated by this evolution equation. By Theorem 8.2 the equilibrium at $x = 0$ is unstable if $\alpha > 0$. If $\alpha < 0$, $x = 0$ is exponentially stable and the estimate $|u(t, x)| \leq e^{-|\alpha| t}|x|$ is valid for all $t \in \mathcal{R}^+$, $x \in \mathcal{R}$; hence, $x = 0$ is globally exponentially stable.

(ii) Consider the dynamical system u generated on $\mathcal{R}$ by

$$\dot{x}(t) = (x(t))^2 - (x(t))^3, \qquad t \in \mathcal{R}^+,$$
$$x(0) = x_0 \in \mathcal{R}.$$

Defining $V(x) = -x$ for $x \in \mathcal{R}$, we see that

$$\dot{V}(x) = \frac{\partial V(x)}{\partial x} \cdot (x^2 - x^3) = -x^2 + x^3$$

and V is a Liapunov function for u on the set $\{x \in \mathcal{R} | x \leq 1\}$. Defining, for example, $\mathcal{G} = \{x \in \mathcal{R} | \cdot |x| < 1\}$, we see by Theorem 8.2 that the equilibrium at $x = 0$ is unstable. To investigate the other equilibrium at $x = +1$, consider $\hat{V}(x) \equiv (x-1)^2$ and note that $\dot{\hat{V}}(x) = -2x^2(1-x)^2$; hence $\hat{V}$ is a Liapunov function for u on all of $\mathcal{R}$. With the definition $\hat{\mathcal{G}} = \{x \in \mathcal{R} | \, |x - 1| < 1\}$, for example, Theorem 8.1 implies that $x = +1$ is asymptotically stable.

(iii) Let $f: \mathcal{R}^n \to \mathcal{R}^n$ be continuous and such that $\langle x, f(x) \rangle \leq 0$ for every

$x \in \mathfrak{R}^n$. Note that this implies that $f(0) = 0 \in \mathfrak{R}^n$, and consider the dynamical system u generated by

$$\dot{x}(t) = f(x(t)), \qquad t \in \mathfrak{R}^+,$$
$$x(0) = x_0 \in \mathfrak{R}^n.$$

Defining $V(x) = \langle x, x \rangle$ and noting that $\dot{V}(x) = 2\langle x, f(x)\rangle \leqslant 0$ for all $x \in \mathfrak{R}^n$, we see that V: $\mathfrak{R}^n \rightarrow \mathfrak{R}$ is a Liapunov function for u on all of $\mathfrak{R}^n$ and, by Theorem 8.1, the equilibrium at $x = 0$ is stable. Moreover, as $V(u(\cdot, x))$: $\mathfrak{R}^+ \rightarrow \mathfrak{R}$ is nonincreasing for each $x \in \mathfrak{R}^n$, it follows that $\|u(t, x)\|^2 = V(u(t, x)) \leqslant V(x) = \|x\|^2$ for every $x \in \mathfrak{R}^n$; hence, all motions of u are bounded. Under the stronger assumption that $\langle x, f(x)\rangle < 0$ for all nonzero x, we can conclude global asymptotic stability of $x = 0$; under the even stronger assumption that $\langle x, f(x)\rangle \leqslant \alpha\langle x, x\rangle$ for some $\alpha < 0$ and all $x \in \mathfrak{R}^n$, we conclude global exponential stability of $x = 0$.

Exercise 8.2. Let A: $\mathfrak{R}^n \rightarrow \mathfrak{R}^n$ be linear and symmetric in terms of some inner product $\langle\ ,\ \rangle$, (i.e., $\langle y, Ax\rangle = \langle Ay, x\rangle$ for all x, $y \in \mathfrak{R}^n$), and let u be the dynamical system generated by (4). Consider the function V: $\mathfrak{R}^n \rightarrow \mathfrak{R}$ defined by $V(x) = -\langle x, Ax\rangle$, $x \in \mathfrak{R}^n$.

(i) Show that V is a Liapunov function for u on $\mathfrak{R}^n$.

(ii) Under what (additional) conditions on A can we be sure that the equilibrium at $x = 0$ is stable? unstable? asymptotically stable? exponentially stable? globally asymptotically stable? globally exponentially stable?

Remark: As A is symmetric, it has only real eigenvalues, and its eigenvectors span $\mathfrak{R}^n$ and are orthogonal under $\langle \cdot, \cdot \rangle$. Hence, $\langle x, Ax\rangle > 0$ (< 0) for all $x \neq 0$ if and only if all eigenvalues of A are positive (negative) (see Ref. 18).

Given a particular dynamical system, it is sometimes quite difficult to construct any useful Liapunov function. Therefore, it is reasonable to ask whether there even exists a useful Liapunov function. This question motivates various "converse theorems" for Liapunov's Direct Method (Refs. 11, 15, 17). We provide only a very special existence result here.

Theorem 8.3. Let u be a linear dynamical system on $\mathfrak{R}^n$, and let the equilibrium at $x = 0$ be stable. Then there exists a Liapunov function V for u on $\mathfrak{R}^n$, with $V(x) > V(0)$ for $x \neq 0$ and $V(x) \rightarrow \infty$ as $\|x\| \rightarrow \infty$.

Proof. As the equilibrium at $x = 0$ is stable and u is linear, all motions of u are stable; moreover, it can be shown that all motions of u are bounded, and therefore a function V: $\mathfrak{R}^n \rightarrow \mathfrak{R}$ is well defined by

$$V(x) \equiv \sup_{t \in \mathfrak{R}^+} \|u(t, x)\|, \qquad x \in \mathfrak{R}^n.$$

As every motion is stable, it is easily verified that V is continuous on $\mathfrak{R}^n$.

Since

$$V(u(t, x)) = \sup_{\tau \in \mathcal{R}^+} \|u(\tau + t, x)\|$$
$$\leqslant \sup_{s \in \mathcal{R}^+} \|u(s, x)\| = V(x)$$

for all $t \in \mathcal{R}^+$, $x \in \mathcal{R}^n$, it follows that $\dot{V}(x) \leqslant 0$ for all $x \in \mathcal{R}^n$. The proof is complete. □

Exercise 8.3. In the proof of Theorem 8.3, verify that V: $\mathcal{R}^n \to \mathcal{R}$ is continuous.

The major difficulty to be overcome in any of the various existence theorems lies in the proof of continuity of V. Given any dynamical system u, it is not difficult to show existence of a variety of real-valued functions that are nonincreasing along motions of u, and this is obviously the useful property of any Liapunov function. For example, define

$$V(x) \equiv \sup_{t \in \mathcal{R}^+} \frac{\|u(t, x)\|}{1 + \|u(t, x)\|}, \qquad x \in \mathcal{R}^n,$$

and note that this function meets all conditions of Definition 8.4 except possibly, that of continuity; it may or may not be continuous. Nevertheless, as this is only one of many possible candidates for a Liapunov function, we can be reasonably sure that useful continuous Liapunov functions exist for every given dynamical system. We also observe that abstract existence proofs are of little help in actually constructing a Liapunov function for a particular application, since u is generally unknown; if the mapping u: $\mathcal{R}^+ \times \mathcal{R}^n \to \mathcal{R}$ were explicitly known, there would be little reason to construct a Liapunov function.

Example 8.2. In many applications, the physical problem itself suggests the form of some Liapunov function. Consider an unforced linear spring–mass–damper system having n degrees of freedom. If the equations of motion are obtained from Lagrange's equations by using a kinetic energy function $T(v)$, an isothermal potential energy function $U(y)$, and a dissipation function $D(v)$, we obtain

$$M\ddot{y}(t) + C\dot{y}(t) + Ky(t) = 0, \qquad t \in \mathcal{R}^+,$$
$$y(0) = y_0 \in \mathcal{R}^n, \qquad \dot{y}(0) = v_0 \in \mathcal{R}^n,$$

where each of the linear operators M: $\mathcal{R}^n \to \mathcal{R}^n$, C:$\mathcal{R}^n \to \mathcal{R}^n$, K: $\mathcal{R}^n \to \mathcal{R}^n$, is symmetric (in terms of $\langle\ ,\ \rangle_{\mathcal{R}^n}$) and nonnegative (e.g., $\langle y, Ky\rangle_{\mathcal{R}^n} \geqslant 0$ for every $y \in \mathcal{R}^n$). Moreover, $T(v) = \frac{1}{2}\langle v, Mv\rangle_{\mathcal{R}^n}$, $U(y) = \frac{1}{2}\langle y, Ky\rangle_{\mathcal{R}^n}$, $D(v) = \frac{1}{2}\langle v, Cv\rangle_{\mathcal{R}^n}$.

Assuming that $M > 0$ (i.e., $\langle v, Mv\rangle_{\mathcal{R}^n} > 0$ for all $v \neq 0$), we consider the dynamical system u generated on $\mathcal{R}^{2n}$ by (4),

$$\dot{x}(t) = Ax(t),\ t \in \mathcal{R}^+,$$

$$x(0) = x_0 \in \mathcal{R}^{2n},$$

with

$$x_0 = \begin{bmatrix} y_0 \\ v_0 \end{bmatrix}, \qquad A = \begin{bmatrix} 0 & I_n \\ -M^{-1}K & -M^{-1}C \end{bmatrix}.$$

By the nature of the physical problem, it seems clear from the laws of thermodynamics that the total mechanical energy cannot increase from its initial value $\frac{1}{2}\langle v_0, Mv_0\rangle_{\mathcal{R}^n} + \frac{1}{2}\langle y_0, Ky_0\rangle_{\mathcal{R}^n}$, and this idea suggests that we define a function $V: \mathcal{R}^{2n} \to \mathcal{R}$ by

$$V(x) \equiv \tfrac{1}{2}\langle v, Mv\rangle_{\mathcal{R}^n} + \tfrac{1}{2}\langle y, Ky\rangle_{\mathcal{R}^n}, \qquad (y, v) = x \in \mathcal{R}^{2n}.$$

Computing $\dot{V}: \mathcal{R}^{2n} \to \mathcal{R}$, we find that

$$\dot{V}(x) = -\langle v, Cv\rangle = -2D(v) \leqslant 0$$

for all $(y, v) = x \in \mathcal{R}^{2n}$, and V is indeed a Liapunov function for u on $\mathcal{R}^{2n}$. If both M and K are positive definite (i.e., if $\langle y, My\rangle > 0$ and $\langle y, Ky\rangle > 0$ for all nonzero $y \in \mathcal{R}^n$), then Theorem 8.1 implies that the equilibrium at $(y, v) = x = 0$ is stable; moreover, as $V(u(\cdot, x)): \mathcal{R}^+ \to \mathcal{R}$ is nonincreasing and there exists $\alpha > 0$ such that $\alpha\|x\|^2_{\mathcal{R}^{2n}} \leqslant V(x)$ for every $x \in \mathcal{R}^n$, we see that

$$\alpha\|u(t, x)\|^2_{\mathcal{R}^{2n}} \leqslant V(u(t, x)) \leqslant V(x)$$

for all $t \in \mathcal{R}^+$, and every motion of u is bounded.

If, in addition, C is positive definite ($C > 0$), we might suspect that $x = 0$ is asymptotically stable, perhaps even globally exponentially stable. Unfortunately, no such conclusion follows from Theorem 8.1 since $\dot{V}(x) = 0$ for every $x = (y, 0)$, not just for $x = (0, 0)$. To validate such conclusions (if true), we will need either a better Liapunov function or a better theorem. We will return to this example in Section 11.

One of the nice features of using physical laws to suggest a Liapunov function is that often a nonlinear problem does not seem significantly more difficult than a linear one.

Exercise 8.4. Consider a nonlinear system of the type described in Example 8.2, with $T(v)$ again given by $\frac{1}{2}\langle v, Mv\rangle_{\mathcal{R}^n}$, symmetric linear $M > 0$. Assume only that U and D are twice continuously differentiable, with $U(y) \to \infty$ for $\|y\| \to \infty$, and $\langle y, \nabla D(y)\rangle \geqslant 0$ for all $y \in \mathcal{R}^n$. Lagrange's equations then lead to

$$M\ddot{y}(t) + \nabla D(\dot{y}(t)) + \nabla U(y(t)) = 0, \qquad t \in \mathcal{R}^+,$$

$$y(0) = y_0 \in \mathcal{R}^n, \qquad \dot{y}(0) = v_0 \in \mathcal{R}^n.$$

When this equation is set in the form (3), a result presented in Section 12 can be used to show that it generates a dynamical system on $\mathfrak{R}^{2n}$ (no finite escape times).

(i) Show that $V(x) \equiv T(v) + U(y)$ remains a Liapunov function on $\mathfrak{R}^{2n}$.

(ii) Show that all motions are bounded.

(iii) Under what conditions on U can we conclude that there is a stable equilibrium at $x = 0$?

(iv) Do there necessarily exist any equilibria? If so, where are they located?

We notice that if, in Example 8.2 or in Exercise 8.4, we have $D(v) \equiv 0$ and we define $x = (y, Mv)$, rather than $x = (y, v)$, then equation (3) becomes a *Hamiltonian equation.* Such equations are currently of considerable interest in applied mathematics; a major question is the determination of conditions on the potential energy U such that one or more motions will be periodic in time.

The construction of Liapunov functions has been an area of continuous research activity for more than a quarter of a century. Despite the numerous special approaches and techniques now available, there remains the discouraging fact that the construction of a useful Liapunov function often involves a great deal of luck and ingenuity. We shall consider some important aspects of the construction problem in the following two sections.

9. Stability of Equilibrium: Linear Dynamical Systems

As we saw in Section 5, a dynamical system u: $\mathfrak{R}^+ \times \mathfrak{R}^n \to \mathfrak{R}^n$ is linear if and only if it is generated by an autonomous equation of the form (4)

$$\dot{x}(t) = Ax(t), \qquad t \in \mathfrak{R}^+,$$
$$x(0) = x_0 \in \mathfrak{R}^n, \tag{4}$$

where A: $\mathfrak{R}^n \to \mathfrak{R}^n$ is linear. Here we intend to investigate the construction of Liapunov functions for u having the quadratic form $V(x) = \langle x, Bx \rangle$, for some linear B: $\mathfrak{R}^n \to \mathfrak{R}^n$ that is symmetric in terms of the Euclidean inner product $\langle \cdot, \cdot \rangle$; that is, if B is represented by a real $n \times n$ matrix, then B is equal to its transpose B^T. For such a V: $\mathfrak{R}^n \to \mathfrak{R}$, we find that $\dot{V}(x) = \langle x, BAx \rangle + \langle Ax, Bx \rangle = -\langle x, Cx \rangle$, where $C = C^T$ is a real $n \times n$ matrix that is related to the matrices B and A by *Liapunov's matrix equation*

$$BA + A^T B = -C. \tag{10}$$

We see that with A represented by a real $n \times n$ matrix, the problem of constructing a quadratic Liapunov function for u on $\mathcal{R}^n$ reduces to the problem of finding a pair of real $n \times n$ matrices $B = B^T$, $C = C^T$, satisfying Liapunov's equation (10), such that $C \geqslant 0$, i.e., in the Euclidean inner product, $\langle x, Cx \rangle \geqslant 0$ for all $x \in \mathcal{R}^n$. Our first result is that a nontrivial function V of this form necessarily exists.

Theorem 9.1. Let u be generated by (4) with A a real matrix. There exist real matrices $B = B^T$, $C = C^T$, $C \geqslant 0$, such that $BA + A^TB = -C$ and $Bx \neq 0$ for $x \neq 0$.

Partial Proof. We view A as $A: \hat{C}^n \to \hat{C}^n$ and, for simplicity, we shall assume that every (complex) eigenvalue $\lambda_p \in \hat{C}$ has geometric multiplicity equal to its algebraic multiplicity; that is, we shall assume that the eigenvectors of A span $\hat{C}^n$. Then there exists a linear $T: \hat{C}^n \to \hat{C}^n$ such that $TAT^{-1} = \Lambda$, where $\Lambda: \hat{C}^n \to \hat{C}^n$ is a diagonal matrix (of eigenvalues) having conjugate transpose Λ^*. Let $\Gamma: \hat{C}^n \to \hat{C}^n$ be diagonal with diagonal elements $\gamma_p = 1$ if $\mathrm{Re}\,(\lambda_p) \leqslant 0$, $\gamma_p = -1$ if $\mathrm{Re}\,(\lambda_p) > 0$, and define $B = T^*\Gamma T$. Then B is a symmetric real matrix having no zero eigenvalue, and

$$BA + A^TB = T^*\Gamma\Lambda T + T^*\Lambda^*\Gamma T = T^*(\Gamma\Lambda + \Lambda^*\Gamma)T.$$

As $\Gamma\Lambda + \Lambda^*\Gamma$ is a real diagonal matrix with diagonal elements $-|\mathrm{Re}\,(\lambda_p)|$, we see that $C \equiv -T^*(\Gamma\Lambda + \Lambda^*\Gamma)T$ is a nonnegative symmetric real matrix. Under the simplifying assumption made, the proof is finished. □

Under certain conditions on A, given *any* positive definite symmetric matrix C, Liapunov's equation (10) does have a symmetric solution B. The following special result is interesting because of the simplicity of its proof. We note that all eigenvalues of A have negative real parts if and only if $x = 0$ is globally exponentially stable, and this in turn is true if and only if $x = 0$ is asymptotically stable (see Exercise 9.2).

Theorem 9.2. If all (complex) eigenvalues of the real matrix A have negative real parts, there exists a real matrix $B = B^T$, $B > 0$, such that $BA + A^TB = -C$, for any given real matrix $C = C^T$, $C > 0$.

Proof. Given $C = C^T$, $C > 0$, and $\max_p \mathrm{Re}\,(\lambda_p) = -\alpha$, $\alpha > 0$, we note that

$$|\langle e^{tA}y, Ce^{tA}x \rangle| \leqslant \|C\|\;\|e^{tA}y\|\;\|e^{tA}x\| \leqslant ce^{-\beta t}\|x\|\;\|y\|,$$

$$\text{some } c > 0,\ 0 < \beta \leqslant 2\alpha,$$

for all $x, y \in \mathfrak{R}^n$, and therefore the improper integral

$$\int_0^\infty \langle e^{tA}y, Ce^{tA}x\rangle dt$$

converges for all $x, y \in \mathfrak{R}^n$, where $\langle \cdot , \cdot \rangle$ denotes the Euclidean inner product. Hence, a symmetric linear $B: \mathfrak{R}^n \to \mathfrak{R}^n$ is well defined by the bilinear form

$$\langle y, Bx\rangle \equiv \int_0^\infty \langle e^{tA}y, Ce^{tA}x\rangle dt, \qquad x, y \in \mathfrak{R}^n,$$

and we see that $\langle x, Bx\rangle > 0$ for all nonzero $x \in \mathfrak{R}^n$. Finally, it follows that

$$\begin{aligned}\langle y, (BA + A^TB)x\rangle &= \int_0^\infty \langle e^{tA}y, Ce^{tA}Ax\rangle dt + \int_0^\infty \langle e^{tA}Ay, Ce^{tA}x\rangle dt \\ &= \int_0^\infty \left(\frac{d}{dt}\langle e^{tA}y, Ce^{tA}x\rangle\right) dt \\ &= -\langle y, Cx\rangle,\end{aligned}$$

for all $x, y \in \mathfrak{R}^n$; hence, $BA + A^TB = -C$, and the proof is complete. □

Exercise 9.1. Let a linear operator $C: \mathfrak{R}^n \to \mathfrak{R}^n$ be symmetric in terms of the inner product $\langle\ ,\ \rangle$. Then all eigenvalues λ_p are real; moreover, the eigenvectors of C span $\mathfrak{R}^n$ and are orthogonal under $\langle \cdot , \cdot \rangle$ (see Ref. 18). Show that $\lambda_{\min}\langle x, x\rangle \leqslant \langle x, Cx\rangle \leqslant \lambda_{\max}\langle x, x\rangle$ for all $x \in \mathfrak{R}^n$.

The most general conditions under which Liapunov's equation (10) has a solution $B = B^T$, given a completely *arbitrary* real $n \times n$ matrix $C = C^T$, $C > 0$, are provided by the following result.

Theorem 9.3. If $\{\lambda_m\}$ is the set of (complex) eigenvalues of the real matrix A, Liapunov's equation $BA + A^TB = -C$ has a unique real solution $B = B^T$ for every given real $C = C^T$, $C > 0$, if and only if $\lambda_m \neq 0$ and $\lambda_m + \lambda_p \neq 0$ for all $m, p = 1, 2, \ldots, n$. Then $Bx \neq 0$ for $x \neq 0$. Moreover, $B > 0$ if and only if the equilibrium at $x = 0$ is asymptotically stable.

A proof of Theorem 9.3 can be found in Ref. 13. We remark that the eigenvalue conditions of this result have to do with guaranteeing a *unique* solution $B = B^T$ for *every* (arbitrary) $C = C^T$, $C > 0$. As all that we require, in a particular application, is *some* solution $B = B^T$ for a *specific* $C = C^T$, $C \geqslant 0$, that we happen to choose, these eigenvalue conditions need not be verified *a priori.* Essentially, then, our approach is to choose some $C = C^T > 0$, and attempt to solve (10); by the symmetry of C and the unknown B, (10) consists of $(n^2 + n)/2$ linear scalar equations in the $(n^2 + n)/2$ unknown real elements of B. Whether or not the conditions of Theorem 9.3 are satisfied, a real solution B may be found; if not, we simply try another $C = C^T > 0$. If, after several attempts, no real solution B has been found, we begin to suspect that A may have a zero or imaginary eigenvalue, since it can be shown that this is the only condition under which *no* solution of (10) exists for *any* $C = C^T > 0$. In this case, we continue our search but choose $C \geqslant 0$ rather than $C > 0$. By Theorem 9.1, we shall, sooner or later, find a nontrivial solution B of (10).

Exercise 9.2. Using only Theorems 9.3 and 8.1, show that the equilibrium at $x = 0$ of the dynamical system generated by (4) is asymptotically stable if and only if it is globally exponentially stable.

For practical application of the preceding theorems, it is fortunate that Sylvester's lemma provides a simple test for positivity of a symmetric real matrix.

Lemma 9.1. A real symmetric $n \times n$ matrix $C = C^T$ is positive definite ($C > 0$) if and only if all principal minors have positive determinants; this is equivalent to the n conditions

$$c_{11} > 0, \qquad \det\begin{vmatrix} c_{11} & c_{12} \\ c_{21} & c_{22} \end{vmatrix} > 0, \qquad \det\begin{vmatrix} c_{11} & c_{12} & c_{13} \\ c_{21} & c_{22} & c_{23} \\ c_{31} & c_{32} & c_{33} \end{vmatrix} > 0,$$

$$\ldots, \qquad \det|C| > 0.$$

For C to be nonnegative ($C \geqslant 0$), it is necessary and sufficient that these n principal minors have nonnegative determinants and that $c_{mm} \geqslant 0$ for $m = 1, 2, \ldots, n$.

Partial Proof. If $C > 0$, all principal minors must also be positive definite, and necessity follows. Sufficiency can be proved by "completion of squares"; for example, consider $n = 2$ and note that

$$\langle x, Cx \rangle = \sum_{i=1}^{2} \sum_{j=1}^{2} c_{ij} x_i x_j$$

$$= c_{11}\left(x_1 + \frac{c_{12}}{c_{11}} x_2\right)^2 + \frac{1}{c_{11}}(c_{11}c_{22} - c_{12}^2)x_2^2.$$

Therefore, $c_{11} > 0$ and $c_{11}c_{22} - c_{12}^2 > 0$ implies that $C > 0$, and the proof is complete for $n = 2$. □

Example 9.1. Consider the evolution equation

$$\dot{x}(t) = Ax(t), \qquad t \in \mathcal{R}^+, A = \begin{bmatrix} 0 & 1 \\ -1 & -1 \end{bmatrix},$$

$$x(0) = x_0 \in \mathcal{R}^2,$$

and choose $C = I$, the identity on $\mathcal{R}^2$. For B a real 2×2 matrix, $B = B^T$, we attempt to solve Liapunov's equation (10), noting that $A^TB = (BA)^T$ and

$$BA = \begin{bmatrix} b_1 & b_2 \\ b_2 & b_3 \end{bmatrix} \begin{bmatrix} 0 & 1 \\ -1 & -1 \end{bmatrix} = \begin{bmatrix} -b_2 & b_1 - b_2 \\ -b_3 & b_2 - b_3 \end{bmatrix}.$$

Therefore, we seek $b_1, b_2, b_3 \in \mathcal{R}$ such that

$$BA + A^TB = \begin{bmatrix} -2b_2 & b_1 - b_2 - b_3 \\ b_1 - b_2 - b_3 & 2b_2 - 2b_3 \end{bmatrix} = -C = \begin{bmatrix} -1 & 0 \\ 0 & -1 \end{bmatrix},$$

and we see that $b_2 = 1/2$, $b_3 = 1$, $b_1 = 3/2$. Hence, by Lemma 9.1,

$$B = \frac{1}{2}\begin{bmatrix} 3 & 1 \\ 1 & 2 \end{bmatrix} > 0,$$

and by Theorem 8.1 we conclude that the equilibrium at $x = 0$ is globally exponentially stable.

Exercise 9.3. Consider the evolution equation (4) on $\mathcal{R}^2$ for the following cases:

$$A = \begin{bmatrix} 1 & 1 \\ 0 & 4 \end{bmatrix}, \quad \begin{bmatrix} -1 & 0 \\ 1 & -1 \end{bmatrix}, \quad \begin{bmatrix} 0 & 1 \\ -4 & 3 \end{bmatrix}, \quad \begin{bmatrix} 0 & -1 \\ 3 & 4 \end{bmatrix}, \quad \begin{bmatrix} 0 & 0 \\ 0 & 0 \end{bmatrix}.$$

In each case, attempt to solve Liapunov's equation (10) with $C = I$, the identity on $\mathcal{R}^2$. If this is not possible, choose $C = C^T$, $C \geqslant 0$, such that a nontrivial solution B of (10) does exist. Using Theorems 8.1 and 8.2, what can you conclude about the stability properties of the equilibrium at $x = 0$ for each case?

For the construction of quadratic Liapunov functions, it is fortunate that when a pair of matrices $B = B^T$, $C = C^T \geqslant 0$ have been found to satisfy Liapunov's equation (10), it is then possible to produce many more such pairs quite easily.

Exercise 9.4. Given real $n \times n$ matrices $B = B^T$, $C = C^T \geqslant 0$ that satisfy $BA + A^TB = -C$, show that this equation is also satisfied by the matrices $\hat{B} = \hat{B}^T$, $\hat{C} = \hat{C}^T \geqslant 0$ defined as

(a)

$$\hat{B} \equiv (\alpha_0 I + \alpha_1 A + \cdots + \alpha_{n-1}A^{n-1})^T B(\alpha_0 I + \alpha_1 A + \cdots + \alpha_{n-1}A^{n-1}),$$

$$\hat{C} \equiv (\alpha_0 I + \alpha_1 A + \cdots + \alpha_{n-1}A^{n-1})^T C(\alpha_0 I + \alpha_1 A + \cdots + \alpha_{n-1}A^{n-1}),$$

for any $\alpha_i \in \mathfrak{R}$, $i = 0, 1, \ldots, n-1$.

(b)

$$\hat{B} \equiv \left[(\alpha_0 I + \alpha_1 A + \cdots + \alpha_{n-1}A^{n-1})^{-1}\right]^T B(\alpha_0 I + \alpha_1 A + \cdots + \alpha_{n-1}A^{n-1})^{-1},$$

$$\hat{C} \equiv \left[(\alpha_0 I + \alpha_1 A + \cdots + \alpha_{n-1}A^{n-1})^{-1}\right]^T C(\alpha_0 I + \alpha_1 A + \cdots + \alpha_{n-1}A^{n-1})^{-1},$$

for any $\alpha_i \in \mathfrak{R}$, $i = 0, 1, \ldots, n-1$, such that

$$\det |\alpha_0 I + \alpha_1 A + \cdots + \alpha_{n-1}A^{n-1}| \neq 0.$$

(c) $\hat{B} \equiv H^T BH$, $\hat{C} \equiv H^T CH$, where H is any real $n \times n$ matrix that commutes with A.

10. Quadratic Liapunov Functions and Linearization

Let us now consider a nonlinear dynamical system u that is generated by (3),

$$\begin{aligned} \dot{x}(t) &= f(x(t)), \qquad t \in \mathfrak{R}^+, \\ x(0) &= x_0 \in \mathfrak{R}^n. \end{aligned} \tag{3}$$

Let us suppose that u has an equilibrium x_e [i.e., $f(x_e) = 0$] and that the matrix $[(\partial/\partial x_j) f_i]$ of partial derivatives is continuous in a neighborhood of x_e. Supposing that one wishes to investigate the stability of x_e, it is common in many branches of engineering (and particularly so in control theory) to linearize the problem, i.e., one defines

$$A = \left[\frac{\partial f_i}{\partial x_j}\right]_{x = x_e}, \qquad g(y) = f(x_e + y) - Ay,$$

and considers the "linearized" dynamical system $\hat{u}$ generated by

$$\dot{y}(t) = Ay(t), \qquad t \in \mathcal{R}^+,$$
$$y(0) = y_0 \in \mathcal{R}. \tag{11}$$

As $\hat{u}$ is linear, it is clearly easier to study the stability properties of the equilibrium of $\hat{u}$ at $y = 0$ than it is to perform a similar study of the equilibrium of u at x_e, and this is what has been done without further question in many areas of engineering. It is not often asked whether or not this is legitimate, i.e., do the equilibria of u and $\hat{u}$ have the *same* stability properties? In general, the answer is "not necessarily." All the linear theory for control system design relies on the following fortunate result.

Theorem 10.1. If u is generated by (3), $\hat{u}$ is generated by the linearized equation (11), and the equilibrium of $\hat{u}$ at $y = 0$ is asymptotically stable, then the equilibrium of u at x_e is asymptotically stable.

Proof. By Theorem 9.3 there exists a matrix solution $B = B^T > 0$ of $A^TB + BA = -I$, since $y = 0$ is an asymptotically stable equilibrium of $\hat{u}$. Define $\hat{V}(y) = \langle y, By \rangle$, using the Euclidean inner product and norm; then $\dot{\hat{V}}_{\hat{u}}(y) = -\langle y, y \rangle$ and $\|B^{-1}\|\ \|y\|^2 \leqslant \langle y, By \rangle \leqslant \|B\|\ \|y\|^2$ for all $y \in \mathcal{R}^n$. Hence $\hat{V}$ is a quadratic Liapunov function for $\hat{u}$ that, by Theorem 8.1, assures exponential stability of $y = 0$.

Now defining $V(x) = \hat{V}(x - x_e)$, we see that for u, as generated by (3),

$$\begin{aligned}\dot{V}_u(x) &= 2\langle x - x_e, Bf(x) \rangle \\ &= 2\langle x - x_e, BA(x - x_e) + Bg(x - x_e) \rangle \\ &\leqslant -\|x - x_e\|^2 + 2\|B\|\ \|x - x_e\|\ \|g(x - x_e)\|\end{aligned}$$

for all $x \in \mathcal{R}^n$. By the assumed continuity of $[(\partial/\partial x_j) f_i]$ in a neighborhood of x_e, it follows that $(\|g(y)\|/\|y\|) \to 0$ as $y \to 0$. Therefore, there exists a neighborhood $\mathcal{N}(x_e)$ of x_e such that $2\|B\|\ \|g(x - x_e)\| < \|x - x_e\|$ for all $x \in \mathcal{N}(x_e)$, and it follows that

$$\dot{V}_u(x) < 0 \qquad \text{for all} \qquad x \in \mathcal{N}(x_e) \setminus \{x_e\}.$$

Consequently, V is a Liapunov function for u on $\mathcal{N}(x_e)$ and, by Theorem

8.1, the equilibrium of u at x_e is asymptotically stable. The proof is complete. □

As linear control theory is, for several reasons, always used to design an asymptotically stable (linearized) control system, Theorem 10.1 is always applicable; consequently, most controllers designed by linearization actually do operate more or less as planned, for sufficiently small initial $\|x_0 - x_e\|$ (and small reference signals). However, were the linearized control system $\hat{u}$ designed only to give stability of $y = 0$, but not asymptotic stability, then the equilibrium x_e of u might very well be unstable!

The proof of the preceding theorem demonstrates a very useful and much used technique for constructing Liapunov functions, i.e., linearization about some equilibrium x_e of the nonlinear dynamical system u, construction of a quadratic Liapunov function for the linearized $\hat{u}$ by Liapunov's equation (10), and use of this function with the original dynamical system u. Often, the results can then be improved by some judicious nonquadratic modification of the function so obtained. Just as a given dynamical system u usually admits many different Liapunov functions, a given function V is usually a Liapunov function for many "similar" dynamical systems. Even though few dynamical systems are linear, the importance of Liapunov's equation (10) should now be apparent.

Exercise 10.1. View the equation

$$\ddot{z}(t) + \dot{z}(t) + z(t) - (z(t))^3 + (z(t))^5 = 0, \qquad t \in \mathfrak{R}^+,$$
$$z(0) = z_0 \in \mathfrak{R}, \qquad \dot{z}(0) = v_0 \in \mathfrak{R},$$

in the form (3),

$$\dot{z}(t) = v(t),$$
$$\dot{v}(t) = -v(t) - z(t) + (z(t))^3 - (z(t))^5, \qquad t \in \mathfrak{R}^+,$$
$$(z(0), v(0)) = (z_0, v_0) \in \mathfrak{R}^2.$$

It can be shown that this equation generates a dynamical system $u: \mathfrak{R}^+ \times \mathfrak{R}^2 \to \mathfrak{R}^2$.

(a) By linearization about the equilibrium at $x = (z, v) = (0, 0)$, obtain a quadratic Liapunov function on some neighborhood $\mathfrak{N}(0)$, and investigate the stability properties of $x = 0$.

(b) Noting that the total energy is a Liapunov function on all of $\mathfrak{R}^2$, are all motions bounded? Using this function, as well as the preceding one, can you find the *domain of attraction* for $x = 0$, $\mathfrak{D} \equiv \{x \in \mathfrak{R}^n \mid u(t, x) \to 0 \text{ as } t \to \infty\}$? Can you locate any subset of $\mathfrak{D}$? Discuss what you suspect, versus what you are now able to prove.

11. The Invariance Principle and Asymptotic Behavior

We have used Liapunov functions to obtain information about bounded motions and stability of equilibria of dynamical systems; a result by J. P. LaSalle (Refs. 5, 12), called the Invariance Principle, often allows us to make much better use of a given Liapunov function. This is very important, for good Liapunov functions are often like good men (hard to find). The Invariance Principle provides information about the asymptotic behavior of motions as $t \to \infty$, whether or not there are equilibria. At the same time, it strengthens our ability to investigate the stability properties of equilibria by allowing us to use Liapunov functions that need not satisfy all hypotheses of Theorem 8.1 or 8.2. A strong form of the Invariance Principle is proved in Chapter IV (see Theorem IV.4.2); here we provide (without proof) a simpler and weaker form of this important and very useful result.

Definition 11.1. Let u be a dynamical system on $\mathcal{R}^n$ and let $\mathcal{M} \subset \mathcal{R}^n$. The subset $\mathcal{M}$ is *positive invariant* under u if $x \in \mathcal{M}$ implies $u(t, x) \in \mathcal{M}$ for all $t \in \mathcal{R}^+$.

We note that both $\mathcal{R}^n$ and the empty set are positive invariant, a set consisting only of equilibria is positive invariant, and $\bigcup_{0 \leqslant t \leqslant T} u(t, x)$ is positive invariant if the motion $u(\cdot, x)$ is periodic with period T. Unions and intersections of positive invariant sets are positive invariant.

Theorem 11.1. (Invariance Principle). Let $V: \mathcal{R}^n \to \mathcal{R}$ be a Liapunov function for the dynamical system $u: \mathcal{R}^+ \times \mathcal{R}^n \to \mathcal{R}$ on a bounded subset $\mathcal{G} \subset \mathcal{R}^n$ with closure $\bar{\mathcal{G}}$. If $x \in \mathcal{G}$, then either there exists $t > 0$ such that $u(t, x) \notin \mathcal{G}$, or $u(t, x) \to \mathcal{M}$ as $t \to \infty$, where $\mathcal{M}$ is the largest positive invariant subset of $\{x \in \bar{\mathcal{G}} \mid \dot{V}(x) = 0\}$. (If $\mathcal{M}$ is empty, the first alternative must hold.)

In Chapter IV we shall prove a stronger version of this result. As an example of how it might be used, suppose that $V: \mathcal{R}^n \to \mathcal{R}$ is a Liapunov function on a bounded subset $\mathcal{G}$. Suppose further that $\mathcal{G}$ contains a disjoint component $\mathcal{G}_\alpha$ of the set $\{x \in \mathcal{R}^n \mid V(x) < \alpha\}$, or of $\{x \in \mathcal{R}^n \mid V(x) \leqslant \alpha\}$, for some $\alpha \in \mathcal{R}$. Then, by the remarks following Definition 8.4, $\mathcal{G}_\alpha$ is positive invariant, and Theorem 11.1 therefore implies that $u(t, x) \to \mathcal{M}_\alpha$ as $t \to \infty$ for every $x \in \mathcal{G}_\alpha$, where $\mathcal{M}_\alpha$ is the largest positive invariant subset of $\{x \in \bar{\mathcal{G}}_\alpha \mid \dot{V}(x) = 0\}$ and $\bar{\mathcal{G}}_\alpha$ is the closure of $\mathcal{G}_\alpha$. If u is generated by (3), and the set $\{x \in \bar{\mathcal{G}}_\alpha \mid \dot{V}(x) = 0\}$ is known, we can then use (3) to determine $\mathcal{M}_\alpha$.

Example 11.1. Consider Exercise 10.1, where

$$\dot{z}(t) = v(t),$$
$$\dot{v}(t) = -v(t) - z(t) + (z(t))^3 - (z(t))^5, \qquad t \in \mathfrak{R}^+,$$
$$(z(0), v(0)) = (z_0, v_0) = x_0 \in \mathfrak{R}^2.$$

Defining $V(x) = v^2 + z^2 - z^4/2 + z^6/3$, $x = (z, v) \in \mathfrak{R}^2$, we see that $\dot{V}(x) = -2v^2$ for the dynamical system u generated on $\mathfrak{R}^2$. Defining $\bar{\mathcal{G}}_\alpha \equiv \{x \in \mathfrak{R}^n \,|\, V(x) \leqslant \alpha\}$, it is apparent that $\bar{\mathcal{G}}_\alpha$ is bounded for every $\alpha \in \mathfrak{R}$, since $V(x) \to \infty$ as $\|x\| \to \infty$. Note that

$$\{x \in \bar{\mathcal{G}}_\alpha \,|\, \dot{V}(x) = 0\} = \{x = (z, v) \in \mathfrak{R}^2 \,|\, v = 0,\, z^2 - z^4/2 + z^6/3 \leqslant \alpha\}.$$

Hence, if $x_0 = (z_0, v_0) \in \mathfrak{M}_\alpha$, the largest positive invariant subset of $\{x \in \bar{\mathcal{G}}_\alpha \,|\, \dot{V}(x) = 0\}$, we must have $v_0 = 0$ and $u(t, x_0) \equiv (z(t), 0)$ for all $t \in \mathfrak{R}^+$, where $z(t)$ satisfies the generating equation with $v(t) \equiv 0$ on $\mathfrak{R}^+$. This implies that $z(t) \equiv z_0$ and $z_0 - z_0^3 + z_0^5 = 0$, for which the only real root is $z_0 = 0$. Therefore, $\mathfrak{M}_\alpha = \{0\}$ for every $\alpha \geqslant 0$, and $u(t, x) \to 0$ for every $x \in \mathfrak{R}^n$, since every x is contained in some $\bar{\mathcal{G}}_\alpha$ for α sufficiently large. We have shown that u has a globally asymptotically stable equilibrium at $x = 0$, a conclusion that could not be obtained from Theorem 8.1 on the basis of this Liapunov function.

Example 11.2. Let us again consider the linear vibration problem of Example 8.2,

$$\dot{x}(t) = Ax(t), \qquad t \in \mathfrak{R}^+, \quad A \equiv \begin{bmatrix} 0 & I \\ -M^{-1}K & -M^{-1}C \end{bmatrix},$$
$$x(0) = (y_0, v_0) \in \mathfrak{R}^{2n},$$

where M: $\mathfrak{R}^n \to \mathfrak{R}^n$, K: $\mathfrak{R}^n \to \mathfrak{R}^n$, are symmetric and positive definite, while C: $\mathfrak{R}^n \to \mathfrak{R}^n$ is symmetric and nonnegative ($C \geqslant 0$). We recall that, even under the assumption $C > 0$, we were able to prove only that $x = 0$ is stable; we suspected but could not prove that $x = 0$ is asymptotically stable.

Using the Liapunov function

$$V(x) = \tfrac{1}{2}\langle v, Mv\rangle_{\mathfrak{R}^n} + \tfrac{1}{2}\langle y, Ky\rangle_{\mathfrak{R}^n}, \qquad x = (y, v) \in \mathfrak{R}^{2n},$$

and recalling that $\dot{V}(x) = -\langle v, Cv\rangle$ for all $x = (y, v) \in \mathfrak{R}^n$, let us now employ the Invariance Principle, assuming only that $C \geqslant 0$. Defining the closed and bounded set $\bar{\mathcal{G}}_\alpha \equiv \{x \in \mathfrak{R}^{2n} \,|\, V(x) \leqslant \alpha\}$, some $\alpha \in \mathfrak{R}^+$, we note that

$$\{x \in \bar{\mathcal{G}}_\alpha \,|\, \dot{V}(x) = 0\} = \{x = (y, v) \in \mathfrak{R}^{2n} \,|\, Cv = 0,\, \langle y, Ky\rangle_{\mathfrak{R}^n} + \langle v, \mathfrak{M}v\rangle_{\mathfrak{R}^n} \leqslant 2\alpha\},$$

where we have used the facts that C is symmetric and nonnegative.

To find the largest positive invariant set $\mathfrak{M}_\alpha$ in $\bar{\mathcal{G}}_\alpha$, we note that if $x_0 = (y_0, v_0) \in \mathfrak{M}_\alpha$, we must have $Cv_0 = 0$ and $u(t, x_0) = (y(t), v(t))$ for all $t \in \mathcal{R}^+$, where $(y(t), v(t))$ satisfies the generating equation with $Cv(t) \equiv 0$ on $\mathcal{R}^+$. Hence, $C\dot{v}(t) \equiv 0$, and this implies that $CM^{-1}Ky(t) + CM^{-1}Cv(t) \equiv 0$, which in turn implies that $CM^{-1}Ky(t) \equiv 0$ on $\mathcal{R}^+$. Continuing in this manner, we find that we must have $C(M^{-1}K)^m v(t) \equiv 0$ for $m = 1, 2, \ldots,$ if $x_0 = (y_0, v_0) \in \mathfrak{M}_\alpha$. Suppose that the $n \times n^2$ matrix $[C, (KM^{-1})C, \ldots, (KM^{-1})^{n-1}C]$ has rank n [i.e., that the pair $(C, M^{-1}K)$ is *completely observable*]; then all these conditions can be satisfied only if $v(t) \equiv 0$, and from the generating equation it follows that $y(t) \equiv 0$. Consequently, if the pair $(C, M^{-1}K)$ is completely observable, $\mathfrak{M}_\alpha = \{0\}$; and from Theorem 11.1 it follows that $x = 0$ is asymptotically stable. By linearity, this implies that $x = 0$ is globally exponentially stable. It is not too difficult to show that this complete observability condition is necessary as well as sufficient for asymptotic stability (Ref. 19). Note that $C > 0$ is sufficient but not necessary for $(C, M^{-1}K)$ to be completely observable.

Exercise 11.1. Consider a system described by

$$\ddot{y}(t) + h(\dot{y}(t)) + g(y(t)) = 0, \qquad t \in \mathcal{R}^+,$$

$$y(0) = y_0 \in \mathcal{R}, \qquad \dot{y}(0) = v_0 \in \mathcal{R}.$$

All that is known about the functions h and g is that they are everywhere locally Lipschitz continuous, $yg(y) > 0$ for every $y \neq 0$, $vh(v) > 0$ for every $v \neq 0$, and

$$\int_0^y g(y)\,dy \to \infty \qquad \text{as} \qquad |y| \to \infty.$$

It can be shown that no solution has finite escape time. Making use of the total energy function, what can you determine about (a) local existence and uniqueness of solutions? (b) equilibrium solutions? (c) generation of a dynamical system u on $\mathcal{R}^2$? (d) unbounded solutions? (e) periodic solutions? (f) behavior of solutions as t increases?

Exercise 11.2. Consider the linear dynamical system u on $\mathcal{R}^3$ that is generated by (4) with

$$A = \begin{bmatrix} 0 & 0 & -1 \\ 0 & 2 & 0 \\ 4 & 0 & -3 \end{bmatrix}.$$

(a) Construct a quadratic Liapunov function V having $\dot{V}(x) \equiv -\langle x, x \rangle$, and show that $x = 0$ is an unstable equilibrium.

(b) Prove that if $V(x_0) < 0$, then $u(t, x_0) \to \infty$ as $t \to \infty$.

(c) If x_0 is such that $V(x_0) > 0$, does this imply that $u(t, x_0) \to 0$ as $t \to \infty$?

Exercise 11.3. Consider the linear dynamical system on $\mathcal{R}^2$ that is generated by (4) with

$$A = \begin{bmatrix} 0 & 1 \\ -1 & 1 \end{bmatrix}.$$

Defining $V(x) = -\|x\|^2$, where $\|\cdot\|$ is the Euclidean norm, show that V is a

Liapunov function on $\mathcal{R}^2$; then using only this function and Theorem 11.1, show that $x = 0$ is unstable. Note that V does *not* satisfy all the conditions of Theorem 8.2.

So far we have considered only a few of the uses of the Invariance Principle; that it has other uses is shown by the following example, admittedly contrived.

Example 11.3. Consider the equation

$$\begin{bmatrix} \dot{v}(t) \\ \dot{y}(t) \end{bmatrix} = \begin{bmatrix} v(t) - [v(t) + y(t)]f(v(t), y(t)) \\ v(t)f(v(t), y(t)) \end{bmatrix}, \qquad t \in \mathcal{R}^+,$$

$$(v(0), y(0)) = (v_0, y_0) \in \mathcal{R}^2,$$

where $f(v, y) \equiv v^2 + 2y^2$. By Theorem 3.1, no solution has finite escape time; hence, by Theorem 2.3, this equation generates a dynamical system u on $\mathcal{R}^n$.

Consider the function V: $\mathcal{R}^2 \to \mathcal{R}$ defined by $V(x) = y^2 + v^2$ for $x = (v, y) \in \mathcal{R}^2$; then $\dot{V}(x) = -2v^2[v^2 + 2y^2 - 1]$ and V is a Liapunov function for u on the set $\{(v, y) \in \mathcal{R}^2 | v^2 + 2y^2 \geqslant 1\}$. Choosing some $\beta > 1$, let us consider the bounded annulus $\mathcal{G}_\beta \equiv \{(v, y) \in \mathcal{R}^2 | \beta > v^2 + y^2 \geqslant 1\}$ and note that V is a Liapunov function on $\mathcal{G}_\beta$ with $\{x \in \bar{\mathcal{G}}_\beta | \dot{V}(x) = 0\} = \{(v, y) \in \bar{\mathcal{G}}_\beta |$ either $v = 0$, or $v^2 + 2y^2 = 1$ and $v^2 + y^2 \geqslant 1\} = \{(v, y) \in \bar{\mathcal{G}}_\beta |$ either $v = 0$, or $y = 0$ and $v^2 = 1\}$. Using the evolution equation, we see that this set contains no (nonempty) positive invariant subset; therefore, for each $x_0 \in \mathcal{G}_\beta$, Theorem 11.1 implies the existence of $T(x_0) > 0$ such that $u(T, x_0) \notin \mathcal{G}_\beta$. As $V(u(\cdot, x_0))$ is nonincreasing so long as $u(t, x_0)$ remains in $\mathcal{G}_\beta$, it follows that the motion can leave $\mathcal{G}_\beta$ only by crossing the boundary defined by $v^2 + y^2 = 1$; moreover, having left $\mathcal{G}_\beta$, it cannot reenter. Noting that $\beta > 1$ was arbitrary, these conclusions apply for every x_0 such that $v_0^2 + y_0^2 \geqslant 1$.

Now consider $V^*(x) \equiv -V(x) = -y^2 - v^2$ for $x = (v, y) \in \mathcal{R}^2$, and note that $\dot{V}^*(x) = -2v^2[1 - v^2 - 2y^2]$. Therefore, V^* is a Liapunov function for u on the set $\{(v, y) \in \mathcal{R}^2 | v^2 + 2y^2 \leqslant 1\}$, and we consider the subset $\mathcal{S} \equiv \{(v, y) \in \mathcal{R}^2 | 2v^2 + 2y^2 \leqslant 1\}$. We see that $\{x \in \bar{\mathcal{S}} | \dot{V}^*(x) = 0\} = \{(v, y) \in \mathcal{S} |$ either $v = 0$, or $v^2 + 2y^2 = 1$ and $2v^2 + 2y^2 \leqslant 1\} = \{(v, y) \in \mathcal{R}^2 | v = 0,\ 2y^2 \leqslant 1\}$ and the largest positive invariant subset is $\{(0, 0)\}$. Employing the Invariance Principle and reasoning as before, we see that $x = 0$ is an unstable equilibrium, and for each nonzero $x_0 \in \mathcal{S}$ there exists $T(x_0)$ such that $u(t, x_0) \notin \mathcal{S}$ for any $t \geqslant T(x_0)$.

Summing up our conclusions, if x_0 is in the annulus $\mathcal{Q} \equiv \{(y, v) \in \mathcal{R}^2 | 2 > 2v^2 + 2y^2 > 1\}$, then $u(t, x_0) \in \mathcal{Q}$ for all $t \in \mathcal{R}^+$; if $x_0 \notin \mathcal{Q}$, there exists $T(x_0) > 0$ such that $u(t, x_0) \in \mathcal{Q}$ for all $t \geqslant T(x_0)$. As $\mathcal{Q}$ contains no equilibria, the Poincaré–Bendixson theorem (Ref. 2) can be used to show that $\mathcal{Q}$ contains at least one periodic motion, and that every motion eventually approaches a periodic motion contained in $\mathcal{Q}$.

Exercise 11.4. In the equation of Example 11.3, replace $f(v, y)$ by $v^2 + \alpha y^2$, arbitrary $\alpha > 0$, $\alpha \neq 1$, and repeat the analysis of that example. State your conclusions in terms of some annulus $\mathcal{Q}_\alpha$. Also do a careful analysis for the case $\alpha = 1$, and discuss the corresponding set $\mathcal{Q}_1$.

12. Comments and Extensions

Under the assumptions that $f: \mathcal{R}^n \to \mathcal{R}^n$ is everywhere locally Lipschitz continuous and that there exists an inner product such that $\|x\| > \beta$ implies $\langle x, f(x)\rangle \leqslant \alpha\langle x, x\rangle$, for some $\alpha < \infty$, $\beta < \infty$, we have been able to use Theorems 2.3 and 3.1 to show that the evolution equation (3),

$$\dot{x}(t) = f(x(t)), \qquad t \in \mathcal{R}^+,$$
$$x(0) = x_0 \in \mathcal{R}^n, \tag{3}$$

does generate a dynamical system on $\mathcal{R}^n$. In Section 2 we noticed that Theorem 2.3 also provides (local) backward existence and uniqueness of solutions, conclusions that are unneeded for a dynamical system and untrue for certain physical systems. We now show that, by strengthening the condition of Theorem 3.1, we can weaken the Lipschitz continuity requirement to simple continuity and still be sure that (3) generates a dynamical system.

Theorem 12.1. Let $f: \mathcal{R}^n \to \mathcal{R}^n$ be everywhere continuous and such that, for some inner product $\langle \cdot, \cdot \rangle$ and some $\omega \in \mathcal{R}$, $\omega I - f$ is *monotone*, i.e.,

$$\langle x - y, (\omega I - f)x - (\omega I - f)y\rangle \geqslant 0 \quad \text{for all} \quad x, y \in \mathcal{R}^n.$$

Then equation (3) generates a dynamical system $u: \mathcal{R}^+ \times \mathcal{R}^n \to \mathcal{R}^n$; further, in terms of the compatible norm, $\|u(t, x) - u(t, y)\| \leqslant e^{\omega t}\|x - y\|$ for all $x, y \in \mathcal{R}^n$, $t \in \mathcal{R}^+$. Furthermore, if $\omega < 0$, then u has at most one equilibrium.

Exercise 12.1. Using Theorems 2.1, 2.2, and 3.1, as well as the fact that for differentiable $g(\cdot)$: $[0, b) \to \mathcal{R}^m$,

$$2\langle g(t), \dot{g}(t)\rangle = \frac{d}{dt}\langle g(t), g(t)\rangle = \frac{d}{dt}\| g(t)\|^2$$

for $t \in [0, b)$, prove Theorem 12.1. Note that $\dot{g}(0)$ denotes the right derivative at $t = 0$.

Compared with our previous method for showing that (3) generates a dynamical system, Theorem 12.1 has advantages and disadvantages.

Example 12.1. Consider (3) on $\mathcal{R}^1$, with $f(x) \equiv -x^{1/2}$ if $x \geqslant 0$, $f(x) \equiv 0$ if $x < 0$. Clearly, f is continuous and $(\omega I - f)$: $\mathcal{R} \to \mathcal{R}$ is monotone for any nonnegative choice of ω, but f is not Lipschitz continuous at $x = 0$. Although Theorem

2.3 does not apply, Theorem 12.1 assures us that (3) generates a dynamical system, i.e., for every $x_0 \in \mathfrak{R}$, there exists a unique differentiable $x(\cdot)$: $\mathfrak{R}^+ \to \mathfrak{R}$ that satisfies (3), and $x(t)$ depends continuously on x_0 at each $t \in \mathfrak{R}^+$. Note that this is Example 2.2 with time reversed.

Exercise 12.2. Consider (3) on $\mathfrak{R}^2$, with

$$\begin{bmatrix} \dot{z}(t) \\ \dot{y}(t) \end{bmatrix} = \begin{bmatrix} -y(t)\left[(z(t))^2 + (y(t))^2\right] \\ z(t)\left[(z(t))^2 + (y(t))^2\right] \end{bmatrix}, \qquad t \in \mathfrak{R}^+,$$

$$(z(0), y(0)) = (z_0, y_0) \in \mathfrak{R}^2.$$

Exercise 3.3 shows that this equation does generate a dynamical system on $\mathfrak{R}^2$. Show that $(\omega I - f)$: $\mathfrak{R}^2 \to \mathfrak{R}^2$ is not monotone for any $\omega \in \mathfrak{R}$, relative to any choice of the inner product on $\mathfrak{R}^2$. Hence, Theorem 12.1 does not apply.

A result similar to Theorem 12.1 can also be obtained for the generation of a process by equation (2),

$$\dot{x}(t) = f(x(t), t), \qquad t \geqslant t_0 \in \mathfrak{R},$$
$$x(t_0) = x_0 \in \mathfrak{R}^n. \tag{2}$$

Theorem 12.2. Let f: $\mathfrak{R}^n \times \mathfrak{R} \to \mathfrak{R}^n$ be everywhere continuous and such that, for some inner product $\langle \cdot, \cdot \rangle$ and some continuous $\omega(\cdot)$: $\mathfrak{R} \to \mathfrak{R}$, $\omega(t)I - f(\cdot, t)$ is monotone for each $t \in \mathfrak{R}$. Then equation (2) generates a process u: $\mathfrak{R}^+ \times \mathfrak{R}^n \times \mathfrak{R} \to \mathfrak{R}$; moreover, in terms of the compatible norm, $\|u(\tau, x, t_0) - u(\tau, y, t_0)\| \leqslant \exp\{h(\tau, t_0)\}\|x - y\|$ for all $x, y \in \mathfrak{R}^n$, $\tau \in \mathfrak{R}^+$, $t_0 \in \mathfrak{R}$, where

$$h(\tau, t_0) \equiv \int_0^\tau \omega(s + t_0)\, ds.$$

Exercise 12.3. Referring to your proof of Theorem 12.1, prove Theorem 12.2.

Theorems 12.1 and 12.2 provide a means of relating an evolution equation to a dynamical system (or process) that is closely related to the C_0-semigroup methods that we shall discuss in Chapter III; there, results similar to Theorem 12.1 will be obtained with $\mathfrak{R}^n$ replaced by more general spaces, and the continuity condition on f will be replaced by a (weaker) condition involving the range of $I - \lambda f$ for $\lambda > 0$, leading to a somewhat weaker relationship between the evolution equation and the corresponding dynamical system.

Exercise 12.4. Let f: $\mathcal{R}^1 \to \mathcal{R}^1$ be such that $\omega I - f$ is monotone for some $\omega \in \mathcal{R}$. Show that f is continuous if and only if $I - \lambda f$ has range $\mathcal{R}^1$ for all positive $\lambda < 1/|\omega|$; that is, given any $y \in \mathcal{R}^1$, the equation $x - \lambda f(x) = y$ has a solution $x \in \mathcal{R}^1$. Do you think this result holds with $\mathcal{R}^1$ replaced by $\mathcal{R}^n$?

The proof of Theorem 12.1 involves showing that the monotonicity condition of Theorem 12.1 implies satisfaction of the condition of Theorem 3.1 for nonexistence of solutions with finite escape times. Unfortunately, the condition of Theorem 3.1 is quite strong; in several of the examples and exercises of Sections 8, 10, 11, we simply asserted without proof that no solutions have finite escape times. To prove such assertions we need a generalization of Theorem 3.1.

Theorem 12.3. Let V: $\mathcal{R}^n \times \mathcal{R} \to \mathcal{R}$ be continuously differentiable on $\mathcal{R}^n \times \mathcal{R}$, $V(x, t) \to +\infty$ as $\|x\| \to \infty$ at each $t \in [t_0, \infty)$, and let there exist continuous functions $\alpha(\cdot)$: $\mathcal{R} \to \mathcal{R}$, $\beta(\cdot)$: $\mathcal{R} \to \mathcal{R}^+$, such that $V(x, t) > \beta(t)$ and $t \geq t_0$ imply

$$\frac{\partial}{\partial t} V(x, t) + \langle \nabla_x V(x, t), f(x, t) \rangle \leq \alpha(t) V(x, t).$$

Then no solution $x(\cdot)$: $[t_0, b) \to \mathcal{R}^n$ of equation (2) has finite escape time.

Proof. As $x(\cdot)$: $[t_0, b) \to \mathcal{R}^n$ is some solution of (2), it is differentiable on $[t_0, b)$ (right-differentiable at $t = t_0$); as V was assumed continuously differentiable, we see that $(d/dt)V(x(t), t)$ exists on $[t_0, b)$. Defining $g(\cdot)$: $[t_0, b) \to \mathcal{R}$ by

$$g(t) \equiv \exp\left\{ -\int_{t_0}^{t} |\alpha(s)|\, ds \right\} V(x(t), t),$$

we see that $g(\cdot)$ is differentiable and, for $t \in [t_0, b)$, either $g(t) \leq \beta(t)$ or

$$\begin{aligned} \frac{d}{dt} g(t) &= -|\alpha(t)|\, g(t) + \exp\left\{ -\int_{t_0}^{t} |\alpha(s)|\, ds \right\} \\ &\qquad \times \left[\frac{\partial}{\partial t} V(x(t), t) + \langle \nabla_x V(x(t), t), \dot{x}(t) \rangle \right] \\ &\leq (\alpha(t) - |\alpha(t)|) g(t) \leq 0. \end{aligned}$$

Therefore, $g(t) \leq \max[g(0), \beta(t)]$ on $[t_0, b)$, and it follows that $V(x(t), t)$ is bounded above on bounded subintervals of $[t_0, b)$. Consequently, given any bounded sequence $\{t_p\} \subset [t_0, b)$, the remaining condition—that

$V(x, t) \to \infty$ as $\|x\| \to \infty$, each $t \in [t_0, \infty)$—shows that $\|x(t_p)\|$ is bounded, and the proof is complete. □

The unproved assertion in each of the aforementioned examples and exercises can now be proved by using the function V mentioned in each specific case. For example, reconsider Example 11.1 and define $V: \mathcal{R}^n \to \mathcal{R}$ as there; then $V(x) \to \infty$ as $\|x\| \to \infty$ and $\langle \nabla V(x), f(x) \rangle = -2v^2 \leq 0$ for all $x \in \mathcal{R}^n$. Therefore, Theorem 12.3 applies and we reach the *a priori* conclusion that no solution of the equation can have finite escape time.

More dramatically than Theorem 3.1, Theorem 12.3 demonstrates the usefulness of *differential inequalities* (Ref. 2) in studying the behavior of solutions of differential equations; the approach is essentially an extension of the Liapunov approach, for we do not require $(d/dt)V(x(t), t) \leq 0$. There are other uses to which such differential inequalities can be put, as Propositions 12.1 and 12.2 demonstrate.

Proposition 12.1. Let all eigenvalues of the $n \times n$ matrix $A: \mathcal{R}^n \to \mathcal{R}^n$ have negative real parts, i.e., let (10) have a solution $B = B^T > 0$ for $C = I$. Let all elements of the real $n \times n$ matrix $\hat{A}(t)$ be continuous on $\mathcal{R}$, and let there exist $\gamma \leq 1$ such that $2\langle x, B\hat{A}(t)x \rangle \leq \gamma \langle x, x \rangle$ for all $t \geq T$, $x \in \mathcal{R}^n$. Then the process generated by

$$\dot{x}(t) = Ax(t) + \hat{A}(t)x(t), \qquad t \geq t_0 \in \mathcal{R},$$
$$x(t_0) = x_0 \in \mathcal{R}^n, \tag{12}$$

has an equilibrium at $x = 0$ that is stable relative to $t_0 \geq T$; if $\gamma < 1$, then $x = 0$ is exponentially stable relative to t_0.

Exercise 12.5. Prove Proposition 12.1. We say that an equilibrium x_e of a process u is exponentially stable relative to t_0 if there exists $\gamma > 0$, $\alpha > 0$, $M < \infty$, such that $\|x - x_e\| < \gamma$ implies $\|u(\tau, x, t_0) - x_e\| \leq Me^{-\alpha\tau}\|x - x_e\|$ for all $\tau \in \mathcal{R}^+$. *Hint:* First show that

$$\frac{d}{dt}[e^{(1-\gamma)t}\langle x(t), Bx(t) \rangle] \leq 0$$

and apply Theorem 12.3.

Proposition 12.1, like the results of Section 10, pertains to the "structural stability" (Ref. 2) of the linear *equation* (4). The following result demonstrates a different but related idea.

Proposition 12.2. Let A, B, $\hat{A}(t)$, and γ be defined as in Proposition 12.1, with $\gamma < 1$, and let $g: \mathcal{R}^n \times \mathcal{R} \to \mathcal{R}^n$ be such that $\|Bg(x, t)\| \leq \delta$ for all $x \in \mathcal{R}^n$, $t \in \mathcal{R}$, and some $\delta < \infty$. If g is continuous, and Lipschitz continuous with respect to x, everywhere on $\mathcal{R}^n \times \mathcal{R}$, then the equation

$$\dot{x}(t) = Ax(t) + \hat{A}(t)x(t) + g(x(t), t), \qquad t \geqslant t_0 \in \mathcal{R},$$

$$x(t_0) = x_0 \in \mathcal{R}^n, \tag{13}$$

generates a process u on $\mathcal{R}^n \times \mathcal{R}$ and all motions are bounded. In addition, defining $\beta \equiv \sup\langle x, Bx\rangle$ over all x in $\mathcal{R}^n$ such that $\|x\| \leqslant 2\delta/(1-\gamma)(1-\varepsilon)$, some $\varepsilon \geqslant 0$, $\varepsilon < 1$, and $\mathcal{S}_\varepsilon \equiv \{x \in \mathcal{R}^n | \langle x, Bx\rangle \leqslant \beta\}$, then $x_0 \in \mathcal{S}_\varepsilon$ implies $u(\tau, x_0, t_0) \in \mathcal{S}_\varepsilon$ for all $\tau \in \mathcal{R}^+$, every $t_0 \in \mathcal{R}$ and $\varepsilon \geqslant 0$. Moreover, for every $x_0 \notin \mathcal{S}_0$ and $t_0 \in \mathcal{R}$, $u(\tau, x_0, t_0) \to \mathcal{S}_0$ as $\tau \to \infty$.

Exercise 12.6. Prove Proposition 12.2. *Hint:* Estimate $(d/dt)\langle x(t), Bx(t)\rangle$ and apply Theorem 12.3.

Conclusions of this type are sometimes described as "total stability" (Ref. 11) or "stability under persistent disturbances" (Ref. 12). Such terminology tends to be misleading; (13) may not even have an equilibrium x_e (necessarily, $x_e \in \mathcal{S}_0$), and there is no assurance that x_e would be stable.

Finally, recall that we have argued that if a physical system is not explosive and is described by a finite number of coordinates, and if all $(x_0, t_0) \in \mathcal{R}^n \times \mathcal{R}$ are physically possible initial conditions, then the motions of the physical system should have essentially those properties required of the motions of a process by Definition 4.1. Unfortunately, one of the caveats of this argument seems unrealistic; few physical systems actually admit *all* (x_0, t_0) in $\mathcal{R}^n \times \mathcal{R}$ as initial conditions. Considering only the autonomous case for simplicity, it may be that only some subset $\mathcal{X} \subset \mathcal{R}^n$ is such that all $x_0 \in \mathcal{X}$ are physically possible; but then this implies that, for the physical motion, $u(t, x_0) \in \mathcal{X}$ for all $t \in \mathcal{R}^+$! Consequently, one is led to consider the idea of a dynamical system u on a subset $\mathcal{X} \subset \mathcal{R}^n$, u: $\mathcal{R}^+ \times \mathcal{X} \to \mathcal{X}$, that otherwise has all the properties of Definition 4.2 (with x restricted to $\mathcal{X}$). If one defines a metric for $\mathcal{X}$ in terms of the norm on $\mathcal{R}^n$,

$$d(x, y) \equiv \|x - y\| \qquad \text{for} \quad x, y \in \mathcal{X},$$

then one has a dynamical system u on the *metric space* $\mathcal{X}$. Interestingly, the methods and results of Sections 8 and 11 remain valid.

A further generalization of this idea will be discussed in Chapter III, wherein the state x is no longer required to be in $\mathcal{R}^n$ and $\mathcal{X}$ is replaced by an abstract metric space. In Chapter IV we shall see that many of the foregoing qualitative methods and results also apply to an abstract dynamical system on a general metric space $\mathcal{X}$. In the next chapter we shall briefly review most of the mathematical concepts and terminology needed for such extensions.

Exercise 12.7. Noting the foregoing description of a dynamical system u on $\mathcal{X} \subset \mathcal{R}^n$, consider the equation discussed in Exercise 3.1,

$$\ddot{y}(t) + y(t) - y^3(t) = 0, \qquad t \in \mathcal{R}^+,$$

$$y(0) = y_0 \in \mathcal{R}, \qquad \dot{y}(0) = v_0 \in \mathcal{R}.$$

Show that this equation generates a dynamical system on

$$\mathcal{X} \equiv \{(y, v) \in \mathcal{R}^2 |\, |y| \leqslant 1, 2v^2 + 2y^2 - y^4 \leqslant 1\}.$$

Remark: It can be shown that for some $(y_0, v_0) \in \mathcal{R}^2$ the solution of this equation has finite escape time, so a dynamical system is *not* generated on all of $\mathcal{R}^2$.

References

1. Coddington, E. A., and Levinson, N., *Theory of Ordinary Differential Equations*, McGraw-Hill, New York, 1955.
2. Hale, J. K., *Ordinary Differential Equations*, Wiley-Interscience, New York, 1969.
3. Lefschetz, S., *Differential Equations: Geometric Theory*, Wiley-Interscience, New York, 1964.
4. Hartman, P., *Ordinary Differential Equations*, John Wiley and Sons, New York, 1963.
5. LaSalle, J. P., *The Stability of Dynamical Systems*, CBMS Regional Conference Series in Applied Mathematics, Society for Industrial and Applied Mathematics, Philadelphia, 1976.
6. Chernoff, P., and Marsden, J., On the continuity and smoothness of group actions, *Bulletin of the American Mathematical Society*, Vol. 76, pp. 1044–1049, 1970.
7. Hille, E., and Phillips, R. S., *Functional Analysis and Semi-groups*, American Mathematical Society Colloquium Publications, Vol. 31, Providence, Rhode Island, 1957.
8. Royden, H. L., *Real Analysis*, MacMillan Company, London, 1968.
9. Cesari, L., *Asymptotic Behavior and Stability Problems in Ordinary Differential Equations*, Springer-Verlag, Berlin, 1963.
10. Coppel, W. A., *Stability and Asymptotic Behavior of Differential Equations*, D. C. Heath and Co., Lexington, Massachusetts, 1965.
11. Hahn, W., *Theory and Application of Liapunov's Direct Method*, Prentice-Hall, Englewood Cliffs, New Jersey, 1963.
12. LaSalle, J. P., and Lefschetz, S., *Stability by Liapunov's Direct Method*, Academic Press, New York, 1961.
13. Lehnigk, S. H., *Stability Theorems for Linear Motions*, Prentice-Hall, Englewood Cliffs, New Jersey, 1966.
14. Liapunov, A. M., Probléme géneral de la stabilité du mouvement, *Annals of Mathematical Studies*, No. 17, Princeton University Press, Princeton, New Jersey, 1949 (reproduction of the French translation in 1907 of a Russian memoir dated 1892).

15. Massera, J. L., Contributions to stability theory, *Annals of Mathematics*, Vol. 64, pp. 182–206, 1956.
16. Yoshizawa, T., *Stability Theory by Liapunov's Second Method*, Mathematical Society of Japan, Tokyo, 1966.
17. Zubov, V. I., *Methods of A. M. Liapunov and Their Applications*, P. Noordhoff, Groningen, 1964.
18. Bellman, R., *Introduction to Matrix Analysis*, McGraw-Hill, New York, 1960.
19. Walker, J. A., and Schmitendorf, W. E., A simple test for asymptotic stability in partially dissipative systems, *ASME Journal of Applied Mechanics*, Vol. 40, pp. 1120–1121, 1973.

II

Preliminaries for Abstract Evolution Equations

In Chapter I we were concerned with a physical system whose time behavior could be described by an ordinary differential equation on the n-dimensional space $\mathcal{X} \equiv \mathcal{R}^n$,

$$\begin{aligned} \dot{x}(t) &= f(x(t), t), \qquad t \geq t_0 \in \mathcal{R}, \\ x(t_0) &= x_0 \in \mathcal{X}. \end{aligned} \tag{1}$$

We now wish to consider this type of equation with $\mathcal{X}$ denoting some type of abstract space. Roughly speaking, we hope to describe the evolution in time of almost *any* physical system by an evolution equation similar to (1); we hope to obtain much information about the time behavior of the system, this information being of the same type as we obtained in Chapter I for physical systems described by differential equations on $\mathcal{R}^n$. In order to do this, it will be necessary to define and briefly illustrate a considerable amount of mathematical language. Hence, it may be helpful to describe some of the problems that we hope to be able to discuss intelligently by using this language.

Heat conduction problems are normally described by partial differential equations of parabolic type; if η denotes position along an insulated rod held to zero temperature at the ends $\eta = 0$, $\eta = 1$, then an example is

$$\frac{\partial}{\partial t}\theta(\eta, t) = \frac{\partial^2}{\partial \eta^2}\theta(\eta, t), \qquad 0 \leq \eta \leq 1, \quad t \geq 0,$$

$$\theta(0, t) = 0 = \theta(1, t), \qquad t \geq 0,$$

with initial temperature data

$$\theta(\eta, 0) = \theta_0(\eta), \qquad 0 \leqslant \eta \leqslant 1.$$

Defining $\mathfrak{X}$ to be some space whose elements are functions $\theta(\cdot)$: $[0, 1] \to \mathfrak{R}$ satisfying $\theta(0) = 0 = \theta(1)$, we see that this equation is of the form (1) with $x(t) \equiv \theta(\cdot, t)$, $x_0 \equiv \theta_0(\cdot)$.

The motions of elastic continua are normally described by partial differential equations of hyperbolic type; if $y(\eta)$ denotes the deflection at position η along a string fixed at $\eta = 0$, $\eta = 1$, then an example is

$$\frac{\partial^2}{\partial t^2} y(\eta, t) - \frac{\partial^2}{\partial \eta^2} y(\eta, t) = 0, \qquad 0 \leqslant \eta \leqslant 1, \quad t \geqslant 0,$$

$$y(0, t) = 0 = y(1, t), \qquad t \geqslant 0,$$

with initial data

$$y(\eta, 0) = y_0(\eta), \quad \frac{\partial}{\partial t} y(\eta, 0) = v_0(\eta), \qquad 0 \leqslant \eta \leqslant 1.$$

If we define $v(\eta, t) \equiv (\partial/\partial t) y(\eta, t)$, the foregoing equation can be written in the form

$$\left(\frac{\partial}{\partial t} y(\eta, t), \frac{\partial}{\partial t} v(\eta, t) \right) = \left(v(\eta, t), \frac{\partial^2}{\partial \eta^2} y(\eta, t) \right), \qquad 0 \leqslant \eta \leqslant 1, \quad t \geqslant 0.$$

If $\mathfrak{X}$ denotes some space of pairs $(y(\cdot), v(\cdot))$ of functions $y(\cdot)$: $[0, 1] \to \mathfrak{R}$, $v(\cdot)$: $[0, 1] \to \mathfrak{R}$, satisfying $y(0) = 0 = y(1)$, $v(0) = 0 = v(1)$, then the equation takes the form (1) with $x_0 \equiv (y_0(\cdot), v_0(\cdot))$.

In certain finite-dimensional feedback control systems there are time delays in the transmission of the output measurements or of the control signal; such systems are described by differential–difference equations on $\mathfrak{R}^n$, of which an example is

$$\dot{y}(t) = Ay(t) + By(t - \tau), \qquad t \geqslant 0,$$

where $y(t) \in \mathfrak{R}^n$, A: $\mathfrak{R}^n \to \mathfrak{R}^n$, B: $\mathfrak{R}^n \to \mathfrak{R}^n$, $\tau > 0$, and the initial data consist of a "history function,"

$$y(-\eta) = u_0(\eta), \qquad -\tau \leqslant -\eta \leqslant 0.$$

Defining $u(\eta, t) \equiv y(t - \eta)$, we see that

$$\frac{\partial}{\partial t} u(\eta, t) = \dot{y}(t - \eta) = -\frac{\partial}{\partial \eta} u(\eta, t),$$

and the equation can be written as

$$\left(\dot{y}(t), \frac{\partial}{\partial t}u(\eta, t)\right) = \left(Ay(t) + Bu(\tau, t), -\frac{\partial}{\partial \eta}u(\eta, t)\right), \quad 0 \leqslant \eta \leqslant \tau, \quad t \geqslant 0,$$

with boundary condition $u(0, t) = y(t)$. Letting $\mathfrak{X}$ denote some space having elements of the form $x = (y, u(\cdot))$, where $y \in \mathfrak{R}^n$ and $u(\cdot)$: $[0, \tau] \to \mathfrak{R}^n$ with $u(0) = y$, the equation takes the form (1) with $x_0 \equiv (u_0(0), u_0(\cdot))$.

Some problems in economics and population dynamics involve "distributed memory" and can be described by functional differential equations on $\mathfrak{R}^n$; for example,

$$\dot{y}(t) = \int_0^\tau A(\eta)y(t - \eta)d\eta, \qquad t \geqslant 0,$$

where $y(t) \in \mathfrak{R}^n$, $A(\eta)$: $\mathfrak{R}^n \to \mathfrak{R}^n$ for $0 < \eta < \tau$, some $\tau > 0$. The initial data consist of some history function,

$$y(-\eta) = u_0(\eta), \qquad 0 \leqslant \eta \leqslant \tau.$$

Defining $u(\eta, t) \equiv y(t - \eta)$, we see that the equation takes the form

$$\left(\dot{y}(t), \frac{\partial}{\partial t}u(\eta, t)\right) = \left(\int_0^\tau A(\eta)u(\eta, t)\, d\eta, -\frac{\partial}{\partial \eta}u(\eta, t)\right), \qquad t \geqslant 0,$$

with boundary condition $u(0, t) = y(t)$. Defining $\mathfrak{X}$ as in the preceding example, the equation takes the form (1) with $x_0 \equiv (u_0(0), u_0(\cdot))$.

Many other such examples can be given (Ref. 1), since the time behavior of almost every physical system can be described by an abstract equation of the form (1); however, four basic questions must be answered: What is $\mathfrak{X}$, precisely? What is f, precisely? What do we mean by $\dot{x}(t)$? What do we mean by $=$? This chapter summarizes the material needed to answer these questions.

1. Abstract Spaces

A space is a set having a certain amount of "structure," topological and/or algebraic. Given a set $\mathfrak{X}$, a topological space is obtained by defining those subsets that are to be called *open sets*, according to certain rules (Refs. 2–4). The simplest way to induce a topology is to define a metric.

Definition 1.1. A *metric space* is a pair $(\mathcal{X}, d)$, $\mathcal{X}$ a set and d a metric. A mapping d: $\mathcal{X} \times \mathcal{X} \to \mathcal{R}$ is a *metric* on $\mathcal{X}$ provided that $d(x, y) = 0$ if and only if $x = y$, and $d(x, y) = d(y, x) \geqslant 0$, $d(x, y) + d(y, z) \geqslant d(x, z)$ for all $x, y, z \in \mathcal{X}$. The last condition is the *triangle inequality*.

For simplicity, a metric space $(\mathcal{X}, d)$ is usually called "the metric space $\mathcal{X}$." For $x \in \mathcal{X}$, $\mathcal{X}$ a metric space, the *open ball* about x with radius $r > 0$ is the set $\mathcal{S}_r(x) \equiv \{y \in \mathcal{X} | d(y, x) < r\}$. If $\mathcal{Q} \subset \mathcal{X}$, then $x \in \mathcal{Q}$ is an *interior point* of $\mathcal{Q}$ if $\mathcal{S}_r(x) \subset \mathcal{Q}$ for some $r > 0$. The subset $\mathcal{Q}$ is *open* if every $x \in \mathcal{Q}$ is an interior point of $\mathcal{Q}$; $\mathcal{Q}$ is *bounded* if $d(x, y) \leqslant \alpha$ for all $x, y \in \mathcal{Q}$, some $\alpha < \infty$. Given $x \in \mathcal{X}$, $\mathcal{N} \subset \mathcal{X}$, $\mathcal{N}$ is a *neighborhood* of x if $\mathcal{S}_r(x) \subset \mathcal{N}$ for some $r > 0$. A *sequence* in $\mathcal{X}$ is an ordered set $\{x_n\} \subset \mathcal{X}$ that can be put in correspondence with the positive real integers, $n = 1, 2, \ldots$. The sequence $\{x_n\}$ *converges* to $x_0 \in \mathcal{X}$ if, for each $\varepsilon > 0$, there exists $N(\varepsilon)$ such that $d(x_n, x_0) < \varepsilon$ for all $n > N(\varepsilon)$; in this case one writes $x_n \to x_0$ as $n \to \infty$, or $\lim_{n\to\infty} x_n = x_0$. A sequence $\{x_n\}$ is *Cauchy* if, for each $\varepsilon > 0$, there exists $N(\varepsilon)$ such that $d(x_m, x_n) < \varepsilon$ for all $m > N(\varepsilon)$, $n > N(\varepsilon)$.

Exercise 1.1. Given a Cauchy sequence $\{x_n\} \subset \mathcal{X}$, $\mathcal{X}$ a metric space, show that $\{x_n\}$ is bounded.

Exercise 1.2. Given a convergent sequence $\{x_n\} \subset \mathcal{X}$, $\mathcal{X}$ a metric space, show that $\{x_n\}$ is Cauchy. *Hint:* The triangle inequality gives $d(x_n, x_m) \leqslant d(x_n, x_0) + d(x_m, x_0)$, where $x_0 = \lim_{p\to\infty} x_p$.

Consider some metric space $\mathcal{X}$ and $\mathcal{Q} \subset \mathcal{X}$. If $\{x_n\} \subset \mathcal{Q}$ and $\lim_{n\to\infty} x_n = x_0 \in \mathcal{X}$, then x_0 is a *limit point* of $\mathcal{Q}$; $\mathcal{Q}$ is said to be *closed* if it contains all its limit points. The *closure* $\bar{\mathcal{Q}}$ of $\mathcal{Q}$ is the set of all limit points of $\mathcal{Q}$; hence, $\mathcal{Q} \subset \bar{\mathcal{Q}} \subset \mathcal{X}$, and $\mathcal{Q}$ is *closed* if and only if $\mathcal{Q} = \bar{\mathcal{Q}}$. The subset $\mathcal{Q}$ is said to be *dense* in $\mathcal{X}$ if $\bar{\mathcal{Q}} = \mathcal{X}$. $\mathcal{Q}$ is *precompact* if every sequence $\{x_n\} \subset \mathcal{Q}$ contains a Cauchy subsequence; $\mathcal{Q}$ is *compact* if every sequence $\{x_n\} \subset \mathcal{Q}$ contains a subsequence converging to a point in $\mathcal{Q}$.

Exercise 1.3. Consider precompact $\mathcal{Q} \subset \mathcal{X}$, $\mathcal{X}$ a metric space. Show that $\mathcal{Q}$ must be bounded.

Definition 1.2. A metric space $\mathcal{X}$ is *complete* if each Cauchy sequence $\{x_n\}$ converges (to some $x \in \mathcal{X}$).

Exercise 1.4. Consider $\mathcal{Q} \subset \mathcal{X}$, $\mathcal{X}$ a complete metric space. Show that $\mathcal{Q}$ is compact if and only if $\mathcal{Q}$ is precompact and closed.

Definition 1.3. A metric space $\mathcal{X}$ is *locally compact* if every closed and bounded subset is compact.

Exercise 1.5. It can be shown that $\mathcal{R}$ is complete and locally compact in terms of the metric induced by the absolute value (Refs. 3, 5–7). Consider $\mathcal{R}^n$ equipped with the metric $d(x, y) \equiv \|x - y\|$. Show that $\mathcal{R}^n$ is complete, and prove that the compact sets *are* the closed and bounded sets.

We now consider the simplest type of space having an algebraic (rather than a topological) structure. By a *field* $\mathcal{F}$ we shall mean either the real line $\mathcal{R}$ or the complex plane $\hat{\mathcal{C}}$, equipped with the usual rules for addition and multiplication.

Definition 1.4. A *linear space* is a quadruple $(\mathcal{X}, \mathcal{F}, +, \cdot)$, $\mathcal{X}$ a set, $\mathcal{F}$ a field, $+: \mathcal{X} \times \mathcal{X} \to \mathcal{X}$, $\cdot: \mathcal{F} \times \mathcal{X} \to \mathcal{X}$, such that

(i) if $x, y, z \in \mathcal{X}$, then $x + y = y + x \in \mathcal{X}$, $x + (y + z) = (x + y) + z$, there exists $0 \in \mathcal{X}$ such that $0 + x = x$, and there exists $(-x) \in \mathcal{X}$ such that $(-x) + x = 0$;

(ii) if $x, y \in \mathcal{X}$ and $\alpha, \beta \in \mathcal{F}$, then $\alpha \cdot (\beta \cdot x) = \alpha\beta \cdot x \in \mathcal{X}$, $1 \cdot x = x$, $\alpha \cdot (x + y) = \alpha x + \alpha y$, $(\alpha + \beta) \cdot x = \alpha \cdot x + \beta \cdot x$.

We say that $(\mathcal{X}, \mathcal{F}, +, \cdot)$ is "the linear space $\mathcal{X}$"; $\mathcal{X}$ is a *real* linear space if $\mathcal{F} = \mathcal{R}$ and a *complex* linear space if $\mathcal{F} = \hat{\mathcal{C}}$. If $\mathcal{Q} \subset \mathcal{X}$, $\mathcal{X}$ a linear space, and if $(\mathcal{Q}, \mathcal{F}, +, \cdot)$ is also a linear space for the same field $\mathcal{F}$, then $\mathcal{Q}$ is a *linear manifold* of $\mathcal{X}$.

Definition 1.5. Consider $x_1, x_2, \ldots, x_n \in \mathcal{X}$, $\mathcal{X}$ a linear space. We say that $x_1, x_2, \ldots, x_n$ are ***linearly independent*** if, for unknown $\alpha_1, \alpha_2, \ldots, \alpha_n \in \mathcal{F}$, the equation $0 = \sum_{i=1}^{n} \alpha_i x_i$ implies $\alpha_i = 0$ for $i = 1, 2, \ldots, n$. If $\mathcal{X}$ contains only a finite number N of linearly independent x_i, then $\mathcal{X}$ is *finite dimensional*, of *dimension* N; otherwise, $\mathcal{X}$ is *infinite dimensional*.

Definition 1.6. A ***norm*** is a mapping $\|\cdot\|: \mathcal{X} \to \mathcal{R}^+$, $\mathcal{X}$ a linear space, such that $\|x\| = 0$ if and only if $x = 0$, and $\|\alpha x\| = |\alpha| \cdot \|x\|$, $\|x + y\| \leqslant \|x\| + \|y\|$ for all $x, y \in \mathcal{X}$, $\alpha \in \mathcal{F}$. Equipped with a particular norm, $\mathcal{X}$ is a *normed linear space*.

Given a linear space $\mathcal{X}$, and some norm $\|\cdot\|_1: \mathcal{X} \to \mathcal{R}^+$, it is possible to define many other norms; for example, $\|x\|_2 \equiv \|\alpha x\|_1$ for some $\alpha > 0$. Two norms on the same linear space $\mathcal{X}$ are said to be ***equivalent*** if there exist $c_2 \geqslant c_1 > 0$ such that $c_2\|x\|_2 \geqslant \|x\|_1 \geqslant c_1\|x\|_2$ for all $x \in \mathcal{X}$. If $\mathcal{X}$ has finite dimension, it can be shown that all possible norms are equivalent (see Refs. 2, 3, 6, 7).

By equipping a linear space $\mathcal{X}$ with a norm $\|\cdot\|$ we obtain a normed linear space $(\mathcal{X}, \|\cdot\|)$, and this induces a metric space $(\mathcal{X}, d)$ upon defining $d(x, y) \equiv \|x - y\|$; this metric space in turn induces a topological space, since a specific meaning has now been assigned to the term "open set." Thus, the choice of the norm defines the ***topology*** of the space. In general, if two normed linear spaces $(\mathcal{X}, \|\cdot\|_1)$ and $(\mathcal{X}, \|\cdot\|_2)$ are defined in terms of the same linear space $\mathcal{X}$, they have different topologies; they have the *same* topology if and only if $\|\cdot\|_1$ and $\|\cdot\|_2$ are ***equivalent*** norms for $\mathcal{X}$. This means, for example, that a finite-dimensional space can be equipped with only one norm-induced topology; however, in general, an infinite-dimensional linear space can be equipped with many different norm-induced topologies, since there may exist many possible norms that are not equivalent.

Definition 1.7. A *Banach space* is a normed linear space that is complete in the metric induced by the norm.

If the linear space $\mathcal{X}$, equipped with the norm $\|\cdot\|_1$, is a Banach space, then $\|\cdot\|_1$ provides the strongest possible norm-induced topology for $\mathcal{X}$; that is, if $\|\cdot\|_2$ is any other norm such that $\|x\|_2 \geqslant c_1\|x\|_1$ for all $x \in \mathcal{X}$, some $c_1 > 0$, then there exists $c_2 > 0$ such that $\|x\|_2 \leqslant c_2\|x\|_1$ for all $x \in \mathcal{X}$ (see Refs. 2, 3).

We have discussed a means of adding topological structure to a linear space $\mathcal{X}$; we may wish instead to increase the algebraic structure of $\mathcal{X}$. One means of doing this is to consider an inner product on $\mathcal{X}$.

Definition 1.8. An *inner product space* is a pair $(\mathcal{X}, \langle\cdot,\cdot\rangle)$, $\mathcal{X}$ a linear space with field $\mathcal{F}$, and $\langle\cdot,\cdot\rangle$ an inner product. A mapping $\langle\cdot,\cdot\rangle: \mathcal{X}\times\mathcal{X}\to\mathcal{F}$ is an *inner product* on $\mathcal{X}$ if $\langle 0, 0\rangle = 0$, $\langle x, x\rangle > 0$ for $x \neq 0$, $\langle x, y\rangle = \overline{\langle y, x\rangle}$, $\langle x+y, z\rangle = \langle x, z\rangle + \langle y, z\rangle$, and $\langle \alpha x, y\rangle = \alpha\langle x, y\rangle$ for all $x, y, z \in \mathcal{X}$, $\alpha \in \mathcal{F}$, where $\bar{\beta}$ denotes the complex conjugate of $\beta \in \mathcal{F}$ when $\mathcal{F} = \hat{\mathcal{C}}$, $\bar{\beta} \equiv \beta$ when $\mathcal{F} = \mathcal{R}$.

Exercise 1.6. For $\mathcal{X}$ an inner product space, show that $\langle x, \alpha y\rangle = \bar{\alpha}\langle x, y\rangle$ for all $x, y \in \mathcal{X}$, $\alpha \in \mathcal{F}$.

Proposition 1.1. If $\mathcal{X}$ is an inner product space, then

$$|\langle x, y\rangle|^2 \leqslant \langle x, x\rangle\langle y, y\rangle \qquad \text{for all} \quad x, y \in \mathcal{X}.$$

Proof. For $x, y \in \mathcal{X}$ and arbitrary $\alpha \in \mathcal{R} \subset \mathcal{F}$,

$$\begin{aligned} 0 &\leqslant \langle x + \alpha\langle x, y\rangle y, x + \alpha\langle x, y\rangle y\rangle \\ &= \langle x, x\rangle + 2\alpha|\langle x, y\rangle|^2 + \alpha^2|\langle x, y\rangle|^2\langle y, y\rangle. \end{aligned}$$

Since a nonnegative quadratic function of a real variable α can be zero for at most one value of α, the discriminant

$$(2|\langle x, y\rangle|^2)^2 - 4\langle x, x\rangle|\langle x, y\rangle|^2\langle y, y\rangle$$

cannot be positive. The proof is complete. □

Exercise 1.7. For $\mathcal{X}$ an inner product space, define $\|x\| \equiv \langle x, x\rangle^{1/2}$ and show that $\|\cdot\|$ is a norm. The pair $(\mathcal{X}, \|\cdot\|)$ is called a *pre-Hilbert space*. Show that $|\langle x, y\rangle| < \|x\|\,\|y\|$ for all $x \in \mathcal{X}$; this is the *Schwarz inequality*.

Definition 1.9. A Hilbert space is a complete pre-Hilbert space.

Exercise 1.8. Given a normed linear space $\mathcal{X}$ (or Banach space $\mathcal{X}$), show that there exists an inner product such that $(\mathcal{X}, \langle\cdot, \cdot\rangle)$ is a pre-Hilbert space (or Hilbert space) if and only if the norm of $\mathcal{X}$ satisfies the *parallelogram law*,

$$\|x + y\|^2 + \|x - y\|^2 = 2\|x\|^2 + 2\|y\|^2$$

for all $x, y \in \mathcal{X}$. *Hint:* To show sufficiency, define

$$\langle x, y\rangle_1 \equiv \frac{1}{4}\|x + y\|^2 - \frac{1}{4}\|x - y\|^2 \qquad \text{if } \mathcal{X} \text{ is real,}$$

and

$$\langle x, y\rangle_2 \equiv \langle x, y\rangle_1 + \sqrt{(-1)}\langle x, \sqrt{(-1)}y\rangle_1 \qquad \text{if } \mathcal{X} \text{ is complex.}$$

Example 1.1. Let $\mathcal{R}^n = \{x = (\eta_1, \eta_2, \ldots, \eta_n) | \eta_k \in \mathcal{R}\}$ with field $\mathcal{F} \equiv \mathcal{R}$; then $\mathcal{R}^n$ is a real linear space. Defining a norm $\|x\| \equiv (\eta_1^2 + \eta_2^2 + \cdots + \eta_n^2)^{1/2}$, $(\mathcal{R}^n, \|\cdot\|)$ is a real normed linear space, implying that it is also a metric space with metric $d(x, y) \equiv \|x - y\|$. Clearly $(\mathcal{R}^n, \|\cdot\|)$ is complete and therefore a Banach space; it becomes a real Hilbert space upon defining an inner product as

$$\langle x, y\rangle \equiv \sum_{k=1}^{n} \eta_k \alpha_k, \qquad x = (\eta_1, \eta_2, \ldots, \eta_n) \in \mathcal{R}^n,$$

$$y = (\alpha_1, \alpha_2, \ldots, \alpha_n) \in \mathcal{R}^n.$$

We referred to $(\mathcal{R}^n, \|\cdot\|, \langle\cdot, \cdot\rangle)$ as a Euclidean space in Chapter I.

If we instead define the norm to be $|x| \equiv \sum_{k=1}^{n} |\eta_k|$ (or $|x| \equiv \max_k |\eta_k|$), then $(\mathcal{R}^n, |\cdot|)$ is a complete real normed linear space (hence, a complete metric space with $d(x, y) = |x - y|$), and it also becomes an inner product space upon defining $\langle x, y\rangle$ as before; however, $(\mathcal{R}^n, |\cdot|, \langle\cdot, \cdot\rangle)$ is not a Hilbert space since the inner product is not compatible with the norm ($\langle x, x\rangle \neq |x|^2$). Moreover, there exists no possible inner product under which $(\mathcal{R}^n, |\cdot|)$ becomes a Hilbert space, since $|\cdot|$

does not satisfy the parallelogram law. However, $(\mathcal{R}^n, |\cdot|)$ is norm-equivalent to the Hilbert space $(\mathcal{R}^n, \|\cdot\|, \langle\cdot,\cdot\rangle)$ since there exist numbers $c_2 \geq c_1 > 0$ such that $c_1\|x\| \leq |x| \leq c_2\|x\|$ for all $x \in \mathcal{R}^n$ (equivalently, $|x|/c_2 \leq \|x\| \leq |x|/c_1$). Hence, if we wish to use an inner product on $\mathcal{R}^n$, we are only inconveniencing ourselves if we do not use the norm that naturally corresponds to this inner product. Unless otherwise stated, $\mathcal{R}^n$ usually denotes the real Hilbert space $(\mathcal{R}^n, \|\cdot\|, \langle\cdot,\cdot\rangle)$.

Example 1.2. Let $\mathcal{R}^\infty$ denote the set of all sequences in $\mathcal{R}$, i.e., $x \in \mathcal{R}^\infty$ means that $x = (\alpha_1, \alpha_2, \ldots, \alpha_p, \ldots)$, $\alpha_p \in \mathcal{R}$, $p = 1, 2, \ldots$. Defining the field $\mathcal{F} \equiv \mathcal{R}$, $\mathcal{R}^\infty$ becomes a linear space. Defining

$$d(x, \hat{x}) \equiv \sum_{n=1}^{\infty} |\alpha_n - \hat{\alpha}_n|/2^n(1 + |\alpha_n - \hat{\alpha}_n|),$$

$\mathcal{R}^\infty$ becomes a complete linear metric space; however, no norm induces this metric and $\mathcal{R}^\infty$ is not a Banach space.

Consider the linear manifold $l_1 \subset \mathcal{R}^\infty$, defined as

$$l_1 \equiv \left\{ x \in \mathcal{R}^\infty \;\middle|\; \lim_{n\to\infty} \sum_{p=1}^{n} |\alpha_p| \text{ exists} \right\}.$$

Equipped with $\|x\|_1 \equiv \sum_{p=1}^{\infty} |\alpha_p|$, l_1 is a normed linear space and can be shown to be complete; hence, l_1 is a Banach space. Similarly, given any positive integer m, equip the linear manifold

$$l_m \equiv \left\{ x \in \mathcal{R}^\infty \;\middle|\; \lim_{n\to\infty} \sum_{p=1}^{n} |\alpha_p|^m \text{ exists} \right\} \qquad \text{with } \|x\|_m \equiv \left(\sum_{p=1}^{\infty} |\alpha_p|^m \right)^{1/m}.$$

Completeness can be demonstrated, and l_m is a Banach space.

Note that l_2 becomes a Hilbert space when equipped with the inner product $\langle x, \hat{x} \rangle \equiv \lim_{n\to\infty} \sum_{p=1}^{n} \alpha_p \hat{\alpha}_p$; however, for $m \neq 2$, $\|\cdot\|_m$ does not satisfy the parallelogram law and there does not exist an inner product under which l_m becomes a Hilbert space. It can be shown that, as sets, $l_1 \supset l_2 \supset l_3 \supset \cdots$, and l_m is a *proper* subset of l_q for $m > q \geq 1$; moreover, for $m > q$, $\|\cdot\|_m$ is not equivalent to $\|\cdot\|_q$ on l_m, but there does exist $c > 0$ such that $\|x\|_m \geq c\|x\|_q$ for all $x \in l_m$.

Example 1.3. Let $\mathcal{C}[0, 1]$ denote the set of continuous real-valued functions defined on the closed interval $[0, 1] \subset \mathcal{R}$ (this implies that these functions are finite valued at $0, 1 \in \mathcal{R}$). Choosing the field $\mathcal{F} \equiv \mathcal{R}$, $\mathcal{C}[0, 1]$ becomes a real linear space; also defining the norm

$$\|x\| \equiv \max|x(s)| \text{ over } s \in [0, 1], \qquad \text{for } x \in \mathcal{C}[0, 1],$$

$(\mathcal{C}, \|\cdot\|)$ is a normed linear space that can be shown to be complete (Refs. 2, 3, 5). Hence, $(\mathcal{C}, \|\cdot\|)$ is a Banach space. The space $(\mathcal{C}, \|\cdot\|)$ cannot be equipped with any inner product for which it becomes a Hilbert space, since $\|\cdot\|$ does not satisfy the parallelogram law. In fact, there exists no equivalent norm that satisfies the parallelogram law. Even so, we could define an inner product for $\mathcal{C}[0, 1]$ by

$$\langle x, y\rangle \equiv \int_0^1 x(\eta)y(\eta)\, d\eta, \qquad x, y \in \mathcal{C}[0, 1],$$

and $(\mathcal{C}, \langle\cdot, \cdot\rangle)$ is an inner product space.

If, instead, we were to equip the real linear space $\mathcal{C}[0, 1]$ with the weaker norm

$$|x| \equiv \left(\int_0^1 (x(\eta))^2\, d\eta\right)^{1/2}, \qquad x \in \mathcal{C}[0, 1],$$

then $(\mathcal{C}, |\cdot|, \langle\cdot, \cdot\rangle)$ is a pre-Hilbert space with $\langle\cdot, \cdot\rangle$ defined as before; however, it is not complete in the metric induced by $|\cdot|$ (i.e., $d(x, y) \equiv |x - y|$) and, hence, it is neither a Hilbert space nor a Banach space. As we prefer to deal with complete spaces, $\mathcal{C}[0, 1]$ will normally refer to the Banach space $(\mathcal{C}, \|\cdot\|)$.

Exercise 1.9. In Example 1.3 show that $\|\cdot\|$ and $|\cdot|$ are norms and that there exists $c > 0$ such that $\|x\| \geqslant c|x|$ for all $x \in \mathcal{C}[0, 1]$; also, show that $\langle\cdot, \cdot\rangle$ is an inner product on $\mathcal{C}[0, 1]$.

Example 1.4. Let $\mathcal{C}^n[0, 1]$ denote the set of continuous real-valued functions having n continuous derivatives on the closed interval $[0, 1] \subset \mathcal{R}$, $0 \leqslant n \leqslant \infty$. Upon defining $\mathcal{F} = \mathcal{R}$ we obtain a real linear space; if $n < \infty$ and we define

$$\|x\| \equiv \sum_{k=0}^{n} \max_{s\in[0, 1]} |x^{(k)}(s)|,$$

we obtain a Banach space that we shall usually also denote by $\mathcal{C}^n[0, 1]$. Clearly, as sets, $\mathcal{C}^n[0, 1] \subset \mathcal{C}^m[0, 1]$ for $0 \leqslant m < n \leqslant \infty$, with $\mathcal{C}^0[0, 1] \equiv \mathcal{C}[0, 1]$, and it is easily shown that the set $\mathcal{C}^n[0, 1]$ is $\|\cdot\|_{\mathcal{C}^m}$-dense in the Banach space $\mathcal{C}^m[0, 1]$ for finite $m \geqslant 0$ and $m < n \leqslant \infty$.

Example 1.5. Let $\mathcal{L}_2(0, 1)$ denote the set of real-valued functions defined on $[0, 1]$ that are Lebesgue square-integrable. Letting $\mathcal{F} = \mathcal{R}$ and

$$\|x\| \equiv \left(\int_0^1 (x(\eta))^2 d\eta\right)^{1/2}, \qquad x \in \mathcal{L}_2(0, 1),$$

$(\mathcal{L}_2, \|\cdot\|)$ is a real normed linear space that can be shown to be complete; hence, it is a Banach space and becomes a Hilbert space upon defining

$$\langle x, y\rangle \equiv \int_0^1 x(\eta)y(\eta)\, d\eta, \qquad x, y \in \mathcal{L}_2(0, 1).$$

Unless otherwise indicated, this Hilbert space will also be denoted by $\mathcal{L}$ ′0, 1). Strictly speaking, $\mathcal{L}_2(0, 1)$ is not a space of functions. An element of $\mathcal{L}_2(0, 1)$ is an *equivalence class* of functions that are equal (pointwise) almost everywhere on (0, 1).

As sets, for any positive integer n,

$$\mathcal{C}^\infty[0, 1] \subset \mathcal{C}^n[0, 1] \subset \mathcal{C}[0, 1] \subset \mathcal{L}_2(0, 1).$$

As it can be shown that the set $\mathcal{C}^\infty[0, 1]$ is $\|\cdot\|_{\mathcal{L}_2}$-dense in the Hilbert space $\mathcal{L}_2(0, 1)$, it follows that the same is true of every $\mathcal{C}^n[0, 1]$, $n = 0, 1, 2, \ldots$. Hence, another way of defining $\mathcal{L}_2(0, 1)$ is to let it be a *completion* of $\mathcal{C}^\infty[0, 1]$ (or $\mathcal{C}^n[0, 1]$, any $n = 0, 1, \ldots$) in $\|\cdot\|_{\mathcal{L}_2}$ (see Refs. 3, 4, 6). Henceforth, the symbol $\int$ will denote Lebesgue integration.

Example 1.6. Consider $x \in \mathcal{L}_2(0, 1)$ and let the function $\partial^n x$ denote the nth *generalized derivative* of x. That is, if $x \in \mathcal{C}^n[0, 1] \subset \mathcal{L}_2(0, 1)$, then $\partial^n x(\eta) = (d^n/d\eta^n)x(\eta)$; if not, then $\partial^n x$ is a function defined on $[0, 1]$ such that

$$\int_0^1 \partial^n x(\eta) \cdot \varphi(\eta)\, d\eta = (-1)^n \int_0^1 x(\eta) \cdot \frac{d^n}{d\eta^n} \varphi(\eta)\, d\eta$$

for every $\varphi \in \mathcal{C}^n[0, 1]$ such that φ and its first $n - 1$ derivatives are zero at $\eta = 0$ and $\eta = 1$. Defining $\mathcal{W}_2^n(0, 1) \equiv \{x \in \mathcal{L}_2(0, 1) | \partial^m x \in \mathcal{L}_2(0, 1),\ m = 1, 2, \ldots, n\}$ for finite n, $\mathcal{F} = \mathcal{R}$, and

$$\|x\| \equiv \left(\int_0^1 \left[(x(\eta))^2 + (\partial x(\eta))^2 + \cdots + (\partial^n x(\eta))^2\right] d\eta\right)^{1/2}, \qquad x \in \mathcal{W}_2^n(0, 1),$$

then $(\mathcal{W}_2^n(0, 1), \|\cdot\|)$ is a real normed linear space known as the *Sobolev space* $\mathcal{W}_2^n(0, 1)$. It is known to be complete and, equipped with the inner product

$$\langle x, y \rangle \equiv \int^1 [x(\eta) \cdot y(\eta) + \cdots + \partial^n x(\eta) \cdot \partial^n y(\eta)]\, d\eta, \qquad x, y \in \mathcal{W}_2^n(0, 1),$$

it is a Hilbert space. As sets, $\mathcal{L}_2(0, 1) = \mathcal{W}_2^0(0, 1) \supset \mathcal{W}_2^1(0, 1) \supset \mathcal{W}_2^2(0, 1) \supset \cdots$, and $\mathcal{W}_2^n(0, 1) \supset \mathcal{C}^n[0, 1]$ for $n = 0, 1, 2, \ldots$. Since it can be shown that $\mathcal{C}^\infty[0, 1]$ is $\|\cdot\|_{\mathcal{W}_2^n}$-dense in $\mathcal{W}_2^n[0, 1]$, it follows that the same is true of $\mathcal{C}^m[0, 1]$ for every $m \geqslant n$.

Finally, we remark that there are some very important results called the Sobolev embedding theorems (Refs. 1, 4, 8); a result of particular interest is that $\mathcal{W}_2^n(0, 1) \subset \mathcal{C}^{n-1}[0, 1]$, which means that for $n \geqslant 1$, every function in (an equivalence class of functions in) $\mathcal{W}_2^n(0, 1)$ is (modulo a set of measure zero in $[0, 1]$) a continuous function that has $n - 1$ continuous derivatives. Later we shall expand this idea by equipping $\mathcal{W}_2^n(0, 1)$ with "boundary conditions"; e.g., we may define

$$\mathcal{X} \equiv \{x \in \mathcal{W}_2^2(0, 1) | x(0) = 0 = \partial x(1)\},$$

by which we will mean that $\mathcal{X}$ is the $\|\cdot\|_{\mathcal{W}_2^2}$-completion of the subset in $\mathcal{C}^2[0, 1]$ that satisfies $|x(0) = 0 = \partial x(1)$.

Exercise 1.10. Let $(\mathcal{X}, \mathcal{F})$ and $(\mathcal{Y}, \mathcal{F})$ be linear spaces having the same field $\mathcal{F}$. Define $\mathcal{X} \times \mathcal{Y} \equiv \{(x, y) | x \in \mathcal{X}, y \in \mathcal{Y}\}$. Show that the *product space* $(\mathcal{X} \times \mathcal{Y}, \mathcal{F})$ is a linear space.

Exercise 1.11. Let $(\mathcal{X}, \mathcal{F}, \|\cdot\|_{\mathcal{X}})$ and $(\mathcal{Y}, \mathcal{F}, \|\cdot\|_{\mathcal{Y}})$ be normed linear spaces having the same field $\mathcal{F}$. Show that both $(\mathcal{X} \times \mathcal{Y}, \mathcal{F}, \|\cdot\|_1)$ and $(\mathcal{X} \times \mathcal{Y}, \mathcal{F}, \|\cdot\|_2)$ are normed linear spaces, where

$$\|(x, y)\|_1 \equiv \|x\|_{\mathcal{X}} + \|y\|_{\mathcal{Y}}, \qquad (x, y) \in \mathcal{X} \times \mathcal{Y},$$

$$\|(x, y)\|_2 \equiv \left(\|x\|_{\mathcal{X}}^2 + \|y\|_{\mathcal{Y}}^2\right)^{1/2}, \qquad (x, y) \in \mathcal{X} \times \mathcal{Y}.$$

Show that these norms are equivalent and, therefore, that they induce the same topology on $\mathcal{X} \times \mathcal{Y}$.

Exercise 1.12. Let $(\mathcal{X}, \mathcal{F}, \|\cdot\|_{\mathcal{X}}, \langle\cdot, \cdot\rangle_{\mathcal{X}})$ and $(\mathcal{Y}, \mathcal{F}, \|\cdot\|_{\mathcal{Y}}, \langle\cdot, \cdot\rangle_{\mathcal{Y}})$ be Hilbert spaces having the same field $\mathcal{F}$. Show that $(\mathcal{X} \times \mathcal{Y}, \mathcal{F}, \|\cdot\|, \langle\cdot, \cdot\rangle)$ is a Hilbert space, where

$$\|(x, y)\| \equiv \left(\|x\|_{\mathcal{X}}^2 + \|y\|_{\mathcal{Y}}^2\right)^{1/2},$$

$$\langle(x, y), (\hat{x}, \hat{y})\rangle \equiv \langle x, \hat{x}\rangle_{\mathcal{X}} + \langle y, \hat{y}\rangle_{\mathcal{Y}},$$

for all pairs $(x, y), (\hat{x}, \hat{y}) \in \mathcal{X} \times \mathcal{Y}$.

We should mention that our metric space definition of compactness is usually called *sequential compactness*, while a subset of a topological space is said to be *compact* if every open covering contains a finite subclass that forms a covering; an open covering of a subset is a class of open sets whose union contains that subset. For metric spaces the two definitions are equivalent (Refs. 3, 4), and this fact is often useful when dealing with functions.

2. Functions

The terms function, mapping, operator, and transformation are synonomous and are used according to taste.

Definition 2.1. Given sets $\mathcal{X}$ and $\mathcal{Y}$, a ***function*** F, from (or on) a subset $\mathcal{D}(F) \subset \mathcal{X}$ to (or into) the set $\mathcal{Y}$, is a rule that assigns to each $x \in \mathcal{D}(F)$ a unique $y \in \mathcal{Y}$. This is denoted by $F: (\mathcal{D}(F) \subset \mathcal{X}) \to \mathcal{Y}$.

Given a function $F: (\mathcal{D}(F) \subset \mathcal{X}) \to \mathcal{Y}$, $\mathcal{X}$ and $\mathcal{Y}$ sets, $\mathcal{D}(F)$ is the ***domain*** of F and $\{y \in \mathcal{Y} \mid y = Fx, \text{ some } x \in \mathcal{D}(F)\}$ is the ***range***, $\mathcal{R}(F) \subset \mathcal{Y}$. If $\mathcal{R}(F) = \mathcal{Y}$, then F is ***onto***. If $Fx = F\hat{x}$ only for $x = \hat{x}$, then F is ***one-to-one***; in this case there exists an ***inverse function*** $F^{-1}: (\mathcal{R}(F) \subset \mathcal{Y}) \to \mathcal{X}$ such that $F^{-1}Fx = x$ for all $x \in \mathcal{D}(F) = \mathcal{R}(F^{-1})$ (and $FF^{-1}y = y$ for all $y \in \mathcal{D}(F^{-1}) = \mathcal{R}(F)$). The set $\{(x, y) \in \mathcal{X} \times \mathcal{Y} \mid x \in \mathcal{D}(F), y = Fx\}$ is the *graph* of F.

The *identity* on $\mathcal{X}$ is the function I: $\mathcal{X} \to \mathcal{X}$ defined by $Ix \equiv x$ for every $x \in \mathcal{X}$. Given two functions F: $(\mathcal{D}(F) \subset \mathcal{X}) \to \mathcal{Y}$, G: $(\mathcal{D}(G) \subset \mathcal{X}) \to \mathcal{Y}$, such that $D(F) \subset \mathcal{D}(G)$ and $Fx = Gx$ for every $x \in \mathcal{D}(F)$, F is a *restriction* of G and G is an *extension* of F. If $\mathcal{X}$ is a linear space with field $\mathcal{F}$, a function F: $\mathcal{X} \to \mathcal{F}$ is a *functional* on $\mathcal{X}$; when $\mathcal{R}(F) \subset \mathcal{R}$, F is a *real functional*.

Consider functions F: $(\mathcal{D}(F) \subset \mathcal{X}) \to \mathcal{Y}$, G: $(\mathcal{D}(G) \subset \mathcal{X}) \to \mathcal{Y}$, $\mathcal{X}$ a set and $\mathcal{Y}$ a linear space; then we define the *sum* $(F + G)$: $(\mathcal{D}(F + G) \subset \mathcal{X}) \to \mathcal{Y}$ by $(F + G)x \equiv Fx + Gx$ for all $x \in \mathcal{D}(F + G) \equiv \mathcal{D}(F) \cap \mathcal{D}(G)$. If F: $(\mathcal{D}(F) \subset \mathcal{X}) \to \mathcal{Y}$ and G: $(\mathcal{D}(G) \subset \mathcal{Y}) \to \mathcal{Z}$, where $\mathcal{X}$, $\mathcal{Y}$, $\mathcal{Z}$ are sets, then the *composite* GF: $(\mathcal{D}(GF) \subset \mathcal{X}) \to \mathcal{Z}$ is defined by $(GF)x \equiv G(Fx)$ for all $x \in \mathcal{D}(GF) \equiv \{x \in \mathcal{D}(F) | Fx \in \mathcal{D}(G)\}$.

Definition 2.2. Let $\mathcal{X}$, $\mathcal{Y}$ be linear spaces having the same field $\mathcal{F}$. A function F: $(\mathcal{D}(F) \subset \mathcal{X}) \to \mathcal{Y}$ is *linear* if $\mathcal{D}(F)$ is a linear manifold and $F(\alpha x + \beta \hat{x}) = \alpha Fx + \beta F\hat{x}$ for all $x, \hat{x} \in \mathcal{D}(F)$ and $\alpha, \beta \in \mathcal{F}$.

Exercise 2.1. Let $\mathcal{X}$, $\mathcal{Y}$ be linear spaces and let F: $(\mathcal{D}(F) \subset \mathcal{X}) \to \mathcal{Y}$ be linear. Show that $\mathcal{R}(F)$ is a linear manifold.

Exercise 2.2. Given linear spaces $\mathcal{X}$, $\mathcal{Y}$, and linear operators F: $(\mathcal{D}(F) \subset \mathcal{X}) \to \mathcal{Y}$, G: $(\mathcal{D}(G) \subset \mathcal{X}) \to \mathcal{Y}$, show that $F + G$ is linear.

Exercise 2.3. Given linear spaces $\mathcal{X}$, $\mathcal{Y}$, $\mathcal{Z}$, and linear operators F: $(\mathcal{D}(F) \subset \mathcal{X}) \to \mathcal{Y}$, G: $(\mathcal{D}(G) \subset \mathcal{Y}) \to \mathcal{Z}$, show that GF is linear.

Definition 2.3. Let F: $(\mathcal{D}(F) \subset \mathcal{X}) \to \mathcal{Y}$, $\mathcal{X}$ and $\mathcal{Y}$ metric spaces. F is *continuous at* $x_0 \in \mathcal{D}(F)$ if, for every $\varepsilon > 0$, there exists $\delta(x_0, \varepsilon)$ such that $d_{\mathcal{X}}(x, x_0) < \delta$ implies $d_{\mathcal{Y}}(Fx, Fx_0) < \varepsilon$ (provided that $x \in \mathcal{D}(F)$); equivalently, F is continuous at x_0 if, for every sequence $\{x_n\} \subset \mathcal{D}(F)$, converging to x_0, the sequence $\{Fx_n\} \subset \mathcal{Y}$ converges to Fx_0.

If a function F: $(\mathcal{D}(F) \subset \mathcal{X}) \to \mathcal{Y}$, $\mathcal{X}$ and $\mathcal{Y}$ metric spaces, is continuous at every $x_0 \in \mathcal{D}(F)$, it is called a *continuous function*; it is *uniformly continuous* on $\mathcal{D}(F)$ if $\delta(\varepsilon, x_0)$ can be chosen independent of $x_0 \in \mathcal{D}(F)$. It can be shown that if $\mathcal{S} \subset \mathcal{D}(F)$ is compact and F is continuous on $\mathcal{S}$, then F is uniformly continuous on $\mathcal{S}$ (Refs. 2, 3, 6). The following result, proved in Ref. 3, is of some interest.

Theorem 2.1. Let F: $(\mathcal{D}(F) \subset \mathcal{X}) \to \mathcal{Y}$ be uniformly continuous on $\mathcal{D}(F)$, $\mathcal{X}$ and $\mathcal{Y}$ metric spaces. If $\mathcal{D}(F)$ is dense and $\mathcal{X}$ is complete, then F admits a unique continuous extension $\tilde{F}$: $\mathcal{X} \to \mathcal{Y}$; furthermore, $\tilde{F}$ is uniformly continuous on $\mathcal{X}$.

If a function has the same domain space and range space ($\mathscr{X} = \mathscr{Y}$, algebraically and topologically), then the following idea is related to (and implies) uniform continuity.

Definition 2.4. A function F: $(\mathscr{D}(F) \subset \mathscr{X}) \to \mathscr{X}$, $\mathscr{X}$ a metric space, is a *contraction* if $\mathscr{R}(F) \subset \mathscr{D}(F)$ and, for some real $\alpha < 1$, $d(Fx, F\hat{x}) \leqslant \alpha d(x, \hat{x})$ for all $x, \hat{x} \in \mathscr{D}(F)$.

The following result, proved in Ref. 3, is called a *fixed-point theorem.*

Theorem 2.2. Let F: $\mathscr{X} \to \mathscr{X}$ be a contraction, $\mathscr{X}$ a complete metric space. Then there exists a unique $x_0 \in \mathscr{X}$ such that $Fx_0 = x_0$.

It is apparent that a contraction is uniformly continuous; in fact, it is uniformly Lipschitz continuous if $\mathscr{X}$ is a normed linear space.

Definition 2.5. A function F: $(\mathscr{D}(F) \subset \mathscr{X}) \to \mathscr{Y}$, $\mathscr{X}$ and $\mathscr{Y}$ normed linear spaces, is ***Lipschitz continuous at*** $x_0 \in \mathscr{D}(F)$ if there exists a real $\alpha(x_0)$ such that $\|Fx - Fx_0\|_{\mathscr{Y}} \leqslant \alpha\|x - x_0\|_{\mathscr{X}}$ for all $x \in \mathbb{S}_r(x_0)$, some $r > 0$.

A function F: $(\mathscr{D}(F) \subset \mathscr{X}) \to \mathscr{Y}$, $\mathscr{X}$ and $\mathscr{Y}$ metric spaces, is said to be *Lipschitz continuous on* $\mathbb{S} \subset \mathscr{D}(F)$ if it is Lipschitz continuous at each $x_0 \in \mathbb{S}$; F is *uniformly Lipschitz continuous on* $\mathbb{S} \subset \mathscr{D}(F)$ if $\alpha(x_0)$ can be chosen independent of $x_0 \in \mathbb{S}$, and in this case it can be shown that $\|Fx - F\hat{x}\|_{\mathscr{Y}} \leqslant \alpha\|x - \hat{x}\|_{\mathscr{X}}$ for all $x, \hat{x} \in \mathbb{S}$. If $\mathbb{S}$ is precompact, Lipschitz continuity on $\mathbb{S}$ implies uniform Lipschitz continuity on $\mathbb{S}$.

Definition 2.6. A function F: $(\mathscr{D}(F) \subset \mathscr{X}) \to \mathscr{Y}$, $\mathscr{X}$ and $\mathscr{Y}$ metric spaces, is ***bounded*** (respectively, *compact*) if F maps bounded sets in $\mathscr{D}(F)$ into bounded sets (respectively, precompact sets) in $\mathscr{Y}$.

Exercise 2.4. Consider a function F: $(\mathscr{D}(F) \subset \mathscr{X}) \to \mathscr{X}$, $\mathscr{X}$ a normed linear space. Show that F is uniformly Lipschitz continuous on $\mathscr{D}(F)$ if and only if βF is a contraction for some $\beta > 0$.

Exercise 2.5. Let F: $(\mathscr{D}(F) \subset \mathscr{X}) \to \mathscr{Y}$ be *linear*, $\mathscr{X}$ and $\mathscr{Y}$ normed linear spaces.

(a) Show that F is bounded if and only if

$$\sup_{\substack{\|x\|_{\mathscr{X}} \leqslant 1 \\ 0 \neq x \in \mathscr{D}(F)}} \|Fx\|_{\mathscr{Y}} / \|x\|_{\mathscr{X}} = \alpha < \infty.$$

(b) Assuming that F is bounded and defining $\|F\|_{\mathscr{B}(\mathscr{X}, \mathscr{Y})} \equiv \alpha$, defined in (a), show that

$$\sup_{0 \neq x \in \mathcal{D}(F)} \|Fx\|_{\mathcal{Y}} / \|x\|_{\mathcal{X}} = \|F\|_{\mathcal{B}(\mathcal{X}, \mathcal{Y})} = \sup_{\substack{\|x\|_{\mathcal{X}} = 1 \\ x \in \mathcal{D}(F)}} \|Fx\|_{\mathcal{Y}}.$$

(c) Show that F is continuous if and only if F is bounded.

(d) Show that if F is continuous at one $x_0 \in \mathcal{D}(F)$, it is continuous, uniformly continuous, and uniformly Lipschitz continuous on $\mathcal{D}(F)$.

For normed linear spaces $\mathcal{X}$ and $\mathcal{Y}$ having the same field $\mathcal{F}$, consider the set $\mathcal{B}(\mathcal{X}, \mathcal{Y})$ of bounded linear operators with domain $\mathcal{X}$ and range in $\mathcal{Y}$. In terms of the field $\mathcal{F}$, $\mathcal{B}(\mathcal{X}, \mathcal{Y})$ is a linear space and, equipped with $\|\cdot\|_{\mathcal{B}(\mathcal{X}, \mathcal{Y})}$ as defined in Exercise 2.5, it becomes a normed linear space. If $\mathcal{Y}$ is a Banach space (i.e., if $\mathcal{Y}$ is complete), then so is $\mathcal{B}(\mathcal{X}, \mathcal{Y})$ (Refs. 2, 3, 6).

Exercise 2.6. Consider a *linear* $F: (\mathcal{D}(F) \subset \mathcal{X}) \to \mathcal{Y}$; $\mathcal{X}$ and $\mathcal{Y}$ are normed linear spaces.

(a) Show that F is one-to-one (invertible) if and only if $\|Fx\|_{\mathcal{Y}} = 0$ implies $x = 0$.

(b) Show that F has a bounded inverse if and only if $\|Fx\|_{\mathcal{Y}} \geqslant \alpha\|x\|_{\mathcal{X}}$ for all $x \in \mathcal{D}(F)$, some $\alpha > 0$. Verify that $\|Fx\|_{\mathcal{Y}} \geqslant \|x\|_{\mathcal{X}} / \|F^{-1}\|_{\mathcal{B}(\mathcal{Y}, \mathcal{X})}$ for all $x \in \mathcal{D}(F)$.

We see that for linear $F: (\mathcal{D}(F) \subset \mathcal{X}) \to \mathcal{Y}$, $\mathcal{X}$ and $\mathcal{Y}$ normed linear spaces, the concepts of boundedness, continuity at a point, continuity on $\mathcal{D}(F)$, uniform continuity, and uniform Lipschitz continuity are equivalent. For general $F: (\mathcal{D}(F) \subset \mathcal{X}) \to \mathcal{Y}$, $\mathcal{X}$ and $\mathcal{Y}$ metric spaces, these ideas have much less in common.

In Chapter III we shall be dealing with some operators that have none of these very nice properties; there we shall use the following idea.

Definition 2.7. A function $F: (\mathcal{D}(F) \subset \mathcal{X}) \to \mathcal{X}$, $\mathcal{X}$ a normed linear space, is *accretive* if, for every real $\lambda > 0$,

$$\|(I + \lambda F)x - (I + \lambda F)\hat{x}\| \geqslant \|x - \hat{x}\| \qquad \text{for all } x, \hat{x} \in \mathcal{D}(F).$$

Exercise 2.7. Show that the condition of Definition 2.7 holds for all $\lambda > 0$ if it holds for all positive $\lambda < \lambda_0$, some $\lambda_0 > 0$. *Hint:* For $0 < \lambda < \lambda_0$ and $\alpha \geqslant 1$, consider $\|(I + \alpha\lambda F)x - (I + \alpha\lambda F)\hat{x}\|$.

If $\mathcal{X}$ is a pre-Hilbert space, there is a parallel concept.

Definition 2.8. A function $F: (\mathcal{D}(F) \subset \mathcal{X}) \to \mathcal{X}$, $\mathcal{X}$ a pre-Hilbert space, is *monotone* if $\mathrm{Re}\langle Fx - F\hat{x}, x - \hat{x}\rangle \geqslant 0$ for all $x, \hat{x} \in \mathcal{D}(F)$.

Proposition 2.1. If $\mathcal{X}$ is a pre-Hilbert space, then $F: (\mathcal{D}(F) \subset \mathcal{X}) \to \mathcal{X}$ is monotone if and only if it is accretive.

Proof. We see that for $\lambda > 0$ and $x, \hat{x} \in \mathcal{D}(F)$,

$$\begin{aligned}\|(I + \lambda F)x - (I + \lambda F)\hat{x}\|^2 &= \langle x - \hat{x} + \lambda(Fx - F\hat{x}), x - \hat{x} + \lambda(Fx - F\hat{x})\rangle \\ &= \|x - \hat{x}\|^2 + 2\lambda \operatorname{Re}\langle Fx - F\hat{x}, x - \hat{x}\rangle \\ &\quad + \lambda^2\|Fx - F\hat{x}\|^2.\end{aligned}$$

Therefore, monotonicity implies accretiveness; conversely, if F is accretive, this equality implies that

$$0 \leqslant 2 \operatorname{Re}\langle Fx - F\hat{x}, x - \hat{x}\rangle + \lambda\|Fx - F\hat{x}\|^2.$$

Therefore, letting $\lambda \searrow 0$, it follows that F is monotone and the proof is complete. □

For accretive $F: (\mathcal{D}(F) \subset \mathcal{X}) \to \mathcal{X}$, $\mathcal{X}$ a normed linear space, it is apparent that $I + \lambda F$ is one-to-one for $\lambda > 0$; therefore, there exists an inverse $(I + \lambda F)^{-1}: (\mathcal{R}(I + \lambda F) \subset \mathcal{X}) \to \mathcal{X}$ for each $\lambda > 0$, and $(\lambda I + F)^{-1}$ is *nonexpansive* since

$$\|y - \hat{y}\| \geqslant \|(I + \lambda F)^{-1}y - (I + \lambda F)^{-1}\hat{y}\| \quad \text{for all } y, \hat{y} \in \mathcal{R}(I + \lambda F) \subset \mathcal{X}.$$

If $\mathcal{X}$ is complete (i.e., a Banach space or a Hilbert space) and if $\mathcal{R}(I + \lambda_0 F) = \mathcal{X}$ for *some* $\lambda_0 > 0$, then it also can be shown that $\mathcal{R}(I + \lambda F) = \mathcal{X}$ for *every* $\lambda > 0$ when F is accretive (Ref. 9).

Exercise 2.8. Let $F: (\mathcal{D}(F) \subset \mathcal{X}) \to \mathcal{X}$ be accretive, $\mathcal{X}$ a normed linear space. Show that $\alpha I + \beta F$ is accretive for every $\alpha \geqslant 0$, $\beta > 0$.

There is another operator property that seemingly has little to do with continuity or boundedness.

Definition 2.9. Let $F: (\mathcal{D}(F) \subset \mathcal{X}) \to \mathcal{Y}$, $\mathcal{X}$ and $\mathcal{Y}$ complete metric spaces. F is a ***closed operator*** if its graph $\{(x, y) | y = Fx, x \in \mathcal{D}(F)\}$ is a closed set in $\mathcal{X} \times \mathcal{Y}$ equipped with the metric $d((x, y), (\hat{x}, \hat{y})) \equiv d_{\mathcal{X}}(x, \hat{x}) + d_{\mathcal{Y}}(y, \hat{y})$. Equivalently, F is closed if, given any $d_{\mathcal{X}}$-Cauchy sequence $\{x_n\} \subset \mathcal{D}(F)$ such that $\{Fx_n\}$ is $d_{\mathcal{Y}}$-Cauchy, then $x_n \to x_0 \in \mathcal{D}(F)$ and $Fx_n \to Fx_0$ as $n \to \infty$.

For a function $F: (\mathcal{D}(F) \subset \mathcal{X}) \to \mathcal{Y}$, $\mathcal{X}$ and $\mathcal{Y}$ complete metric spaces, it is easily shown that F is closed if F is continuous and $\mathcal{D}(F)$ is a closed set in $\mathcal{X}$; also, F is closed if F has a closed inverse.

Exercise 2.9. For a linear operator F: $(\mathfrak{D}(F) \subset \mathfrak{X}) \to \mathfrak{Y}$, $\mathfrak{X}$ and $\mathfrak{Y}$ Banach spaces, show that F is closed if any of the following hold.

(a) F is bounded and $\mathfrak{D}(F)$ is closed.
(b) F is bounded and $\mathfrak{D}(F) = \mathfrak{X}$.
(c) $\mathfrak{R}(F$ is closed and a bounded inverse exists.
(d) $\mathfrak{R}(F) = \mathfrak{Y}$ and a bounded inverse exists.
(e) F has a closed inverse.
(f) $\alpha I + F$ is closed for some $\alpha \in \mathfrak{R}$.
(g) $\alpha I + F$ has a closed inverse for some $\alpha \in \mathfrak{R}$.

It is interesting that any function F: $(\mathfrak{D}(F) \subset \mathfrak{X}) \to \mathfrak{Y}$, $\mathfrak{X}$ and $\mathfrak{Y}$ metric spaces, can be viewed in a way that makes it "appear to be" uniformly continuous and bounded! To this end, define a new metric space $\mathfrak{Z}$ to be $\mathfrak{D}(F)$ equipped with the metric defined by $d_g(x, \hat{x}) \equiv d_{\mathfrak{X}}(x, \hat{x}) + d_{\mathfrak{Y}}(Fx, F\hat{x})$; clearly, viewed as F: $\mathfrak{Z} \to \mathfrak{Y}$, F is uniformly continuous and bounded, with $d_{\mathfrak{Y}}(Fx, F\hat{x}) \leqslant d_g(x, \hat{x})$ for all $x, \hat{x} \in \mathfrak{D}(F) = \mathfrak{Z}$. This trick is occasionally useful; for example, assuming $\mathfrak{X}$ and $\mathfrak{Y}$ to be complete, it shows that F: $\mathfrak{Z} \to \mathfrak{Y}$ is a closed operator if and only if $\mathfrak{Z} \equiv (\mathfrak{D}(F), d_g)$ is complete, and from this it follows that F: $(\mathfrak{D}(F) \subset \mathfrak{X}) \to \mathfrak{Y}$ is a closed operator if and only if $\mathfrak{Z}$ is complete. Of course, this is exactly what Definition 2.9 says as well.

Example 2.1. For a given $k \in \mathcal{C}[0, 1]$, define a linear functional F: $\mathcal{C}[0, 1] \to \mathfrak{R}$ by

$$Fx \equiv \int_0^1 k(s)x(s)\, ds, \qquad x \in \mathcal{C}[0, 1].$$

Noting that for $x \in \mathcal{C}[0, 1]$

$$\|Fx\|_{\mathfrak{R}} = |Fx| = \left|\int_0^1 k(s)x(s)\, ds\right|$$

$$\leqslant \left(\max_s |k(s)|\right)\left(\max_s |x(s)|\right) = \|k\|_{\mathcal{C}}\|x\|_{\mathcal{C}},$$

we see that F is bounded with $\|F\|_{\mathfrak{B}(\mathcal{C}, \mathfrak{R})} \leqslant \|k\|_{\mathcal{C}}$; the lowest possible upper bound (in terms of $\|x\|_{\mathcal{C}}$) can be shown to follow from the estimate

$$|Fx| \leqslant \left(\int_0^1 |k(s)|\, ds\right)\left(\max_s |x(s)|\right), \qquad x \in \mathcal{C}[0, 1],$$

and we can conclude that $\|F\|_{\mathfrak{B}(\mathcal{C}, \mathfrak{R})} = \int_0^1 |k(s)|\, ds$.

As F is linear, we see that

$$\|Fx - F\hat{x}\|_{\mathfrak{R}} \leqslant \|F\|_{\mathfrak{B}(\mathcal{C}, \mathfrak{R})}\|x - \hat{x}\|_{\mathcal{C}}$$

for all $x, \hat{x} \in \mathcal{C}[0, 1]$; hence, F is uniformly Lipschitz continuous on $\mathcal{C}[0, 1]$.

Moreover, $\mathcal{C}[0, 1]$ and $\mathcal{R}$ are complete, and we conclude that F is a closed operator.

If, instead, we view F as $F: (\mathcal{D}(F) \subset \mathcal{L}_2(0, 1)) \to \mathcal{R}$, with $\mathcal{D}(F) = \mathcal{C}[0, 1]$, then

$$\begin{aligned} \|Fx\|_{\mathcal{R}} &= |Fx| = |\langle k, x\rangle_{\mathcal{L}_2}| \\ &\leqslant \|k\|_{\mathcal{L}_2}\|x\|_{\mathcal{L}_2} = \left(\int_0^1 (k(s))^2\, ds\right)^{1/2} \|x\|_{\mathcal{L}_2} \end{aligned}$$

for all $x \in \mathcal{D}(F)$. As it can be shown that this is the best possible estimate in terms of $\|x\|_{\mathcal{L}_2}$ (consider $x(s) \equiv k(s)$), it follows that $\|F\|_{\mathcal{B}(\mathcal{L}_2, \mathcal{R})} = \|k\|_{\mathcal{L}_2}$.

Finally, noting that $\mathcal{D}(F)$ is dense in $\mathcal{L}_2(0, 1)$ with $\mathcal{L}_2$-closure equal to $\mathcal{L}_2(0, 1)$, we see that $\mathcal{D}(F)$ is not a closed subset of $\mathcal{L}_2(0, 1)$; as $F: (\mathcal{D}(F) \subset \mathcal{L}_2(0, 1)) \to \mathcal{R}$ is bounded and linear (hence, continuous), we see that F is not a closed operator when viewed in this way. However, by Theorem 2.1, F admits a unique continuous extension $\tilde{F}: \mathcal{L}_2(0, 1) \to \mathcal{R}$ since $\mathcal{L}_2(0, 1)$ is complete; it follows that $\tilde{F}$ is closed, and it is apparent that $\tilde{F}$ is the linear operator defined by

$$\tilde{F}x \equiv \int_0^1 k(s)x(s)\, ds, \qquad x \in \mathcal{L}_2(0, 1).$$

Example 2.2. Let $k: (\mathcal{S} \subset \mathcal{R}^2) \to \mathcal{R}$ be continuous on the compact set

$$\mathcal{S} \equiv \{(\eta, s) \in \mathcal{R}^2 | 0 \leqslant \eta \leqslant 1, 0 \leqslant s \leqslant 1\},$$

and define a linear function $F: \mathcal{C}[0, 1] \to \mathcal{C}[0, 1]$ by

$$Fx \equiv \int_0^1 k(\cdot, s)x(s)\, ds, \qquad x \in \mathcal{C}[0, 1].$$

This function is bounded, since

$$\begin{aligned} \|Fx\|_{\mathcal{C}} &= \max_{\eta} |(Fx)(\eta)| \\ &\leqslant \max_{\eta}\left(\int_0^1 |k(\eta, s)|\, ds\right)\left(\max_{s} |x(s)|\right) \\ &= \max_{\eta}\left(\int_0^1 |k(\eta, s)|\, ds\right)\|x\|_{\mathcal{C}} \end{aligned}$$

for all $x \in \mathcal{C}[0, 1]$; as this can be shown to be the best possible estimate in terms of $\|x\|_{\mathcal{C}}$, we conclude that

$$\|F\|_{\mathcal{B}(\mathcal{C}, \mathcal{C})} = \max_{\eta} \int_0^1 |k(\eta, s)|\, ds.$$

Moreover, $F: \mathcal{C}[0, 1] \to \mathcal{C}[0, 1]$ is closed, since linearity and boundedness imply continuity and $\mathcal{C}[0, 1]$ is complete.

We may also view F as $F: (\mathcal{D}(F) \subset \mathcal{L}_2(0, 1)) \to \mathcal{L}_2(0, 1)$, with $\mathcal{D}(F) \equiv \mathcal{C}[0, 1]$ dense in $\mathcal{L}_2(0, 1)$. The lowest possible upper bound for $\|Fx\|_{\mathcal{L}_2}$, in terms of $\|x\|_{\mathcal{L}_2}$, is obtained from the estimate

$$\|Fx\|_{\mathcal{L}_2}^2 = \int_0^1 \left[\int_0^1 k(\eta, s)x(s)\, ds \right]^2 d\eta = \int_0^1 \langle k(\eta, \cdot), x \rangle_{\mathcal{L}_2}^2\, d\eta$$

$$\leqslant \int_0^1 \|k(\eta, \cdot)\|_{\mathcal{L}_2}^2 \|x\|_{\mathcal{L}_2}^2\, d\eta = \int_0^1 \left[\int_0^1 (k(\eta, s))^2\, ds \right] d\eta \cdot \|x\|_{\mathcal{L}_2}^2$$

for $x \in \mathfrak{D}(F) \subset \mathcal{L}_2(0, 1)$; hence,

$$\|F\|_{\mathfrak{B}(\mathcal{L}_2, \mathcal{L}_2)} = \left\{ \int_0^1 \left[\int_0^1 (k(\eta, s))^2 ds \right] d\eta \right\}^{1/2}.$$

Although F is not closed when viewed in this way, F is uniformly continuous, $\mathfrak{D}(F)$ is $\mathcal{L}_2$-dense in $\mathcal{L}_2[0, 1]$, and there exists a unique continuous extension $\tilde{F} : \mathcal{L}_2(0, 1) \to \mathcal{L}_2(0, 1)$; obviously,

$$\tilde{F}x \equiv \int_0^1 k(\eta, s)x(s)\, ds, \qquad x \in \mathcal{L}_2(0, 1),$$

with $\|\tilde{F}\|_{\mathfrak{B}(\mathcal{L}_2, \mathcal{L}_2)} = \|F\|_{\mathfrak{B}(\mathcal{L}_2, \mathcal{L}_2)}$, and $\tilde{F}$ is a closed operator from $\mathcal{L}_2(0, 1)$ to $\mathcal{L}_2(0, 1)$.

Example 2.3. Consider $x \in \mathfrak{X} \equiv \mathcal{C}[0, 1]$ and its generalized derivative ∂x. It is apparent that we do not have $\partial x \in \mathfrak{X}$ for every $x \in \mathfrak{X}$; clearly, $\partial x \in \mathfrak{X}$ only if $x \in \mathcal{C}^1[0, 1]$. If we define a linear operator $F : \mathcal{C}^1[0, 1] \to \mathfrak{X}$ by $Fx = \partial x$, $x \in \mathcal{C}^1[0, 1]$, we see that F is bounded since $\|\partial x\|_{\mathcal{C}} = \max_\eta |\partial x(\eta)| \leqslant \|x\|_{\mathcal{C}^1}$.

However, if we choose to view this operator entirely in the topology of $\mathfrak{X} \equiv \mathcal{C}[0, 1]$, we have $F : (\mathfrak{D}(F) \subset \mathfrak{X}) \to \mathfrak{X}$ defined by $Fx = \partial x$, $x \in \mathfrak{D}(F) \equiv \mathcal{C}^1[0, 1]$. Viewed in this way, F is an unbounded operator, but we have reduced the number of topologies that need to be considered. It is apparent that $\mathfrak{D}(F)$ is dense in $\mathfrak{X}$, since $\mathcal{C}^1[0, 1]$ is $\|\cdot\|_{\mathcal{C}}$-dense in $\mathcal{C}[0, 1]$. Defining $\|x\|_g \equiv \|x\|_{\mathfrak{X}} + \|Fx\|_{\mathfrak{X}}$, we see that $(\mathfrak{D}(F), d_g)$ is a complete metric space with $d_g(x, y) \equiv \|x - y\|_g$, since $\|\cdot\|_g = \|\cdot\|_{\mathcal{C}^1}$ and $\mathcal{C}^1[0, 1]$ is a Banach space; hence, $F : (\mathfrak{D}(F) \subset \mathfrak{X}) \to \mathfrak{X}$ is a closed operator.

Let us investigate the range of F; $y \in \mathfrak{R}(F)$ if and only if the equation $Fx = y$ has at least one solution $x \in \mathfrak{D}(F)$. If $y \in \mathfrak{X} \equiv \mathcal{C}[0, 1]$ and if we define

$$x(\eta) = \alpha + \int_0^\eta y(\xi)\, d\xi, \qquad \text{any } \alpha \in \mathfrak{R},$$

we see that $Fx(\eta) = (d/d\eta)x(\eta) = y(\eta)$, $\eta \in [0, 1]$. As $y \in \mathfrak{X}$ was arbitrary, we conclude that $\mathfrak{R}(F) = \mathfrak{X}$. As α was arbitrary, we also see that F is not one-to-one and does not have an inverse function.

Example 2.4. Defining a real normed linear space $\mathfrak{X} \equiv \{x \in \mathcal{C}[0, 1] | x(0) = 0 = x(1)\}$ with $\|x\| \equiv \|x\|_{\mathcal{C}} = \max_\eta |x(\eta)|$, we easily verify that $\mathfrak{X}$ is a Banach space (complete) and define $F : (\mathfrak{D}(F) \subset \mathfrak{X}) \to \mathfrak{X}$ as

$$Fx \equiv -\partial^2 x, \qquad x \in \mathfrak{D}(F) \equiv \{x \in \mathcal{C}^2[0, 1] | x \in \mathfrak{X}, \partial^2 x \in \mathfrak{X}\}.$$

We claim that F is accretive. To verify this, notice that if $x \in \mathfrak{D}(F)$ and $|x(\eta)|$ achieves its maximum at $\eta = \eta_0 \in [0, 1]$, then the boundary conditions imply that $\partial^2 x(\eta_0) \leqslant 0$ if $x(\eta_0) \geqslant 0$, and $\partial^2 x(\eta_0) \geqslant 0$ if $x(\eta_0) \leqslant 0$; therefore, for $\lambda > 0$,

$$\|x + \lambda F x\| \geqslant |x(\eta_0) - \lambda \partial^2(\eta_0)| \geqslant |x(\eta_0)| = \|x\|$$

and the linearity of F shows it to be accretive. It also follows that $I + \lambda F$ has a continuous inverse for $\lambda > 0$; in fact, $(I + \lambda F)^{-1} : (\mathfrak{R}(I + \lambda F) \subset \mathfrak{X}) \to \mathfrak{X}$ is nonexpansive for $\lambda > 0$.

We also claim that $\mathfrak{D}(F)$ is dense; to see this, we note that every $x \in \mathfrak{X} \cap \mathcal{C}^1[0, 1]$ admits a $\|\cdot\|_{\mathcal{C}}$-convergent Fourier series expansion of the form $x(\eta) = \sum_{n=1}^{\infty} \alpha_n \sin n\pi x$, which means that if $e_n(\eta) \equiv \sin n\pi\eta$, then

$$\left\| x - \sum_{n=1}^{N} \alpha_n e_n \right\| = \max_{\eta} \left| x(\eta) - \sum_{n=1}^{N} \alpha_n \sin n\pi\eta \right| \to 0 \qquad \text{as } N \to \infty.$$

As $\mathfrak{X} \cap \mathcal{C}^1[0, 1]$ is dense in $\mathfrak{X}$, it follows that the set $\mathfrak{E}$ of finite real linear sums,

$$\mathfrak{E} \equiv \left\{ x = \sum_{n=1}^{N} \alpha_n e_n \,\middle|\, \text{any } \alpha_1, \alpha_2, \ldots, \alpha_N \in \mathfrak{R};\ \text{any } N = 1, 2, \ldots \right\},$$

is dense in $\mathfrak{X}$ (the set $\{e_n\}$ is said to *span* $\mathfrak{X}$, $e_n(\eta) \equiv \sin n\pi\eta$). It is apparent that $\mathfrak{E} \subset \mathfrak{D}(F)$, and it follows that $\mathfrak{D}(F)$ is dense.

Let us now investigate the range of F, or, more generally, the range of $\alpha I + \beta F$ for some $\alpha, \beta \in \mathfrak{R}$. Therefore, we wish to find those $y \in \mathfrak{X}$ for which the equation $\alpha x - \beta \partial^2 x = y$ has a solution $x \in \mathfrak{D}(F)$. Suppose $y = e_n$; then a solution is $x = (\alpha + \beta_n^2 \pi^2)^{-1} e_n$ if $\alpha \neq -\beta n^2 \pi^2$. Assuming that $\alpha \neq -\beta n^2 \pi^2$ for all $n = 1, 2, \ldots,$ and noting that F is linear, we conclude that this equation has a solution $x \in \mathfrak{D}(F)$ for every $y \in \mathfrak{E}$, which implies that $\mathfrak{R}(\alpha I + \beta F) \supset \mathfrak{E}$. Since $\mathfrak{E}$ is dense, $\mathfrak{R}(\alpha I + \beta F)$ must be dense for $\alpha \neq -n^2 \pi^2 \beta$, $n = 1, 2, \ldots$. Equipping $\mathfrak{D}(F)$ with the norm $\|x\|_g \equiv \|x\| + \|Fx\|$, $x \in \mathfrak{D}(F)$, and noting that

$$\left| |\partial x(\eta)| - |\partial x(\eta_0)| \right| \leqslant \left| \int_{\eta_0}^{\eta} |\partial^2 x(\xi)| \, d\xi \right|$$

implies

$$\max_{\eta} |\partial x(\eta)| \leqslant \tfrac{1}{2} \max_{\eta} |\partial^2 x(\eta)| \qquad \text{for } x \in \mathfrak{D}(F),$$

we see that $\|\cdot\|_g$ is equivalent to $\|\cdot\|_{\mathcal{C}^2}$ on $\mathfrak{D}(F)$. As $\mathcal{C}^2[0, 1]$ is complete, it is easy to verify that $(\mathfrak{D}(F), \|\cdot\|_{\mathcal{C}^2})$ is complete, and this means that $(\mathfrak{D}(F), \|\cdot\|_g)$ is complete; hence F is a closed operator. For $\lambda > 0$, we also see that $I + \lambda F$ is closed, $\mathfrak{R}(I + \lambda F)$ is dense, and there exists a closed and bounded inverse $(I + \lambda F)^{-1}$ on $\mathfrak{R}(I + \lambda F)$; it follows that $\mathfrak{R}(I + \lambda F) = \mathfrak{D}((I + \lambda F)^{-1})$ must be all of $\mathfrak{X}$ for every $\lambda > 0$.

Exercise 2.10. Show that, in Example 2.4, $\mathfrak{R}(\alpha I + \beta F) = \mathfrak{X}$ for all $\alpha, \beta \in \mathfrak{R}$ such that $\alpha\beta > 0$. *Hint:* $\mathfrak{X}$ is a linear space.

Example 2.5. Consider the real Hilbert space $\mathcal{X} = \mathcal{L}_2(0, 1)$ and a linear operator $F : (\mathcal{D}(F) \subset \mathcal{X}) \to \mathcal{X}$ by $Fx \equiv -\partial^2 x$, $x \in \mathcal{D}(F)$, where $\mathcal{D}(F)$ is defined to be the $\|\cdot\|_{\mathcal{W}_2^2}$-completion of $\{x \in \mathcal{C}^2[0, 1] | x(0) = 0 = x(1)\}$ and is denoted by $\mathcal{D}(F) \equiv \{x \in \mathcal{W}_2^2(0, 1) | x(0) = 0 = x(1)\}$.

We claim that $\mathcal{D}(F)$ is dense in $\mathcal{X}$; to verify this, consider $e_n \in \mathcal{X}$, $e_n(\eta) \equiv \sin n\pi\eta$, $n = 1, 2, \ldots$, and let $\mathcal{E}$ denote the set of finite real linear sums.

$$\mathcal{E} \equiv \left\{ x = \sum_{n=1}^{N} \alpha_n e_n \,\middle|\, \text{any } \alpha_1, \alpha_2, \ldots, \alpha_N \in \mathcal{R};\ \text{any } N = 1, 2, \ldots \right\}.$$

It is known that every $x \in \mathcal{L}_2(0, 1)$ admits a $\|\cdot\|_{\mathcal{L}_2}$-convergent Fourier expansion of the form $x = \sum_{n=1}^{\infty} \alpha_n e_n$, some sequence $\{\alpha_n\} \subset \mathcal{R}$ (Ref. 10); hence $\mathcal{E}$ is $\|\cdot\|_{\mathcal{L}_2}$-dense in $\mathcal{L}_2(0, 1)$ and, as $\mathcal{E} \subset \mathcal{D}(F)$, $\mathcal{D}(F)$ must be dense in $\mathcal{L}_2(0, 1)$.

We also claim that $F : (\mathcal{D}(F) \subset \mathcal{X}) \to \mathcal{X}$ is a closed operator; we see that this is the case if $(\mathcal{D}(F), \|\cdot\|_g)$ is complete, with $\|x\|_g \equiv (\|x\|_{\mathcal{X}}^2 + \|Fx\|_{\mathcal{X}}^2)^{1/2}$, $x \in \mathcal{D}(F)$. It is easily verified that $(\mathcal{D}(F), \|\cdot\|_{\mathcal{W}_2^2})$ is complete, so we need only show that $\|\cdot\|_g$ is equivalent to $\|\cdot\|_{\mathcal{W}_2^2}$. Noting that $\|x\|_{\mathcal{W}_2^2}^2 = \|x\|_{\mathcal{L}_2}^2 + \|\partial x\|_{\mathcal{L}_2}^2 + \|\partial^2 x\|_{\mathcal{L}_2}^2$ and $\|x\|_g^2 = \|x\|_{\mathcal{L}_2}^2 + \|\partial^2 x\|_{\mathcal{L}_2}^2$, we see that we wish to prove the existence of $c > 0$ such that $\|x\|_{\mathcal{L}_2}^2 + \|\partial^2 x\|_{\mathcal{L}_2}^2 \geqslant c\|\partial x\|_{\mathcal{L}_2}^2$ for all $x \in \mathcal{D}(F)$. To this end, note that for $x \in \mathcal{E}$ we have $x(\eta) = \sum_{n=1}^{N} \alpha_n \sin n\pi\eta$, some $\alpha_1, \alpha_2, \ldots, \alpha_N \in \mathcal{R}$, some $N = 1, 2, \ldots$, and

$$\|\partial^2 x\|_{\mathcal{L}_2}^2 = \int_0^1 |\partial^2 x(\eta)|^2 \, d\eta = \sum_{n=1}^{N} \sum_{p=1}^{N} \alpha_n \alpha_p n^2 p^2 \pi^4 \int_0^1 (\sin n\pi\eta)(\sin p\pi\eta) \, d\eta$$

$$= \sum_{n=1}^{N} \alpha_n^2 n^4 \pi^4 \int_0^1 |\sin n\pi\eta|^2 \, d\eta = \sum_{n=1}^{N} \alpha_n^2 n^4 \pi^4 \int_0^1 |\cos n\pi\eta|^2 \, d\eta$$

$$\geqslant \pi^2 \sum_{n=1}^{N} \alpha_n^2 n^2 \pi^2 \int_0^1 |\cos n\pi\eta|^2 \, d\eta = \pi^2 \sum_{n=1}^{N} \sum_{p=1}^{N} \alpha_n \alpha_p n p \pi^2 \int_0^1 (\cos n\pi\eta)(\cos p\pi\eta) \, d\eta$$

$$= \pi^2 \int_0^1 |\partial x(\eta)|^2 \, d\eta = \pi^2 \|\partial x\|_{\mathcal{L}_2}^2.$$

Denoting this inequality by $f_1(x) \geqslant f_2(x)$ for all $x \in \mathcal{E} \subset \mathcal{D}(F) \subset \mathcal{W}_2^2(0, 1)$, we see that the mappings $f_1 : (\mathcal{D}(F) \subset \mathcal{W}_2^2(0, 1)) \to \mathcal{R}$ and $f_2 : (\mathcal{D}(F) \subset \mathcal{W}_2^2(0, 1)) \to \mathcal{R}$ are continuous; since $\mathcal{E}$ is $\|\cdot\|_{\mathcal{W}_2^2}$-dense in $\mathcal{D}(F)$, it follows from continuity that $f_1(x) \geqslant f_2(x)$ for all $x \in \mathcal{D}(F)$. We conclude that $\|\cdot\|_g$ is equivalent to $\|\cdot\|_{\mathcal{W}_2^2}$ on $\mathcal{D}(F)$, $(\mathcal{D}(F), \|\cdot\|_g)$ is complete, and F is a closed operator.

By exactly the same type of argument we see that, for $x \in \mathcal{E} \subset \mathcal{D}(F)$,

$$\langle x, Fx \rangle \equiv \langle x, Fx \rangle_{\mathcal{L}_2} = -\int_0^1 x(\eta) \cdot \partial^2 x(\eta) \, d\eta$$

$$= \int_0^1 |\partial x(\eta)|^2 \, d\eta$$

$$\geqslant \pi^2 \int_0^1 |x(\eta)|^2 \, d\eta = \pi^2 \|x\|_{\mathcal{L}_2}^2 = \pi^2 \|x\|_{\mathcal{X}}^2,$$

and, by the $\|\cdot\|_{\mathfrak{W}_2^1}$= denseness of $\mathcal{E}$ in $\mathcal{D}(F)$ and the continuity of all terms as functions from $(\mathcal{D}(F), \|\cdot\|_{\mathfrak{W}_2^1})$ to $\mathcal{R}$, we see that

$$\langle x, Fx\rangle_{\mathfrak{X}} = \|\partial x\|_{\mathfrak{L}2}^2 \geq \pi^2\|x\|_{\mathfrak{L}_2}^2 = \pi^2\|x\|_{\mathfrak{X}}^2$$

for every $x \in \mathcal{D}(F)$. By the linearity of F, we see that $\omega I + F$ is monotone (accretive) for every real $\omega \geq -\pi^2$.

Finally, let us investigate the range of $\alpha I + \beta F$, some $\alpha, \beta \in \mathcal{R}$. We note that $y \in \mathcal{R}(\alpha I + \beta F)$ if and only if the equation $\alpha x + \beta Fx = y$ has a solution $x \in \mathcal{D}(F)$, not necessarily unique. Considering $y = e_n$, we see that there exists a solution $x = (\alpha + \beta n^2\pi^2)^{-1}e_n$ provided that $\alpha \neq -\beta n^2\pi^2$. Assuming that $\alpha \neq -\beta n^2\pi^2$ for all $n = 1, 2, \ldots,$ and noting that F is linear, we see that this equation has a solution for every $y \in \mathcal{E}$; therefore, $\mathcal{R}(\alpha I + \beta F) \supset \mathcal{E}$ and $\mathcal{R}(\alpha I + \beta F)$ is dense in $\mathfrak{X}$. Recalling that F is closed and that $\omega I + F$ is accretive for every $\omega \geq -\pi^2$, it follows that $(1 + \lambda\omega)I + \lambda F$ has a closed and bounded inverse for every $\lambda > 0$, $\omega \geq -\pi^2$, and $[(1 + \lambda\omega)I + \lambda F]^{-1}$ has domain $\mathcal{R}[(1 + \lambda\omega)I + \lambda F]$ dense in $\mathfrak{X}$; consequently, $\mathcal{R}[(1 + \lambda\omega)I + \lambda F] = \mathfrak{X}$ for every $\lambda > 0$, $\omega \geq -\pi^2$. As $\mathfrak{X}$ is a linear space, we see that $\mathcal{R}(\alpha I + \beta F) = \mathfrak{X}$ for all $\beta \neq 0$, $\alpha/\beta > -\pi^2$.

Example 2.6. Let $\mathfrak{X}$ be a Banach space with field $\mathfrak{F}$. The space $\mathfrak{B}(\mathfrak{X}, \mathfrak{F})$, of bounded linear functionals taking $\mathfrak{X}$ into $\mathfrak{F}$, is called the *dual space* of $\mathfrak{X}$ and is denoted by $\mathfrak{X}^*$. Obviously, $\mathfrak{X}^*$ is itself a Banach space with field $\mathfrak{F}$; therefore, it also has a dual space $\mathfrak{B}(\mathfrak{X}^*, \mathfrak{F})$ denoted by $\mathfrak{X}^{**}$. Each element x_0 of $\mathfrak{X}$ defines a unique element x_0^{**} of $\mathfrak{X}^{**}$ by $x_0^{**}(x^*) \equiv x^*(x_0)$ for all $x^* \in \mathfrak{X}^*$; denoting this association by $x_0^{**} = Jx_0$, it follows that $J\mathfrak{X} \subset \mathfrak{X}^{**}$. If $\mathfrak{X}^{**} = J\mathfrak{X}$, the Banach space $\mathfrak{X}^{**}$ is said to be *reflexive* (Refs. 2, 4, 6).

It can be shown that every Hilbert space $\mathfrak{X}$ is reflexive (Refs. 2, 4, 6). If $\mathfrak{X}$ is a Hilbert space, the Riesz theorem (Refs. 2, 4, 6) states that, to every $x^* \in \mathfrak{X}^*$, there corresponds a unique $y \in \mathfrak{X}$ such that

$$x^*(x) = \langle x, y\rangle_{\mathfrak{X}}$$

for all $x \in \mathfrak{X}$; moreover, $\|x^*\|_{\mathfrak{X}^*} \equiv \|x^*\|_{\mathfrak{B}(\mathfrak{X}, \mathfrak{F})} = \|y\|_{\mathfrak{X}}$. The last conclusion is easily obtained, since

$$\|x^*\|_{\mathfrak{X}^*} \equiv \|x^*\|_{\mathfrak{B}(\mathfrak{X}, \mathfrak{F})} \equiv \sup_{\|x\|_{\mathfrak{X}}=1} |\langle x, y\rangle_{\mathfrak{X}}| \leq 1 \cdot \|y\|_{\mathfrak{X}},$$

and equality is achieved if $y = 0$, or if $x \equiv \|y\|_{\mathfrak{X}}^{-1}y$, $y \neq 0$. Therefore, each $x^* \in \mathfrak{X}^*$ can be represented by some linear operator $\langle \cdot, y\rangle : \mathfrak{X} \to \mathfrak{F}$, $y \in \mathfrak{X}$, and we can "identify" $\mathfrak{X}$ with its dual $\mathfrak{X}^*$ when $\mathfrak{X}$ is a Hilbert space.

If $\mathfrak{X}$ is a Banach space that is not a Hilbert space, the specification of $\mathfrak{X}^*$ is much more complicated. In this case, suppose that there is a Hilbert space $\mathcal{H}$, having the same field $\mathfrak{F}$, such that $\mathfrak{X} \subset \mathcal{H}$ and $\|x\|_{\mathfrak{X}} \geq c\|x\|_{\mathcal{H}}$ for all $x \in \mathfrak{X}$, some $c > 0$. Then, given any $h \in \mathcal{H}$, the linear operator $\langle \cdot, h\rangle_{\mathcal{H}} : \mathfrak{X} \to \mathfrak{F}$ is continuous; hence, $\langle \cdot, h\rangle_{\mathcal{H}} \in \mathfrak{X}^*$, and we can conclude that $\mathcal{H}^* \subset \mathfrak{X}^*$. We see that the linear operators $\langle \cdot, h\rangle_{\mathcal{H}} : \mathfrak{X} \to \mathfrak{F}$, $h \in \mathcal{H}$, are some of the elements x^* of $\mathfrak{X}^*$, but possibly not all.

Using the dual space $\mathfrak{X}^*$ of a Banach space $\mathfrak{X}$, one can produce analogues of many ideas that were originally restricted to Hilbert spaces; duality theory is a large part of functional analysis. For example, we note that if $x \in \mathfrak{X}$, $\mathfrak{X}$ a Banach space, and $x^* \in \mathcal{X}^*$, then $x^*(x)$ is very much like an inner product; moreover, since $\mathcal{X}^* = \mathcal{B}(\mathcal{X}, \mathcal{F})$, we see that $|x^*(x)| \leq \|x^*\|_{\mathcal{X}^*}\|x\|_{\mathcal{X}}$, which is very much like the Schwarz inequality. The special properties of Hilbert spaces have been found to be very useful, and great efforts have been made to deduce analogous properties for Banach spaces. However, we shall not pursue the subject of duality further.

Exercise 2.11. Let $\mathcal{X} \equiv \mathcal{C}[0, 1]$, with the maximum norm, and consider $f : \mathcal{C}[0, 1] \to \mathcal{R}$ defined by $f(x) \equiv x(1/2)$, $x \in \mathcal{C}[0, 1]$. Show that $f \in \mathcal{X}^* \equiv \mathcal{B}(\mathcal{X}, \mathcal{R})$. In physics and engineering, this functional f is denoted by

$$f(x) = \int_0^1 \delta\left(\eta - \tfrac{1}{2}\right)x(\eta)\, d\eta,$$

and δ is called the "unit impulse function at $\eta = 1/2$"; however, if $\int_0^1$ denotes any known type of integral, δ cannot be a *function*, and therefore the symbols $\int_0^1$ and δ are individually meaningless. Taken together, they mean f as defined here; occasionally, f is called a *distribution* (Refs. 1, 4, 5).

3. Linear Functions

There are several ideas that are appropriate only for linear operators; for example, in Section 2 we saw that the set $\mathcal{B}(\mathcal{X}, \mathcal{Y})$ of all bounded linear operators taking $\mathcal{X}$ into $\mathcal{Y}$, $\mathcal{X}$ and $\mathcal{Y}$ Banach spaces, was itself a Banach space equipped with the norm

$$\|F\|_{\mathcal{B}(\mathcal{X}, \mathcal{Y})} \equiv \sup_{\|x\|_{\mathcal{X}} = 1} \|Fx\|_{\mathcal{Y}}, \qquad F \in \mathcal{B}(\mathcal{X}, \mathcal{Y}).$$

In Chapter I we noted that a linear operator $F : \mathcal{R}^n \to \mathcal{R}^n$ could be represented by an $n \times n$ matrix, and it follows that every linear $F : \mathcal{R}^n \to \mathcal{R}^n$ is bounded, i.e., $F \in \mathcal{B}(\mathcal{R}^n, \mathcal{R}^n)$. This is a special case of the following result.

Proposition 3.1. Let $F : (\mathcal{D}(F) \subset \mathcal{X}) \to \mathcal{Y}$ be linear, $\mathcal{X}$ and $\mathcal{Y}$ normed linear spaces. If the linear manifold $\mathcal{D}(F)$ has finite dimension, then F is bounded.

Proof. Consider a set of n linearly independent $x_p \in \mathcal{D}(F)$, $p = 1, 2, \ldots, n$, where n is the dimension of $\mathcal{D}(F)$, and note that any $x \in \mathcal{D}(F)$ can be expressed as $x = \sum_{p=1}^n \alpha_p(x)x_p$ for some $\alpha_p(x) \in \mathcal{F}$, $p = 1, 2, \ldots, n$. Since $\|x\| \to 0$ implies $\alpha_p(x) \to 0$, $p = 1, 2, \ldots, n$ and as $Fx = \sum_{p=1}^n \alpha_p(x)Fx_p$, it follows that F is continuous at $x = 0$; by linearity, this implies that F is bounded. □

From the remarks of Section 2, we see that a linear operator $F : (\mathfrak{D}(F) \subset \mathfrak{X}) \to \mathfrak{Y}$, $\mathfrak{X}$ and $\mathfrak{Y}$ Banach spaces, is closed if and only if $(\mathfrak{D}(F), \|\cdot\|_g)$ is a Banach space, where $\|\cdot\|_g$ is a *graph norm*; a graph norm is any norm that is equivalent to $\|\cdot\|_{\mathfrak{X}} + \|F\cdot\|_{\mathfrak{Y}}$ on $\mathfrak{D}(F)$. We employed this idea in several examples in Section 2.

In many applications we are interested in functions having a range space that is also the domain space, i.e., $F : (\mathfrak{D}(F) \subset \mathfrak{X}) \to \mathfrak{X}$. If, in fact, $F \in \mathfrak{B}(\mathfrak{X}, \mathfrak{X})$, many nice conclusions follow.

Exercise 3.1. Let $F \in \mathfrak{B}(\mathfrak{X}, \mathfrak{X})$, $\mathfrak{X}$ a Banach space. Defining $F^0 \equiv I$, $F^n \equiv F(F^{n-1})$ for $n = 1, 2, \ldots,$ show that $F^n \in \mathfrak{B}(\mathfrak{X}, \mathfrak{X})$ and $\|F^n\| \leqslant \|F\|^n$.

Proposition 3.2. Let $F \in \mathfrak{B}(\mathfrak{X}, \mathfrak{X})$, $\mathfrak{X}$ a Banach space, with $\|F\| < 1$. Then $I - F$ has a bounded inverse defined on $\mathfrak{R}(I - F) = \mathfrak{X}$, $\|(I - F)^{-1}\| \leqslant 1/(1 - \|F\|)$ and $(I - F)^{-1} = \Sigma_{p=0}^{\infty} F^p$.

Proof. Since $\|(I - F)x\| \geqslant \|x\| - \|Fx\| \geqslant (1 - \|F\|)\|x\|$ for all $x \in \mathfrak{X}$, $I - F$ has a bounded inverse defined on $\mathfrak{R}(I - F)$ and $\|y\| \geqslant (1 - \|F\|)\|(I - F)^{-1}y\|$ for all $y \in \mathfrak{R}(I - F)$; it follows that $\|(I - F)^{-1}\| \leqslant 1/(1 - \|F\|)$. To show that $\mathfrak{R}(I - F) = \mathfrak{X}$, choose arbitrary $y_0 \in \mathfrak{X}$ and define $G : \mathfrak{X} \to \mathfrak{X}$ as $Gx \equiv y_0 + Fx$, $x \in \mathfrak{X}$. As F is a contraction, G is a contraction, and Theorem 2.2 implies that G has a fixed point x_0; hence, $(I - F)x_0 = y_0$ and we have shown that $\mathfrak{R}(I - F) = \mathfrak{X}$.

Noting that $I - F^{n+1} = \Sigma_{p=0}^{n} F^p(I - F)$, it follows that

$$\left\|(I - F)^{-1} - \sum_{p=0}^{n} F^p\right\| \leqslant \|F^{n+1}\|\|(I - F)^{-1}\| \leqslant \|F\|^{n+1}/(1 - \|F\|) \to 0$$

$$\text{as } n \to \infty.$$

Therefore, $\Sigma_{p=0}^{n} F^p$ is $\|\cdot\|_{\mathfrak{B}(\mathfrak{X}, \mathfrak{X})}$-convergent to $(I - F)^{-1}$; this is called "convergence in the *uniform operator topology*." □

Exercise 3.2. Consider a linear operator $F : (\mathfrak{D}(F) \subset \mathfrak{X}) \to \mathfrak{X}$, $\mathfrak{X}$ a Banach space, such that $\omega I - F$ is accretive for some $\omega \in \mathfrak{R}$ and $\mathfrak{R}(\mu_0 I - F) = \mathfrak{X}$ for some real $\mu_0 > \omega$. Show that, for every $\mu > \omega$, $\mu I - F$ is one-to-one, $\mathfrak{R}(\mu I - F) = \mathfrak{X}$, $(\mu I - F)^{-1} \in \mathfrak{B}(\mathfrak{X}, \mathfrak{X})$, $\|(\mu I - F)^{-1}\| \leqslant 1/(\mu - \omega)$, and F is a closed operator.

Definition 3.1. For a linear operator $F : (\mathfrak{D}(F) \subset \mathfrak{X}) \to \mathfrak{X}$, $\mathfrak{X}$ a complex Banach space, the ***resolvent set*** $\rho(F)$ consists of those $\xi \in \hat{\mathfrak{C}}$ such that $\xi I - F$ has dense range and bounded inverse. For $\xi \in \rho(F)$, let $R(\xi, F)$ denote the (unique) extension of $(\xi I - F)^{-1}$ to all of $\mathfrak{X}$; the ***resolvent*** is the mapping $R(\cdot, F) : (\rho(F) \subset \hat{\mathcal{C}}) \to \mathcal{B}(\mathcal{X}, \mathcal{X})$. The *spectrum* of F is $\sigma(F) \equiv \{\xi \in \hat{\mathcal{C}} | \xi \notin \rho(F)\}$.

We notice that if $\xi \in \rho(F)$, then F is a closed operator if and only if $\mathcal{R}(\xi I - F) = \mathcal{X}$, since $(\xi I - F)^{-1}$ is then closed, bounded, and densely defined; in this case, $R(\xi, F) = (\xi I - F)^{-1}$.

Theorem 3.1. For a linear operator $F : (\mathcal{D}(F) \subset \mathcal{X}) \to \mathcal{X}$, $\mathcal{X}$ a complex Banach space, the resolvent set $\rho(F)$ is open, $R(\xi, F)$ is holomorphic for $\xi \in \rho(F)$, and

$$R(\mu, F) - R(\xi, F) = (\xi - \mu) R(\mu, F) R(\xi, F)$$

for u, $\xi \in \rho(F)$.

Proof. For $\xi_0 \in \rho(F)$, $\xi \in \hat{\mathcal{C}}$, we see that

$$\xi I - F = (\xi_0 I - F)\left[I - (\xi_0 - \xi) R(\xi_0, F)\right].$$

If $|\xi - \xi_0| < 1/\|R(\xi_0, F)\|$, then Proposition 3.2 shows that $\xi I - F$ has a bounded inverse defined on $\mathcal{R}(\xi I - F) = \mathcal{R}(\xi_0 I - F)$ and, for $x \in \mathcal{R}(\xi I - F)$,

$$(\xi I - F)^{-1} x = \left[\sum_{n=0}^{\infty} (\xi_0 - \xi)^n (R(\xi_0, F))^n\right] R(\xi_0, F) x,$$

the sum converging in the uniform operator topology. We see that $\xi \in \rho(F)$ and, therefore, that $\rho(F)$ is open; moreover, $R(\cdot, F) : (\rho(F) \subset \hat{\mathcal{C}}) \to \mathcal{X}$ is holomorphic.

For μ, $\xi \in \rho(F)$ we also see that

$$\begin{aligned} R(\mu, F) - R(\xi, F) &= R(\mu, F)\left[(\xi I - F) - (\mu I - F)\right] R(\xi, F) \\ &= (\xi - \mu) R(\mu, F) R(\xi, F), \end{aligned}$$

and the proof is complete. □

The spectrum $\sigma(F)$ of a linear operator $F : (\mathcal{D}(F) \subset \mathcal{X}) \to \mathcal{X}$ is seen to consist of three disjoint subsets. The *point spectrum* $\sigma_p(F)$ consists of those $\xi \in \hat{\mathcal{C}}$ such that $\xi I - F$ is not one-to-one. That is, $\xi \in \sigma_p(F)$ if $(\xi I - F) g = 0$ for some nonzero $g \in \mathcal{D}(F)$; then ξ is an *eigenvalue* and g is an *eigenvector*. The *residual spectrum* $\sigma_r(F)$ consists of those $\xi \in \hat{\mathcal{C}}$ such that $\xi I - F$ is invertible but $\mathcal{R}(\xi I - A)$ is not dense. The *continuous spectrum* $\sigma_c(F)$ consists of those $\xi \in \hat{\mathcal{C}}$ such that $\xi I - F$ is invertible with dense range, but $(\xi I - F)^{-1}$ is not bounded.

We notice that Definition 3.1 assumes a complex Banach space. Suppose, however, that we wish to consider a linear operator $F : (\mathfrak{D}(F) \subset \mathfrak{X}) \to \mathfrak{X}$, $\mathfrak{X}$ a real linear space ($\mathfrak{F} \equiv \mathfrak{R}$); in order to discuss the spectral properties of F, it is logically necessary to "complexify" both the space and the operator. To this end, we first define a complex Banach space ($\mathfrak{F} \equiv \hat{\mathcal{C}}$) as

$$\mathfrak{X}^c \equiv \{z = x + iy | x, y \in \mathfrak{X}, i = \sqrt{(-1)}\}$$

$$\text{with } \|z\|_{\mathfrak{X}^c} \equiv \left(\|x\|_{\mathfrak{X}}^2 + \|y\|_{\mathfrak{X}}^2\right)^{1/2};$$

if $\mathfrak{X}$ happens to be a Hilbert space, then so is $\mathfrak{X}^c$ equipped with

$$\langle z, \hat{z} \rangle_{\mathfrak{X}^c} \equiv \langle x, \hat{x} \rangle_{\mathfrak{X}} + \langle y, \hat{y} \rangle_{\mathfrak{X}} + i(\langle y, \hat{x} \rangle_{\mathfrak{X}} - \langle x, \hat{y} \rangle_{\mathfrak{X}}).$$

F can now be complexified by defining $F^c : (\mathfrak{D}(F^c) \subset \mathfrak{X}^c) \to \mathfrak{X}^c$ as

$$F^c Z \equiv Fx + iFy, \qquad z \in \mathfrak{D}(F^c) \equiv \{z = x + iy | x, y \in \mathfrak{D}(F), i = \sqrt{(-1)}\},$$

and Definition 3.1 now applies to F^c.

Example 3.1. As just described, let us complexify the problem of Example 2.5; thus, we consider $F : (\mathfrak{D}(F) \subset \mathfrak{X}) \to \mathfrak{X}$ with $\mathfrak{X} \equiv \mathfrak{L}_2^c(0, 1)$, $Fx \equiv -\partial^2 x$ for $x \in \mathfrak{D}(F) \equiv \{x \in \mathfrak{W}_2^{c2}(0, 1) | x(0) = 0 = x(1)\}$, and

$$\langle \hat{x}, x \rangle_{\mathfrak{X}} \equiv \int_0^1 \hat{x}(\eta)\, \overline{x(\eta)}\, d\eta, \qquad \hat{x}, x \in \mathfrak{X}.$$

We also define

$$\mathfrak{E} \equiv \left\{ x = \sum_{n=1}^{N} \alpha_n e_n \middle| e_n \equiv \sin n\pi\eta; \text{ any } \alpha_1, \alpha_2, \ldots, \alpha_N \in \hat{\mathcal{C}}; \text{ any } N = 1, 2, \ldots \right\}.$$

Just as in Example 2.5, we can use this set to see that $\mathfrak{D}(F)$ is dense and, for every $\alpha, \beta \in \hat{\mathcal{C}}$, $\alpha \neq -\beta n^2\pi^2$ for all $n = 1, 2, \ldots$, $\mathfrak{R}(\alpha I + \beta F)$ is dense and $\alpha I + \beta F$ is a closed operator. It follows that, for every $\xi \in \hat{\mathcal{C}}$ with $\xi \neq n^2\pi^2$ for all $n = 1, 2, \ldots$, $\xi I - F$ is closed with dense range.

If $x = \sum_{n=1}^{N} \alpha_n e_n \in \mathfrak{E} \subset \mathfrak{D}(F)$ and $\xi \in \hat{\mathcal{C}}$, then by the methods used in Example 2.5 we easily obtain

$$\begin{aligned}
\|(\xi I - F)x\|_{\mathfrak{X}}^2 &= \langle (\xi I - F)x, (\xi I - F)x \rangle_{\mathfrak{X}} \\
&= \sum_{n=1}^{N} |\xi - n^2\pi^2|^2 \langle e_n, e_n \rangle_{\mathfrak{X}} \\
&\geqslant \inf_{n=1,2,\ldots} |\xi - n^2\pi^2|^2 \|x\|_{\mathfrak{X}}^2.
\end{aligned}$$

Using the continuity of the first and last terms as mappings from $(\mathcal{D}(F), \|\cdot\|_{\mathcal{W}_\xi^2})$ into $\mathcal{R}$, we conclude that

$$\|(\xi I - F)x\|_{\mathcal{X}} \geq \inf_{n=1,2,\ldots} |\xi - n^2\pi^2| \|x\|_{\mathcal{X}}$$

for all $x \in \mathcal{D}(F)$. Assuming that $\xi \neq n^2\pi^2$ for all $n = 1, 2, \ldots,$ we see that $\xi I - F$ has a bounded inverse defined on $\mathcal{R}(\xi I - F)$; as $\mathcal{R}(\xi I - F)$ is dense and $(\xi I - F)^{-1}$ is closed, we must have $\mathcal{R}(\xi I - A) = \mathcal{X}$. It follows if $\xi \neq n^2\pi^2$ for all $n = 1, 2, \ldots,$ then ξ is in the resolvent set $\rho(F)$; hence, $\rho(F) \supset \hat{\mathcal{C}} \setminus \{n^2\pi^2\}$ and $\sigma(F) \subset \{n^2\pi^2\}$.

Considering $\xi_n I - F$ for $\xi_n \equiv n^2\pi^2$, we see that the equation $(\xi_n I - F)x = 0$ admits the solution $x = e_n \in \mathcal{E} \subset \mathcal{D}(F)$, $n = 1, 2, \ldots$; hence, e_n is an eigenvector corresponding to the eigenvalue $\xi_n = n^2\pi^2$, $n = 1, 2, \ldots,$ and the point spectrum $\sigma_p(F) \supset \{n^2\pi^2\}$. As $\{n^2\pi^2\} \supset \sigma(F) \supset \sigma_p(F) \supset \{n^2\pi^2\}$, $\rho(F) = \hat{\mathcal{C}} \setminus \{n^2\pi^2\}$, and both $\sigma_r(F)$ and $\sigma_c(F)$ are empty.

Exercise 3.3. In Example 3.1, show that

$$\|R(\xi, F)\|_{\mathcal{B}(\mathcal{X}, \mathcal{X})} = 1/\Big(\inf_{n=1,2,\ldots} |\xi - n^2\pi^2|\Big)$$

for every $\xi \in \rho(F)$.

Exercise 3.4. For $\mathcal{X} \equiv \mathcal{C}^c[0, 1]$, let $Fx \equiv \partial x$ for $x \in \mathcal{D}(F) \subset \mathcal{X}$. With $F : (\mathcal{D}(F) \subset \mathcal{X}) \to \mathcal{X}$, show that

(a) $\rho(F) = \hat{\mathcal{C}}$ and $\sigma(F)$ is empty if $\mathcal{D}(F) \equiv \{x \in \mathcal{C}^{c1}[0, 1] | x(0) = 0\}$;
(b) $\sigma(F) = \sigma_P(F) = \hat{\mathcal{C}}$ and $\rho(F)$ is empty if $\mathcal{D}(F) \equiv \mathcal{C}^{c1}[0, 1]$;
(c) $\sigma(F) = \sigma_p(F) = \{\pm 2n\pi\sqrt{(-1)} | n = 0, 1, 2, \ldots\}$ if

$$\mathcal{D}(F) \equiv \{x \in \mathcal{C}^{c1}[0, 1] | x(0) = x(1)\}.$$

Hint: If the equation $\xi x - \partial x = y$ has a solution x for given $y \in \mathcal{C}^c[0, 1]$, it must satisfy

$$\frac{d}{d\eta}(e^{-\xi\eta}x(\eta)) = -e^{\xi\eta}y(\eta);$$

therefore, $\mathcal{R}(F) = \mathcal{X}$ if and only if $\mathcal{R}(\xi I - F) = \mathcal{X}$ for all $\xi \in \hat{\mathcal{C}}$.

In Example 3.1, the simplicity of the conclusions reached about $\rho(F)$ and $\sigma(F)$ was due largely to the fact that F was a symmetric operator.

Definition 3.2. A linear operator $F : (\mathcal{D}(F) \subset \mathcal{X}) \to \mathcal{X}$, $\mathcal{X}$ a Hilbert space, is *symmetric* if $\langle \hat{x}, Ax \rangle = \langle A\hat{x}, x \rangle$ for all $x, \hat{x} \in \mathcal{D}(F)$.

Exercise 3.5. Consider a symmetric linear $F : (\mathcal{D}(F) \subset \mathcal{X}) \to \mathcal{X}$, $\mathcal{X}$ a complex Hilbert space. Show that all eigenvalues are real ($\sigma_p \subset \mathcal{R}$) and that the eigenvectors are orthogonal ($\langle g_n, g_m \rangle = 0$ for $\xi_n \neq \xi_m$). Also, if m independent eigenvectors ($m > 1$) correspond to the same eigenvalue $\xi_n \in \sigma_p(F)$ (i.e., if the *null space* of $\xi_n I - F$ has dimension m greater than one), show that we can assume these m eigenvectors to be orthogonal (to each other) without loss of generality.

Exercise 3.6. Given a compact symmetric operator $F \in \mathcal{B}(\mathcal{X}, \mathcal{X})$, $\mathcal{X}$ a complex Hilbert space, the set of all eigenvectors is known to span $\mathcal{X}$ (Refs. 2, 3). Show that σ_p is a bounded subset of $\mathcal{R} \subset \hat{\mathcal{C}}$ and, for every $x \in \mathcal{X}$,

$$\alpha\langle x, x\rangle \leqslant \langle x, Fx\rangle \leqslant \beta\langle x, x\rangle$$

$$\alpha\langle x, F^n x\rangle \leqslant \langle x, F^{n+1} x\rangle \leqslant \beta\langle x, F^n x\rangle, \qquad n = 1, 2, \ldots,$$

where

$$\alpha \equiv \inf_{\xi \in \sigma_p} \xi, \qquad \beta \equiv \sup_{\xi \in \sigma_p} \xi.$$

Hint: By Exercise 3.5 we know that $\sigma_p \subset \mathcal{R}$ and we can assume that all eigenvectors are mutually orthogonal.

Exercise 3.7. Consider the unbounded operator $F : (\mathcal{D}(F) \subset \mathcal{X}) \to \mathcal{X}$ of Example 3.1. Show that F is symmetric and that $\langle x, Fx\rangle_{\mathcal{X}} \geqslant \pi^2\langle x, x\rangle_{\mathcal{X}}$ for all $x \in \mathcal{D}(F)$.

4. Differentiation of Functions

The following notion of differentiation is basic to the study of abstract evolution equations.

Definition 4.1. Let $x(\cdot) : (\mathcal{I} \subset \mathcal{R}) \to \mathcal{X}$, $\mathcal{X}$ a normed linear space. If $[t, t + \varepsilon) \subset \mathcal{I}$, some $\varepsilon > 0$, the ***right derivative*** at t is

$$\dot{x}^+(t) \equiv \lim_{h \searrow 0} \frac{1}{h}\left[x(t + h) - x(t)\right], \qquad h > 0,$$

provided that this limit exists. If $(t - \varepsilon, t] \subset \mathcal{I}$, some $\varepsilon > 0$, the *left derivative* at t is

$$\dot{x}^-(t) \equiv \lim_{h \nearrow 0} \frac{1}{h}\left[x(t + h) - x(t)\right], \qquad h < 0,$$

provided that this limit exists. If $t \in \mathcal{I}$, then $x(\cdot)$ is *differentiable* at t with *derivative* $\dot{x}(t)$ if

(a) t is a left limit point of $\mathcal{I}$, $\dot{x}^+(t)$ exists, and $\dot{x}(t) \equiv \dot{x}^+(t)$,
(b) t is a right limit point of $\mathcal{I}$, $\dot{x}(t)$ exists, and $\dot{x}(t) \equiv \dot{x}^-(t)$,

(c) t is an interior point of $\mathcal{I}$, $\dot{x}^+(t)$ and $\dot{x}^-(t)$ exist and are equal, and $\dot{x}(t) \equiv \dot{x}^+(t) = \dot{x}^-(t)$.

We occasionally write

$$\frac{d^+}{dt}x(t), \qquad \frac{d^-}{dt}x(t), \qquad \frac{d}{dt}x(t)$$

in place of $\dot{x}^+(t)$, $\dot{x}^-(t)$, $\dot{x}(t)$, respectively. It is important to notice that a derivative of this type is a limit taken in $\mathcal{X}$, and its existence depends on the topology of $\mathcal{X}$ (norm of $\mathcal{X}$). For example, we see that $\dot{x}(t)$ exists at an interior $t \in \mathcal{I}$ if and only if there exists $y \in \mathcal{X}$ such that

$$\lim_{h\to 0} \frac{1}{h} \|x(t+h) - x(t) - hy\|_{\mathcal{X}} = 0, \qquad h \in \mathcal{R},$$

and then $\dot{x}(t) \equiv y$. This alternative condition can be generalized, as in the following definition (Ref. 11).

Definition 4.2. Let $F : \mathcal{X} \to \mathcal{Y}$, where $\mathcal{X}$ and $\mathcal{Y}$ are normed linear spaces. Then F is *Fréchet differentiable at* $x_0 \in \mathcal{X}$ if there exists a (bounded linear) operator $F'_{x_0} \in \mathcal{B}(\mathcal{X}, \mathcal{Y})$ such that

$$\lim_{\|x\|_{\mathcal{X}}\to 0} \|F(x_0 + x) - F(x_0) - F'_{x_0}x\|_{\mathcal{Y}} / \|x\|_{\mathcal{X}} = 0.$$

Then the operator $F'_{x_0} \in \mathcal{B}(\mathcal{X}, \mathcal{Y})$ is the *Fréchet derivative* of F at x_0.

Example 4.1. Consider a bounded linear operator $B : \mathcal{X} \to \mathcal{X}$, $\mathcal{X}$ a real Hilbert space, and let $F : \mathcal{X} \to \mathcal{R}$ be given by $Fx = \langle x, Bx \rangle$, $x \in \mathcal{X}$. Note that

$$\begin{aligned} F(x_0 + x) - Fx_0 &= \langle x_0 + x, B(x_0 + x) \rangle - \langle x_0, Bx_0 \rangle \\ &= \langle x, Bx_0 \rangle + \langle x_0, Bx \rangle + \langle x, Bx \rangle \\ &= \langle x, Bx_0 \rangle + \langle Bx, x_0 \rangle + \langle x, Bx \rangle. \end{aligned}$$

Noting that $|\langle x, Bx \rangle| \leqslant \|B\| \|x\|^2$, we see that

$$\lim_{\|x\|\to 0} \frac{|\langle x, Bx \rangle|}{\|x\|} = 0.$$

Therefore, F is Fréchet differentiable at every $x_0 \in \mathcal{H}$, with $F'_{x_0} \in \mathcal{B}(\mathcal{X}, \mathcal{R})$ defined by

$$F'_{x_0} \equiv \langle x, Bx_0 \rangle + \langle Bx, x_0 \rangle, \qquad x \in \mathcal{X}.$$

Example 4.2. Let $f : \mathfrak{R} \to \mathfrak{R}$ be given by $f(x) = x^3$. Then

$$f(x_0 + x) - f(x_0) = 3x_0^2 x + 3x_0 x^2 + x^3.$$

Noting that

$$\lim_{|x| \to 0} \frac{|3x_0 x^2 + x^3|}{|x|} = 0,$$

we see that f is Fréchet differentiable at every $x_0 \in \mathfrak{R}$, with $f'_{x_0} \in \mathfrak{B}(\mathfrak{R}, \mathfrak{R})$ given by

$$f'_{x_0} = 3x_0^2.$$

Note that $f'_{x_0}(x) = 3x_0^2 x$, which is linear in x, although not in x_0.

In the case of a function $f : \mathfrak{R} \to \mathfrak{Y}$, $\mathfrak{Y}$ a Banach space, the preceding definition takes the form

$$\lim_{|\tau| \to 0} \frac{\|f(t+\tau) - f(t) - f'_t(\tau)\|}{|\tau|} = 0,$$

which is equivalent to

$$\lim_{\tau \to 0} \|\tau^{-1}[f(t+\tau) - f(t) - f'_t(\tau)]\| = 0.$$

If we define $\dot{f}(t) = f'_t(1)$, then $f'_t(\tau) = \tau \dot{f}(t)$ and our condition becomes

$$\lim_{\tau \to 0} \|\tau^{-1}[f(t+\tau) - f(t)] - \dot{f}(t)\| = 0.$$

That is,

$$\dot{f}(t) = \lim_{\tau \to 0} \frac{1}{\tau}[f(t+\tau) - f(t)],$$

which in Definition 4.1 is called *the derivative* of f at $t \in \mathfrak{R}$. We note that $\dot{f}(t) \in \mathfrak{Y}$ whereas $f'_t \in \mathfrak{B}(\mathfrak{R}, \mathfrak{Y})$; they are related by the *linear isometry* $\tau \cdot \dot{f}(t) \equiv f'_t(\tau)$, $\tau \in \mathfrak{R}$.

Example 4.3. Let $f : \mathfrak{R}^n \to \mathfrak{R}^m$, with the function f represented by $f = (f_1(x), f_2(x), \ldots, f_m(x))$, $f_i(x) \in \mathfrak{R}^1$ and x represented by $x = (\eta_1, \eta_2, \ldots, \eta_n)$, $\eta_j \in \mathfrak{R}^1$. If the partial derivatives $\partial f_i(x)/\partial \eta_j$ exist and are continuous at $x_0 \in \mathfrak{R}^n$ for all i, j, then f is Fréchet differentiable at x_0 and f'_{x_0} can be represented by the $m \times n$ Jacobian matrix at x_0, i.e.,

$$f'_{x_0} = \left[\frac{\partial}{\partial \eta_j} f_i(x_0)\right], \qquad i = 1, 2, \ldots, m, \quad j = 1, 2, \ldots, n.$$

Clearly, $F'_{x_0} \in \mathfrak{B}(\mathfrak{R}^n, \mathfrak{R}^m)$. If $m = 1$, then $\mathfrak{R}^m = \mathfrak{R}^1 = \mathfrak{R}$, the field of $\mathfrak{R}^n$, and the Jacobian is a $1 \times n$ matrix $f'_{x_0} \in \mathfrak{B}(\mathfrak{R}^n, \mathfrak{R}) = \mathfrak{R}^{n*}$, and by using the inner product on $\mathfrak{R}^n$, we can identify f'_{x_0} with an n-tuple $\nabla f(x_0)$ (the *gradient* of f at x_0) that belongs to $\mathfrak{R}^n$; this is the idea of Riesz's theorem, mentioned in Section 3. On the other hand, if $n = 1$ and $m \geqslant 1$, the Jacobian is an $m \times 1$ matrix, $f'_{x_0} \in \mathscr{B}(\mathscr{R}, \mathscr{R}^m)$, that we can identify with $\dot{f}(x_0) \in \mathscr{R}^m$ by $f'_{x_0}x = x\dot{f}(x_0)$, $x \in \mathscr{R}$.

Exercise 4.1. Let $F \in \mathscr{B}(\mathscr{X}, \mathscr{Y})$, $\mathscr{X}$ and $\mathscr{Y}$ Banach spaces. Show that F is everywhere Fréchet differentiable, and $F'_{x_0} = F$ for every $x_0 \in \mathscr{X}$.

The idea behind Fréchet differentiation is "linear approximation"; if $F : \mathscr{X} \to \mathscr{Y}$ is Fréchet differentiable at x_0, with Fréchet derivative $F'_{x_0} \in \mathscr{B}(\mathscr{X}, \mathscr{Y})$, then we see that we have a local approximation of the form $F(x_0 + x) = F(x_0) + F'_{x_0}(x) + \Gamma(x)$ where $(\|\Gamma(x)\|_{\mathscr{Y}}/\|x\|_{\mathscr{X}}) \to 0$ as $\|x\|_{\mathscr{X}} \to 0$.

Exercise 4.2. Consider $F : \mathscr{X} \to \mathscr{Y}$, $\mathscr{X}$ and $\mathscr{Y}$ Banach spaces, such that F *is* Fréchet differentiable at some $x_0 \in \mathscr{X}$. Show that $F'_{x_0} \in \mathscr{B}(\mathscr{X}, \mathscr{Y})$ can be computed by the formula

$$F'_{x_0}x = \lim_{h \to 0} \frac{1}{h}[F(x_0 + hx) - F(x_0)], \qquad x \in \mathscr{X}, h \in \mathscr{R},$$

the limit being taken in $\mathscr{Y}$. *Note:* This is *not* an alternative definition (Ref. 11).

Exercise 4.3. Let the function $x(\cdot) : (\mathscr{I} \subset \mathscr{R}) \to \mathscr{X}$, $\mathscr{X}$ a Banach space, be differentiable at some $t \in \mathscr{I}$. Show that $x(\cdot)$ is *weakly differentiable* at t, i.e., for all $x^* \in \mathscr{X}^* \equiv \mathscr{B}(\mathscr{X}, \mathscr{F})$, the derivative $(d/dt)[x^*(x(t))]$ exists and equals $x^*(y)$ for some $y \in \mathscr{X}$. Show that $(d/dt)[x^*(x(t))] = x^*(\dot{x}(t))$. *Remark:* If $\mathscr{X}$ has finite dimension, we can easily verify that weak differentiability is equivalent to differentiability; in general, however, this is not true.

In Chapter IV we have use for a *chain rule* of the following form.

Proposition 4.1. For $\mathscr{X}$ and $\mathscr{Y}$ Banach spaces, let $x(\cdot) : (\mathscr{I} \subset \mathscr{R}) \to \mathscr{X}$ be right-differentiable at some $t \in \mathscr{I}$, and let $F : \mathscr{X} \to \mathscr{Y}$ be Fréchet differentiable at $x_0 = x(t)$. Then the function $F(x(\cdot)) : (\mathscr{I} \subset \mathscr{R}) \to \mathscr{Y}$ is right-differentiable at t and $(d^+/dt)F(x(t)) = F'_{x(t)}\dot{x}^+(t)$.

Proof. Define $y(\cdot) \equiv F(x(\cdot))$, $x_0 = x(t)$, and $\Delta x \equiv x(t + h) - x(t)$, small $h > 0$, and note that $F'_{x_0} \in \mathscr{B}(\mathscr{X}, \mathscr{Y})$ and

$$\begin{aligned}\frac{1}{h}\big[y(t+h) - y(t)\big] &= \frac{1}{h}\big[F(x(t+h)) - F(x(t)) + F'_{x_0}(\Delta x) - F'_{x_0}(\Delta x)\big]\\ &= F'_{x_0}\Big(\frac{1}{h}\big[x(t+h) - x(t)\big]\Big)\\ &\quad + \Big(\frac{1}{\|\Delta x\|}\big[F(x_0 + \Delta x) - F(x_0) - F'_{x_0}(\Delta x)\big]\Big)\\ &\quad \times \Big(\frac{1}{h}\big[x(t+h) - x(t)\big]\Big).\end{aligned}$$

Therefore,

$$\left\| F'_{x_0}\dot{x}^+(t) - \frac{1}{h}[y(t+h) - y(t)] \right\|_{\mathcal{Y}}$$

$$\leqslant \|F'_{x_0}\|_{\mathcal{B}(\mathcal{X},\mathcal{Y})} \left\| \dot{x}^+(t) - \frac{1}{h}[x(t+h) - x(t)] \right\|_{\mathcal{X}} + \big(\|F(x_0 + \Delta x) - F(x_0)$$

$$- F'_{x_0}(\Delta x)\|_{\mathcal{Y}} / \|\Delta x\|_{\mathcal{X}}\big)\left(\frac{1}{h}\|x(t+h) - x(t)\|_{\mathcal{X}}\right),$$

and, as $h \searrow 0$,

$$\left\| \dot{x}(t) - \frac{1}{h}[x(t+h) - x(t)] \right\|_{\mathcal{X}} \to 0,$$

$$\|\Delta x\|_{\mathcal{X}} \equiv \|x(t+h) - x(t)\|_{\mathcal{X}} \to 0.$$

Moreover,

$$\|F(x_0 + \Delta x) - F(x_0) - F'_{x_0}(\Delta x)\|_{\mathcal{Y}} / \|\Delta x\|_{\mathcal{X}} \to 0 \qquad \text{as } \|\Delta x\|_{\mathcal{X}} \to 0.$$

This implies that $y(\cdot)$ is right differentiable at t with $\dot{y}^+(t) = F'_{x_0}\dot{x}^+(t)$, and the proof is complete. □

From the proof of Proposition 4.1, it is clear that "right-differentiable" could be replaced throughout by "left-differentiable" or "differentiable." Unfortunately, the Fréchet differentiability requirement on F is quite strong; for certain specific functions F, not necessarily Fréchet differentiable, a chain rule is still valid.

Proposition 4.2. For $\mathcal{X}$ a normed linear space, let $x(\cdot) : (\mathcal{I} \subset \mathcal{R}) \to \mathcal{X}$ be right-differentiable at some $t \in \mathcal{I}$. Then $\|x(\cdot)\|$ is right-differentiable at t, and $|(d^+/dt)\|x(t)\|| \leqslant \|\dot{x}^+(t)\|$.

Proof. Define $x_0 = x(t)$, $v = \dot{x}^+(t)$, and note that for any real $h > 0$, $0 < \theta \leqslant 1$,

$$\|x_0 + \theta h v\| \leqslant (1 - \theta)\|x_0\| + \theta\|x_0 + hv\|,$$

which implies that

$$\frac{1}{\theta h}(\|x_0 + \theta h v\| - \|x_0\|) \leqslant \frac{1}{h}(\|x_0 + hv\| - \|x_0\|) \geqslant -\|v\|.$$

Hence, as a function of h, $h \in (0, \infty)$, $h^{-1}(\|x_0 + hv\| - x_0\|)$ is nondecreasing (monotone) and bounded below. It follows (Rf. 12) that

$\lim_{h\searrow 0} h^{-1}(\|x_0 + hv\| - \|x_0\|)$ exists (and is finite); as $|\|x_0 + hv\| - \|x_0\||$ $\leqslant h\|v\|$, we see that

$$\lim_{h\searrow 0} \frac{1}{h}|(\|x_0 + hv\| - \|x_0\|)| \leqslant \|v\|.$$

We now note that

$$\left|\frac{1}{h}(\|x(t+h)\| - \|x(t)\|) - \frac{1}{h}(\|x_0 + hv\| - \|x_0\|)\right|$$

$$= \left|\frac{1}{h}(\|x(t+h)\| - \|x(t) + h\dot{x}^+(t)\|)\right|$$

$$\leqslant \frac{1}{h}\|x(t+h) - x(t) - h\dot{x}^+(t)\| \to 0 \qquad \text{as } h \searrow 0,$$

which implies that $(d^+/dt)\|x(t)\|$ exists and

$$\left|\frac{d^+}{dt}\|x(t)\|\right| = \lim_{h\searrow 0} \frac{1}{h}(\|x_0 + hv\| - \|x_0\|) \leqslant \|\dot{x}^+(t)\|.$$

The proof is complete. □

From the proof of Proposition 4.2, it is clear that "right-differentiable" could be replaced throughout by "left-differentiable" but not by "differentiable" (consider $\mathcal{I} = \mathfrak{R} = \mathfrak{X}$, $x(t) \equiv t$, with $\|x(t)\| \equiv |t|$).

Exercise 4.4. For $\mathfrak{X}$ a Banach space let $x(\cdot) : (\mathcal{I} \subset \mathfrak{R}) \to \mathfrak{X}$ be right-differentiable at some $t \in \mathcal{I}$. Using Propositions 4.1 to 4.2, show that $g(\cdot) \equiv \|x(\cdot)\|^2 : (\mathcal{I} \subset \mathfrak{R}) \to \mathfrak{R}$ is right-differentiable at t, and $|\dot{g}^+(t)| \leqslant 2\|x(t)\|\|\dot{x}^+(t)\|$. *Hint:* Consider $F : \mathfrak{R} \to \mathfrak{R}$ defined by $F\alpha \equiv \alpha^2$, $\alpha \in \mathfrak{R}$, in Proposition 4.1.

It is also possible to discuss integration of $\mathfrak{X}$-valued functions defined on $\mathfrak{R}$. A *partition* of $(a, b] \subset \mathfrak{R}$ is a finite set $P \equiv \{(\hat{t}_1, A_1), \ldots, (\hat{t}_n, A_n)\}$ of pairs $(\hat{t}_k, A_k)$ such that $\hat{t}_k \in \bar{A}_k \equiv [t_k, t_{k+1}]$, $A_k \equiv (t_k, t_{k+1}]$, for some selection of t_k, $a \equiv t_0 < t_1 < t_2 < \cdots < t_n \equiv b$; $\gamma = \max_k|t_{k+1} - t_k|$ is the *gauge* of P. Given some partition P of $(a, b]$ and a function $f : ([a, b] \subset \mathfrak{R}) \to \mathfrak{X}$, $\mathfrak{X}$ a Banach space, the corresponding *Riemann sum* is

$$S(P, f) \equiv \sum_{k=1}^{n} f(\hat{t}_k)|t_{k+1} - t_k|.$$

Definition 4.3. Let $f : [a, b] \subset \mathfrak{R} \to \mathfrak{X}$, $\mathfrak{X}$ a Banach space, and consider $x \in \mathfrak{X}$. If for each $\varepsilon > 0$ there exists $\delta(\varepsilon) > 0$ such that $\|x - S(P, f)\|$ $< \varepsilon$ for every partition P of $(a, b]$ having gauge $\gamma < \delta$, then f is *Riemann* *integrable* on $[a, b]$ with *Riemann integral* x, denoted by

$$x = \int_a^b f(t)\, dt.$$

It is easily shown that

$$\left\| \int_a^b f(t)\, dt \right\| \leqslant \int_a^b \|f(t)\|\, dt$$

if f is Riemann integrable on $[a, b] \subset \mathcal{R}$ (Ref. 11). If $-\infty < a < b < \infty$ and if $f : ([a, b] \subset \mathcal{R}) \to \mathcal{X}$ is continuous, $\mathcal{X}$ a Banach space, then f is uniformly continuous on $[a, b]$ and it follows that f is Riemann integrable on $[a, t]$, for every $t \in [a, b]$; moreover, the Riemann integral

$$x(t) \equiv \int_a^t f(\tau)\, d\tau, \qquad t \in [a, b],$$

is differentiable on $[a, b]$, with $\dot{x}(t) = f(t)$ (Ref. 11). In fact, most results on Riemann integration of real-valued functions carry over to Riemann integration of $\mathcal{X}$-valued functions defined in $\mathcal{R}$, $\mathcal{X}$ a Banach space.

For $f : ([a, b] \subset \mathcal{R}) \to \mathcal{X}$, $\mathcal{X}$ a Banach space, more general integrals have also been defined. For example, the *Bochner integral* (Ref. 4) is one extension of the idea of Lebesgue integration to $\mathcal{X}$-valued functions. We shall not go further into this subject here.

5. Abstract Evolution Equations

At the beginning of this chapter we suggested that the time behavior of almost any physical system can be described by an abstract evolution equation of the form (1), although the "classical" (or "formal") description may be a partial differential equation, functional differential equation, etc. To see how this idea can be made precise, consider the formal heat conduction problem described earlier.

$$\frac{\partial}{\partial t}\theta(\eta, t) = \frac{\partial^2}{\partial \eta^2}\theta(\eta, t), \qquad 0 \leqslant \eta \leqslant 1, \qquad t \geqslant 0,$$

$$\theta(0, t) = 0 = \theta(1, t), \qquad t \geqslant 0, \tag{2}$$

with initial data $\theta(\eta, 0) = \theta_0(\eta)$, $0 \leqslant \eta \leqslant 1$. If this P.D.E. is satisfied by some $\theta(\cdot, \cdot) : [0, 1] \times \mathcal{R}^+ \to \mathcal{R}$, with $(\partial/\partial t)\theta$ and $(\partial^2/\partial\eta^2)\theta$ continuous on $[0, 1] \times \mathcal{R}^+$, we must have $\theta(\cdot, t) \in \mathcal{C}^2[0, 1]$, $(\partial/\partial t)\theta(\cdot, t) \in \mathcal{C}[0, 1]$, and $\theta(0, t) = 0 = \theta(1, t)$ for every $t \in \mathcal{R}^+$; moreover, the initial data must be such that $\theta(\cdot, 0) = \theta_0(\cdot) \in \mathcal{C}^2[0, 1]$, with $\theta_0(0) = 0 = \theta_0(1)$. Assuming this, define $x(t) \equiv \theta(\cdot, t) : [0, 1] \to \mathcal{R}$ for each $t \in \mathcal{R}^+$; then $x(\cdot)$ maps $\mathcal{R}^+$ into $\mathcal{X}$, where $\mathcal{X}$ is the Banach space

$$\mathcal{X} \equiv \{x \in \mathcal{C}[0, 1] | x(\eta) = 0 \text{ at } \eta = 0, 1\},$$

$$\|x\| \equiv \max_{\eta} |x(\eta)|.$$

In fact, $x(\cdot) : \mathcal{R}^+ \to \mathcal{X}$ is continuously differentiable, since $(\partial/\partial t)\theta$ was assumed continuous on $[0, 1] \times \mathcal{R}^+$ and $[0, 1]$ is compact; moreover, $x(t) \in \mathcal{D}(F)$ for every $t \in \mathcal{R}^+$, where

$$\mathcal{D}(F) \equiv \{x \in \mathcal{C}^2[0, 1] | x(\eta) = 0 = \partial^2 x(\eta) \text{ at } \eta = 0, 1\}.$$

Defining $F : (\mathcal{D}(F) \subset \mathcal{X}) \to \mathcal{X}$ by $Fx \equiv \partial^2 x$, $x \in \mathcal{D}(F)$, and $x_0 \equiv \theta_0(\cdot)$, we see that our formal equation is completely equivalent to the abstract evolution equation

$$\dot{x}(t) = Fx(t), \qquad t \in \mathcal{R}^+,$$

$$x(0) = x_0 \in \mathcal{D}(F) \subset \mathcal{X}. \tag{3}$$

Although this particular abstract formulation of the heat conduction problem is completely equivalent to the original formal equation, there are several good reasons that it actually is preferable, not the least of which is the possibility of extending the qualitative methods of Chapter I to our abstract equation. Another reason becomes apparent when we realize that, in the physical system, there is no reason to assume that the initial temperature field x_0 belongs to $\mathcal{D}(F)$. There are several methods we might use to consider initial data $x_0 \notin \mathcal{D}(F)$. Suppose that (3) has a solution $x_n(t)$ for initial data $x_n \in \mathcal{D}(F)$, the sequence $\{x_n\}$ converging to $x_0 \in \mathcal{X}$, $x_0 \notin \mathcal{D}(F)$, and also suppose that this sequence of solutions $x_n(\cdot)$ converges in some sense to a continuous function $y(\cdot) : \mathcal{R}^+ \to \mathcal{X}$, $y(0) = x_0$. If the motions of the physical system depend continuously on the initial data (in $\mathcal{X}$), and if x_0 is physically possible, then $y(\cdot)$ must be the physical motion corresponding to initial data x_0. This is the basic idea involved in relating an abstract evolution equation in $\mathcal{X}$ to an abstract dynamical system on $\mathcal{X}$, which will be discussed in Chapter III.

Another way that we might generalize our view of the formal equation (2), and thereby obtain another and possibly more general description of the physical system, would be to set up the abstract equation in a different space; (3) is not the only abstract formulation of the heat conduction problem. For example, suppose we define the Hilbert space $\tilde{\mathfrak{X}} \equiv \mathcal{L}_2(0, 1)$ and $\tilde{F} : (\mathfrak{D}(\tilde{F}) \subset \tilde{\mathfrak{X}}) \to \tilde{\mathfrak{X}}$ as

$$\tilde{F}x \equiv \partial^2 x, \qquad x \in \mathfrak{D}(\tilde{F}) \equiv \{x \in \mathfrak{W}_2^2(0, 1) | x(\eta) = 0 \text{ at } \eta = 0, 1\},$$

and consider

$$\dot{x}(t) = \tilde{F}x(t), \qquad t \in \mathfrak{R}^+,$$
$$x(0) = x_0 \in \mathfrak{D}(\tilde{F}) \subset \tilde{\mathfrak{X}}. \tag{4}$$

Then every solution of (2) (equivalently, (3)), is a solution of (4), but not conversely. $\mathfrak{D}(\tilde{F})$ is strictly larger than $\mathfrak{D}(F)$, and $\dot{x}(t)$ now has a different and weaker meaning, since $\dot{x}(t)$ now represents a limit in $\mathcal{L}_2(0, 1)$ rather than in $\mathcal{C}[0, 1]$.

A given physical system can often be described by abstract evolution equations on a variety of spaces; if there is a "most appropriate" space (there may be several), the proper choice follows from the physics of the system and not from any "formal" equation. In our search for an appropriate space, our primary clues are supplied by a firm belief that a physical problem has unique "motions" for all physically possible initial data, that physical motions exist on $\mathfrak{R}^+$ if the physical system is not explosive, and that there is some sense in which physical motions depend continuously on time. We will look further into this question in Chapter III.

References

1. Carroll, R. W., *Abstract Methods in Partial Differential Equations*, Harper and Row, New York, 1969.
2. Dunford, N., and Schwarz, J., *Linear Operators, Part I*, Wiley-Interscience, New York, 1958.
3. Friedman, A., *Foundations of Modern Analysis*, Holt, Rinehart and Winston, New York, 1970.
4. Yosida, K., *Functional Analysis*, Springer-Verlag, New York, 1971.
5. Kolmogorov, A. N., and Fomin, S. V., *Functional Analysis*, Graylock Press, Rochester, New York, 1957.
6. Rudin, W., *Functional Analysis*, McGraw-Hill, New York, 1973.
7. Royden, H. L., *Real Analysis*, MacMillan Company, London, 1968.

8. Volevich, L. R., and Paneyakh, B. P., Certain spaces of generalized functions and embedding theorems, *Russian Mathematical Surveys*, Vol. 20, pp. 1092–1153, 1965.
9. Komura, Y., Nonlinear semi-groups in Hilbert space, *Journal of the Mathematical Society of Japan*, Vol. 13, pp. 493–507, 1967.
10. Stakgold, I., *Boundary Value Problems of Mathematical Physics*, MacMillan Company, London, 1967.
11. Tapia, R. A., The differentiation and integration of nonlinear operators, *Nonlinear Functional Analysis and Applications*. Edited by L. B. Rall, Academic Press, New York, pp. 45–101, 1971.
12. McShane, E. J., *Integration*, Princeton University Press, Princeton, New Jersey, 1947.

III

Abstract Dynamical Systems and Evolution Equations

When a physical system is modeled by a "formal" equation of evolution (e.g., a P.D.E. or F.D.E.), often assumptions are made that have little to do with the physics of the system but are made for the convenience of the modeler; that is, they are made so that the formal equation "makes sense." To a mathematician not interested in applications, such artificial assumptions are as good as any others, and their occasionally disruptive effect on existence, uniqueness, and continuity of solutions (of the formal equation) is an interesting phenomenon, worthy of considerable study. However, to someone interested in applications, such phenomena are irritating rather than interesting, indicating only that a better model is needed. Experience indicates that a physical system does have a unique "motion" for any physically possible initial data and that physical motions are usually in some sense continuously dependent on time; hence, a good mathematical model should also have these properties.

This modeling difficulty has been quite frustrating for people interested in studying finer properties of the motions of physical systems, such as boundedness, stability, and asymptotic behavior. Beginning essentially with the work of V. I. Zubov (Ref. 1), one approach taken to the modeling problem has been to ignore it; by this approach one simply postulates an abstract mathematical object called a dynamical system, flow, or process, whose motions have certain properties that can be reasonably expected of the motions of physical systems; the modeling problem is left for others to solve. This rather cavalier approach has led to a large body of theory regarding the behavior of motions of these mathematical objects. We shall discuss and apply some of this theory in Chapter IV.

The key to solving the modeling problem seems to be provided by certain results in what might appear to be a very different area, the theory of abstract groups and semigroups. The linear autonomous modeling problem was essentially solved by Hille, Phillips (Ref. 2), and Yosida (Ref. 3). During the past decade, great advances have also been made on the modeling problem for nonlinear and nonautonomous systems. In this section we shall discuss some of these results and demonstrate their application to autonomous systems.

1. Dynamical Systems and C_0-Semigroups

Our definition of an abstract dynamical system is an obvious extension of Definition I.4.2; the terms "semiflow" and "semidynamical system" are often used instead.

Definition 1.1. A *dynamical system* on a metric space $\mathfrak{X}$ is a mapping $u: \mathfrak{R}^+ \times \mathfrak{X} \to \mathfrak{X}$ such that

(i) $u(\cdot, x): \mathfrak{R}^+ \to \mathfrak{X}$ is continuous (right-continuous at $t = 0$),
(ii) $u(t, \cdot): \mathfrak{X} \to \mathfrak{X}$ is continuous,
(iii) $u(0, x) = x$,
(iv) $u(t + s, x) = u(t, u(s, x))$,

for all $t, s \in \mathfrak{R}^+$, $x \in \mathfrak{X}$.

It has been shown that the conditions of this definition are sufficient to ensure (joint) continuity of $u: \mathfrak{R}^+ \times \mathfrak{X} \to \mathfrak{X}$ (Ref. 4).

For u a dynamical system on a metric space $\mathfrak{X}$, the mapping $u(\cdot, x): \mathfrak{R}^+ \to \mathfrak{X}$ is called the *motion* originating at $x \in \mathfrak{X}$, and x_e is an *equilibrium* if $u(t, x_e) \equiv x_e$ for all $t \in \mathfrak{R}^+$. A set $\mathfrak{S} \subset \mathfrak{X}$ is *positive invariant* under u if $x \in \mathfrak{S}$ implies that $u(t, x) \in \mathfrak{S}$ for all $t \in \mathfrak{R}^+$. We refer to $\mathfrak{X}$ as the *state space*, and $u(t, x)$ is the *state* at time $t \geqslant 0$.

Definition 1.2. For $\mathfrak{X}$ a metric space, a family $\{S(t)\}_{t \geqslant 0}$ of continuous operators, $S(t): \mathfrak{X} \to \mathfrak{X}$, is a *strongly continuous semigroup* of continuous operators if

(i) $S(\cdot)x: \mathfrak{R}^+ \to \mathfrak{X}$ is continuous (right-continuous at $t = 0$),
(ii) $S(0) = I$,
(iii) $S(t + \tau) = S(t)S(\tau)$,

for all $t, \tau \in \mathfrak{R}^+$, $x \in \mathfrak{X}$.

A family $\{S(t)\}_{t \geqslant 0}$ of continuous operators, satisfying Definition 1.2, is usually called a C_0-semigroup, where C_0 refers to (i) and conditions (ii) and (iii) are called the semigroup property. Clearly, every C_0-semigroup $\{S(t)\}_{t\geqslant 0}$ determines a dynamical system, and conversely, by the definition $u(t, x) \equiv S(t)x$, $t \in \mathcal{R}^+$, $x \in \mathcal{X}$. Hence, these concepts are equivalent. The semigroup notation is often less confusing, but the dynamical system terminology is usually more concise, so we shall henceforth refer to the "dynamical system" $\{S(t)\}_{t\geqslant 0}$ when $\{S(t)\}_{t\geqslant 0}$ is a C_0-semigroup of continuous operators, $S(t)$: $\mathcal{X} \to \mathcal{X}$ for $t \geqslant 0$.

2. Linear Dynamical Systems

Our definiton of a linear abstract dynamical system is a simple extension of Definition I.5.1, although here we shall convert to semigroup notation for the sake of convenience.

Definition 2.1. For $\mathcal{X}$ a Banach space, a dynamical system $\{S(t)\}_{t\geqslant 0}$ is linear if $S(t) \in \mathcal{B}(\mathcal{X}, \mathcal{X})$ for every $t \in \mathcal{R}^+$, i.e., if $S(t)$ is a bounded linear operator for each $t \in \mathcal{R}^+$.

Proposition 2.1. If $\{S(t)\}_{t\geqslant 0}$ is a linear dynamical system on a Banach space $\mathcal{X}$, there exist real numbers M, ω, $M \geqslant 1$, such that

$$\|S(t)\|_{\mathcal{B}(\mathcal{X}, \mathcal{X})} \leqslant Me^{\omega t} \qquad \text{for all} \quad t \in \mathcal{R}^+.$$

Proof. We first show that $\|S(t)\| \leqslant M$ for all $t \in [0, T]$, some $T > 0$, some $M \geqslant 1$. If not, there exists a sequence $\{t_n\} \subset \mathcal{R}^+$, $t_n \searrow 0$ as $n \to \infty$, such that $\|S(t_n)\| \geqslant n$; but then the resonance theorem (Ref. 3) implies that the sequence $\{S(t_n)x\}$ is unbounded for some $x \in \mathcal{X}$, and this contradicts (i) of Definition 1.2. Hence there exist $T > 0$ and M such that $\|S(t)\| \leqslant M$ for all $t \in [0, T]$, and $M \geqslant 1$ since $S(0) = I$.

Given $t > 0$, choose an integer $n \geqslant 0$ and $s \geqslant 0$, $s < T$, such that $t = nT + s$; then, by the semigroup property,

$$\|S(t)\| = \|S(s)(S(T))^n\| \leqslant M^{n+1} \leqslant MM^{t/T} = Me^{\omega t},$$

where $\omega \equiv T^{-1} \ln M$. The proof is complete. □

The result of Proposition 2.1 is not only interesting but will also turn out to be very useful to us. It also is very important that with each linear dynamical system on $\mathcal{X}$, a Banach space, there is associated a certain operator A: $(\mathcal{D}(A) \subset \mathcal{X}) \to \mathcal{X}$, called the infinitesimal generator.

Definition 2.2. For $\{S(t)\}_{t\geq 0}$ a linear dynamical system on a Banach space $\mathfrak{X}$, we define $\mathfrak{D}(A)$ to be the set of $x \in \mathfrak{X}$ for which the limit

$$\lim_{t \searrow 0} \frac{1}{t}[S(t)x - x]$$

exists, and we define Ax to be this limit for $x \in \mathfrak{D}(A)$. Then $A: (\mathfrak{D}(A) \subset \mathfrak{X}) \to \mathfrak{X}$ is called the *infinitesimal generator* of $\{S(t)\}_{t\geq 0}$.

Exercise 2.1. For $\mathfrak{X}$ a Banach space, suppose that $A: (\mathfrak{D}(A) \subset \mathfrak{X}) \to \mathfrak{X}$ is the infinitesimal generator of a linear dynamical system $\{S(t)\}_{t\geq 0}$. Show that for any given $\alpha \in \mathfrak{R}$, $\{e^{\alpha t}S(t)\}_{t\geq 0}$ is a linear dynamical system with infinitesimal generator $\alpha I + A$.

The importance of the infinitesimal generator is due to our wish to associate linear dynamical systems with linear autonomous evolution equations, and conversely.

Theorem 2.1. Let $A: (\mathfrak{D}(A) \subset \mathfrak{X}) \to \mathfrak{X}$, $\mathfrak{X}$ a Banach space, be the infinitesimal generator of a linear dynamical system $\{S(t)\}_{t\geq 0}$ on $\mathfrak{X}$. Then

(i) A is a closed linear operator and $\mathfrak{D}(A)$ is dense;

(ii) $\mathfrak{D}(A)$ is positive invariant and, for every $x \in \mathfrak{D}(A)$,

$$\frac{d}{dt}S(t)x = AS(t)x = S(t)Ax, \qquad t \in \mathfrak{R}^+.$$

Proof. By the properties of the Riemann integral, we see that for $h > 0$, $t > 0$, $x \in \mathfrak{X}$,

$$\begin{aligned}\frac{1}{h}[S(h) - I]\int_0^t S(s)x\,ds &= \frac{1}{h}\int_0^t [S(s+h)x - S(s)x]\,ds\\ &= \frac{1}{h}\int_t^{t+h} S(s)x\,ds - \frac{1}{h}\int_0^h S(s)x\,ds\\ &\to S(t)x - x \qquad \text{as } h \searrow 0,\end{aligned}$$

and this shows that

$$\int_0^t S(s)x\,ds \in \mathfrak{D}(A), \qquad A\int_0^t S(s)x\,ds = S(t)x - x,$$

for all $x \in \mathfrak{X}$, $t > 0$. Noting that $t^{-1}\int_0^t S(s)x\,ds \to x$ as $t \searrow 0$, for any given $x \in \mathfrak{X}$, it follows that $\mathfrak{D}(A)$ is dense.

For $t \geqslant 0$, $h > 0$, and $x \in \mathfrak{D}(A)$, we see that

$$\frac{1}{h}[S(h) - I]S(t)x = S(t)\left(\frac{1}{h}[S(h)x - x]\right) \to S(t)Ax \qquad \text{as } h \searrow 0.$$

Hence, $S(t)x \in \mathfrak{D}(A)$ for all $t \in \mathfrak{R}^+$ if $x \in \mathfrak{D}(A)$; moreover, $(d^+/dt)S(t)x = AS(t)x = S(t)Ax$ for all $t \in \mathfrak{R}^+$ if $x \in \mathfrak{D}(A)$. We also see that if $h < 0$, $t + h \geqslant 0$, $x \in \mathfrak{D}(A)$, then $S(t)Ax - S(t+h)Ax \to 0$ as $h \nearrow 0$ and

$$\begin{aligned}\frac{1}{h}[S(t+h)x - S(t)x] - S(t)Ax + S(t)Ax - S(t+h)Ax \\ = S(t+h)\left(\frac{1}{-h}[S(-h)x - x] - Ax\right) \\ \to 0 \qquad \text{as } h \nearrow 0,\end{aligned}$$

since $\|S(t+h)\| \leqslant Me^{\omega(t+h)}$ by Proposition 2.1. It follows that $(d^-/dt)S(t)x = S(t)Ax$, and we conclude that $(d/dt)S(t)x = S(t)Ax = AS(t)x$ for all $t \in \mathfrak{R}^+$, $x \in \mathfrak{D}(A)$.

As linearity of A is apparent from Definition 2.2, it only remains to show that A is a closed operator. For $x \in \mathfrak{D}(A)$, we see that $(d/dt)S(\cdot)x$ is continuous on $\mathfrak{R}^+$, and

$$S(t)x - x = \int_0^t \left(\frac{d}{ds}S(s)x\right) ds = \int_0^t S(s)Ax\, ds, \qquad t \in \mathfrak{R}^+.$$

Hence, consider $x \in \mathfrak{X}$ and a sequence $\{x_n\} \subset \mathfrak{D}(A)$ with $x_n \to x$ and $Ax_n \to y \in \mathfrak{X}$ as $n \to \infty$. Then

$$\begin{aligned}S(t)x_n - x_n &= \int_0^t S(s)Ax_n\, ds, \qquad t > 0,\\ &\to \int_0^t S(s)y\, ds \qquad \text{as } n \to \infty.\end{aligned}$$

Hence,

$$S(t)x - x = \int_0^t S(s)y\, ds$$

and, for $t > 0$,

$$\frac{1}{t}[S(t)x - x] = \frac{1}{t}\int_0^t S(s)y\, ds \to y \qquad \text{as } t \searrow 0.$$

It follows that $x \in \mathfrak{D}(A)$ and $Ax = y$; therefore, A is closed and the proof is complete. □

Theorem 2.1 is a very useful result; if $A: (\mathfrak{D}(A) \subset \mathfrak{X}) \to \mathfrak{X}$ is an infinitesimal generator, $\mathfrak{X}$ a Banach space, then Theorem 2.1 implies that the motion $S(\cdot)x_0: \mathfrak{R}^+ \to \mathfrak{X}$ is a strong solution of the linear autonomous evolution equation

$$\dot{x}(t) = Ax(t), \qquad t \in \mathfrak{R}^+,$$

$$x(0) = x_0 \in \mathfrak{D}(A) \subset \mathfrak{X}, \tag{1}$$

for each $x_0 \in \mathfrak{D}(A)$, and $\mathfrak{D}(A)$ is dense. By a *strong solution* of (1), we mean a differentiable function $x(\cdot): \mathfrak{R}^+ \to \mathfrak{X}$ such that $x(0) = x_0 \in \mathfrak{D}(A)$, $x(t) \in \mathfrak{D}(A)$ for all $t \in \mathfrak{R}^+$, and $\dot{x}(t) = Ax(t)$ for all $t \in \mathfrak{R}^+$. Under the assumption that A is an infinitesimal generator, we shall be able to show later that solutions of (1) are unique; assuming this for now, we see that every solution of (1) is a motion of $\{S(t)\}_{t \geq 0}$, and it follows from the estimate $\|S(t)x - S(t)x_0\| \leq Me^{\omega t}\|x - x_0\|$that the solutions of (1) depend continuously on the initial data $x_0 \in \mathfrak{D}(A) \subset \mathfrak{X}$.

These ideas are apparently quite useful for the solution of linear autonomous abstract evolution equations, provided that A in (1) happens to be an infinitesimal generator. We now ask which types of linear operators happen to be infinitesimal generators; obviously, by Theorem 2.1, A must be closed, linear, and densely defined in the Banach space $\mathfrak{X}$.

In Section I.5 we noted that every linear $A: \mathfrak{R}^n \to \mathfrak{R}^n$, represented by a real $n \times n$ matrix, is the infinitesimal generator of a linear dynamical system $\{S(t)\}_{t \geq 0}$ defined by e^{tA}. The following proposition generalizes this result.

Proposition 2.2. If $A \in \mathfrak{B}(\mathfrak{X}, \mathfrak{X})$, $\mathfrak{X}$ a Banach space, then both A and $-A$ are infinitesimal generators.

Proof. Note that $\Sigma_{n=0}^{N}(t^n/n!)A^n \in \mathfrak{B}(\mathfrak{X}, \mathfrak{X})$ for finite N and every $t \in \mathfrak{R}^+$; moreover,

$$\left\| \sum_{n=m}^{P} (t^n/n!)A^n \right\| \leq \sum_{n=m}^{P} (t^n/n!)\|A\|^n,$$

which implies that $S(t) \equiv \Sigma_{n=0}^{\infty}(t^n/n!)A^n$ exists, converging in the uniform operator topology (topology of $\mathfrak{B}(\mathfrak{X}, \mathfrak{X})$), and $\|S(t)\| \leq e^{t\|A\|}$. The semigroup property follows just as in Section I.5, and we note that

$$\|S(t)-I\| = \left\| \sum_{n=1}^{\infty} (t^n/n!)A^n \right\| \leqslant t\|A\|e^{t\|A\|}.$$

Hence, for $h>0$, $t \geqslant 0$,

$$\|S(t+h)x - S(t)x\| \leqslant \|S(t)\|\ \|S(h)x - x\|$$
$$\leqslant \|S(t)\|h\|A\|e^{hA}\|x\| \to 0 \qquad \text{as } h \searrow 0,$$

while for $h<0$, $t+h>0$,

$$\|S(t+h)x - S(t)x\| \leqslant \|S(t+h)\|\ \|x - S(-h)x\|$$
$$\leqslant e^{t\|A\|}|h|\ \|A\|e^{|h|\|A\|}\|x\| \to 0 \qquad \text{as } h \nearrow 0.$$

It follows that $\{S(t)\}_{t\geqslant 0}$ is a linear dynamical system on $\mathfrak{X}$; moreover, since

$$\left\| \frac{1}{t}[S(t)-I] - A \right\| = \left\| \sum_{n=2}^{\infty} (t^{n-1}/n!)A^n \right\| \leqslant \|A\|\ \|S(t)-I\|,$$

A is the infinitesimal generator of $\{S(t)\}_{t\geqslant 0}$. As the same reasoning applies to $-A$, the proof is complete. □

When both A and $-A$ are infinitesimal generators, as was concluded in Proposition 2.2, the C_0-semigroup $\{S(t)\}_{t\geqslant 0}$ generated by A must be invertible, and $-A$ generates $\{S^{-1}(t)\}_{t\geqslant 0}$. Then, with the definition $S(-t) \equiv S^{-1}(t)$, $t \geqslant 0$, A is said to generate the *C_0-group* $\{S(t)\}_{t\in\mathfrak{R}}$, and the "directionality" of time is of no importance.

Unfortunately, if a linear autonomous physical system requires an infinite-dimensional state space, then often the infinitesimal generator is not bounded. Although this complicates things considerably, results presented in the following section completely characterize the linear operators that are infinitesimal generators.

3. Generation of Linear Dynamical Systems

Proposition 2.1 states that every linear dynamical system $\{S(t)\}_{t\geqslant 0}$ satisfies an estimate of the form $\|S(t)x\| \leqslant Me^{\omega t}\|x\|$ for some $\omega, M \in \mathfrak{R}$, $M \geqslant 1$. The proof of Proposition 2.2 shows that this estimate holds for $M=1$, some $\omega \leqslant \|A\|$, if the infinitesimal generator A is bounded. If A is unbounded, such an estimate may only hold for some $M > 1$, and this is inconvenient for several reasons. The following result shows that there always exists an equivalent norm such that the estimate holds for $M = 1$. This is very important for applications.

Theorem 3.1. If $\{S(t)\}_{t\geq 0}$ is a linear dynamical system on a Banach space $\mathcal{X}$, there exists $\omega \in \mathcal{R}$ and an equivalent norm $\|\cdot\|_e$ for $\mathcal{X}$ such that $\|S(t)x\|_e \leq e^{\omega t}\|x\|_e$ for all $t \in \mathcal{R}^+$, $x \in \mathcal{X}$.

Proof. By Proposition 2.1 there exist $M, \omega \in \mathcal{R}$, $M \geq 1$, such that $\|S(t)x\| \leq Me^{\omega t}\|x\|$ for all $t \in \mathcal{R}^+$, $x \in \mathcal{X}$. Defining

$$\|x\|_e \equiv \sup_{t\geq 0} e^{-\omega t}\|S(t)x\|, \qquad x \in \mathcal{X},$$

we note that $\|x\| \leq \|x\|_e \leq M\|x\|$, and $\|\alpha x\|_e = |\alpha|\,\|x\|_e$ for all $x \in \mathcal{X}$, $\alpha \in \mathcal{F}$. Moreover,

$$\begin{aligned}\|x+y\|_e &= \sup_{t\geq 0} e^{-\omega t}\|S(t)x + S(t)y\| \\ &\leq \sup_{t\geq 0} e^{-\omega t}\|S(t)x\| + \sup_{t\geq 0} e^{-\omega t}\|S(t)y\| \\ &= \|x\|_e + \|y\|_e, \qquad x, y \in \mathcal{X},\end{aligned}$$

and it follows that $\|\cdot\|_e$ is an equivalent norm for $\mathcal{X}$. Finally,

$$\begin{aligned}\|S(t)x\|_e &= \sup_{\tau\geq 0} e^{-\omega\tau}\|S(t+\tau)x\| = \sup_{s\geq t} e^{-\omega(s-t)}\|S(s)x\| \\ &\leq e^{\omega t}\sup_{s\geq 0} e^{-\omega s}\|S(s)x\| = e^{\omega t}\|x\|_e, \qquad t \in \mathcal{R}^+, x \in \mathcal{X},\end{aligned}$$

and the proof is complete. □

In Section 4 we shall discuss how such an equivalent norm can be found. Its existence leads to the following characterization of infinitesimal generators.

Theorem 3.2. A linear operator $A\colon (\mathcal{D}(A) \subset \mathcal{X}) \to \mathcal{X}$, $\mathcal{X}$ a Banach space, is the infinitesimal generator of a linear dynamical system $\{S(t)\}_{t\geq 0}$, satisfying $\|S(t)\|_{\mathcal{B}(\mathcal{X},\mathcal{X})} \leq Me^{\omega t}$ for all $t \in \mathcal{R}^+$, if and only if $\mathcal{D}(A)$ is dense, $\mathcal{R}(I - \lambda A) = \mathcal{X}$ for all sufficiently small $\lambda > 0$, and there exists an equivalent norm $\|\cdot\|_e$ such that $\omega I - A$ is $\|\cdot\|_e$-accretive. Then $\|S(t)x\|_e \leq e^{\omega t}\|x\|_e$ for all $t \geq 0$, $x \in \mathcal{X}$.

This theorem can be proved in various ways; we defer the proof until the end of this section, as the proof we have chosen is quite difficult. Our reason for using this method of proof is that it permits a partial extension to nonlinear dynamical systems, as we shall see in Section 5.

Theorem 3.2 can be shown to be completely equivalent to the famous Hille–Yosida–Phillips theorem (Refs. 2, 3, 5, 6).

Theorem 3.3. A linear operator $A: (\mathfrak{D}(A) \subset \mathfrak{X}) \to \mathfrak{X}$, $\mathfrak{X}$ a complex Banach space, is the infinitesimal generator of a linear dynamical system $\{S(t)\}_{t \geq 0}$, satisfying $\|S(t)\|_{\mathfrak{B}(\mathfrak{X}, \mathfrak{X})} \leq Me^{\omega t}$ for all $t \in \mathfrak{R}^+$, if and only if

(i) A is closed and $\mathfrak{D}(A)$ is dense;

(ii) every real $\mu > \omega$ is in the resolvent set $\rho(A)$;

(iii) $\|(R(\mu, A))^n\| \leq M/(\mu - \omega)^n$ for all $\mu > \omega$, $n = 1, 2, \ldots,$

where $R(\cdot, A): (\rho(A) \subset \hat{\mathcal{C}}) \to \mathfrak{B}(\mathfrak{X}, \mathfrak{X})$ is the resolvent.

Various proofs of this theorem can be found in Refs. 2, 3, 5, 6. We shall find Theorem 3.2 to be much more useful for applications. The disadvantage of Theorem 3.2 is that a suitable norm $\|\cdot\|_e$ must be found. Aside from the nuisance of insisting that $\mathfrak{X}$ be complex, the principal disadvantage of Theorem 3.3 is that condition (iii) represents an infinite number of conditions. Condition (iii) is "easy" only if $\|R(\mu, A)\| \leq (\mu - \omega)^{-1}$ for all $\mu > \omega$; however, if this is true, the choice $\|\cdot\|_e \equiv \|\cdot\|_{\mathfrak{X}}$ works in Theorem 3.2 (and conversely). For applications, Theorem 3.2 seems by far the easier to use. We shall discuss the problem of finding a suitable equivalent norm in Section 4; often we can use the original norm on $\mathfrak{X}$, as when $A \in \mathfrak{B}(\mathfrak{X}, \mathfrak{X})$.

Exercise 3.1. If $A \in \mathfrak{B}(\mathfrak{X}, \mathfrak{X})$, $\mathfrak{X}$ a Banach space, show that A meets all conditions of Theorem 3.2 with $\|\cdot\|_e \equiv \|\cdot\|_{\mathfrak{X}}$, for some $\omega \in \mathfrak{R}$.

Exercise 3.2. Suppose that $A: (\mathfrak{D}(A) \subset \mathfrak{X}) \to \mathfrak{X}$ meets all conditions of Theorem 3.2 for some $\omega \in \mathfrak{R}$, and some $\|\cdot\|_e \neq \|\cdot\|_{\mathfrak{X}}$, $\mathfrak{X}$ a complex Banach space. Show that A meets conditions (i), (ii), (iii) of Theorem 3.3 (without using Theorem 3.3).

Example 3.1. As discussed in Section II.5, consider an abstract heat conduction problem framed in the Banach space $\mathfrak{X} \equiv \{x \in \mathcal{C}[0, 1] | x(\eta) = 0 \text{ at } \eta = 0, 1\}$, $\|x\| \equiv \max_\eta |x(\eta)|$, and described by

$$\dot{x}(t) = Ax(t), \qquad t \in \mathfrak{R}^+,$$

$$x(0) = x_0 \in \mathfrak{D}(A) \subset \mathfrak{X},$$

where

$$Ax \equiv \partial^2 x, \qquad x \in \mathfrak{D}(A) \equiv \{x \in \mathcal{C}^2[0, 1] | x(\eta) = 0 = \partial^2(\eta) \text{ at } \eta = 0, 1\}.$$

From Example II.2.4 it follows that $\mathfrak{D}(A)$ is dense, $\mathfrak{R}(I - \lambda A) = \mathfrak{X}$ for every $\lambda > 0$, and $-A$ is accretive in the original norm on $\mathfrak{X}$.

Choosing $\omega = 0$ and applying Theorem 3.2, we see that $A: (\mathfrak{D}(A) \subset \mathfrak{X}) \to \mathfrak{X}$ is the infinitesimal generator of a linear dynamical system $\{S(t)\}_{t \geq 0}$ on $\mathfrak{X}$, and $\|S(t)x\| \leq \|x\|$ for all $t \in \mathfrak{R}^+$, every $x \in \mathfrak{X}$. Also, $S(\cdot)x_0$: $\mathfrak{R}^+ \to \mathfrak{X}$ is the (unique) strong solution of the evolution equation for every $x_0 \in \mathfrak{D}(A)$.

Example 3.2. As also discussed in Section II.5, consider an abstract heat conduction problem framed in the Hilbert space $\mathfrak{X} \equiv \mathfrak{L}_2(0, 1)$ with

$$\langle x, y \rangle \equiv \int_0^1 x(\eta) y(\eta) d\eta,$$

and described by

$$\dot{x}(t) = Ax(t), \qquad t \in \mathfrak{R}^+,$$
$$x(0) = x_0 \in \mathfrak{D}(A) \subset \mathfrak{X},$$

where

$$Ax \equiv \partial^2 x, \qquad x \in \mathfrak{D}(A) \equiv \{x \in \mathfrak{W}_2^2(0, 1) | x(\eta) = 0 \text{ at } \eta = 0, 1\}.$$

From Example II.2.5 it follows that $\mathfrak{D}(A)$ is dense, $\mathfrak{R}(I - \lambda A) = \mathfrak{X}$ for all $\lambda > 0$, and $\omega I - A$ is monotone (accretive) in the original norm for every $\omega \geq -\pi^2$.

Choosing $\omega = -\pi^2$ and applying Theorem 3.2, we see that $A: (\mathfrak{D}(A) \subset \mathfrak{X}) \to \mathfrak{X}$ is the infinitesimal generator of a linear dynamical system $\{S(t)\}_{t \geq 0}$ on $\mathfrak{X}$, and $\|S(t)x\| \leq \exp\{-\pi^2 t\} \|x\|$ for all $t \in \mathfrak{R}^+$, every $x \in \mathfrak{X}$. Also, $S(\cdot)x_0$ is the (unique) strong solution of the evolution equation for every $x_0 \in \mathfrak{D}(A)$.

Exercise 3.3. Set up the heat conduction problem in the Hilbert space $\mathfrak{X} \equiv \{x \in \mathfrak{W}_2^1(0, 1) | x(\eta) = 0 \text{ at } \eta = 0, 1\}$, with

$$\langle x, y \rangle \equiv \int_0^1 [x(\eta) \cdot y(\eta) + \partial x(\eta) \cdot \partial y(\eta)]\, d\eta.$$

It can be shown that the set $\mathfrak{S}$ of Example II.2.5 is dense in $\mathfrak{X}$. Show that A is the infinitesimal generator of a linear dynamical system $\{S(t)\}_{t \geq 0}$ satisfying $\|S(t)x\| \leq \exp\{-\pi^2 t\} \|x\|$, $t \in \mathfrak{R}^+$, $x \in \mathfrak{X}$.

Exercise 3.4. Set up the heat conduction problem in the Banach space $\mathfrak{X} \equiv \{x \in \mathcal{C}^1[0, 1] | x(\eta) = 0 \text{ at } \eta = 0, 1\}$ with $\|x\| \equiv \max_\eta |x(\eta)| + \max_\eta |\partial x(\eta)|$. It can be shown that the set $\mathfrak{S}$ of Example II.2.4 is dense in $\mathfrak{X}$. Show that A is the infinitesimal generator of a linear dynamical system $\{S(t)\}_{t \geq 0}$ satisfying $\|S(t)x\| \leq \|x\|$, $t \in \mathfrak{R}^+$, $x \in \mathfrak{X}$. Note that $|\partial x(\eta)|$ might assume its maximum anywhere on $[0, 1]$, including $\eta = 0, 1$.

In the preceding examples and exercises we were fortunate that in terms of the original norm, $\omega I - A$ was found to be accretive for some

$\omega \in \mathcal{R}$; often, this is not the case and adjustments must be made in the norm in order to obtain accretiveness. In fact, we may wish to adjust the norm in any case, to obtain a lower value for ω.

Example 3.3. Consider the scalar version of the differential-difference equation mentioned in Chapter II,

$$\dot{y}(t) = \alpha y(t) + \beta y(t-\tau), \qquad t \geqslant 0,$$

$$y(-\eta) \equiv u_0(\eta), \qquad 0 \leqslant \eta \leqslant \tau,$$

for given $\alpha, \beta, \tau \in \mathcal{R}$, $\tau > 0$, and $u_0(\cdot)$: $[0, \tau] \to \mathcal{R}$. As in Chapter II, we note that this equation is equivalent to

$$\dot{y}(t) = \alpha y(t) + \beta u(\tau, t),$$

$$\frac{\partial}{\partial t} u(\eta, t) = -\frac{\partial}{\partial \eta} u(\eta, t), \qquad t \geqslant 0, 0 \leqslant \eta \leqslant \tau,$$

with initial data $y(0) = u_0(0)$, $u(\eta, 0) = u_0(\eta)$; here we have merely defined $u(\eta, t) \equiv y(t-\eta)$ for $\eta \in [0, \tau]$, $t \in \mathcal{R}^+$.

To place this equation in the abstract form (1), suppose that we define a Hilbert space $\mathcal{X} \equiv \mathcal{R} \times \mathcal{L}_2(0, \tau)$ with

$$\langle x_1, x_2 \rangle \equiv y_1 y_2 + \gamma \int_0^\tau u_1(\eta) u_2(\eta)\, d\eta, \qquad x_i \equiv (y_i, u_i),$$

but leave the value of $\gamma > 0$ arbitrary for the moment. We note that all norms induced by differing values of γ are equivalent. Now defining $Ax \equiv (\alpha y + \beta u(\tau), -\partial u)$ for $x = (y, u) \in \mathcal{D}(A)$,

$$\mathcal{D}(A) \equiv \{(y, u) \in \mathcal{R} \times \mathcal{W}_2^1(0, \tau) | u(0) = y\},$$

it is easily verified that $\mathcal{D}(A)$ is dense, because $\{u \in \mathcal{W}_2^1(0, \tau) | u(0) = y\}$ is $\|\cdot\|_{\mathcal{L}_2}$-dense in $\mathcal{L}_2(0, \tau)$ for each fixed $y \in \mathcal{R}$.

Let us investigate the range of $I - \lambda A$, $\lambda > 0$, by considering the equations

$$y - \lambda\alpha y - \lambda\beta u(\tau) = \hat{y},$$

$$u(\eta) + \lambda \partial u(\eta) = \hat{u}(\eta), \qquad 0 \leqslant \eta \leqslant \tau,$$

for given $(\hat{y}, \hat{u}) \in \mathcal{R} \times \mathcal{L}_2(0, \tau)$. The second equation has a solution

$$u(\eta) = e^{-\eta/\lambda}\left(y + \int_0^\eta \hat{u}(\xi)\, d\xi\right), \qquad 0 \leqslant \eta \leqslant \tau,$$

for every arbitrary $y \in \mathcal{R}$. The first equation now becomes

$$(1 - \lambda\alpha - \lambda\beta e^{-\tau/\lambda})y = \hat{y} + \lambda\beta e^{-\tau/\lambda} \int_0^\tau \hat{u}(\xi)\, d\xi,$$

and, by choosing small enough $\lambda > 0$, we see that $1 - \lambda\alpha - \lambda\beta e^{-\tau/\lambda} \neq 0$; hence, this equation also has a solution y. Noting that $u \in \mathcal{W}_2^1(0, \tau)$ because $\hat{u} \in \mathcal{L}_2(0, 1)$, and that $u(0) = y$, we conclude that the equation $(I - \lambda A)x = \hat{x}$ has a solution $x \in \mathcal{D}(A)$ for every $\hat{x} \in \mathcal{X}$; therefore, $\mathcal{R}(I - \lambda A) = \mathcal{X}$ for sufficiently small $\lambda > 0$.

We now ask whether $\omega I - A$ is accretive (monotone) for some $\omega \in \mathcal{R}$. We note that for $x = (y, u) \in \mathcal{D}(A)$,

$$\begin{aligned}\langle x, Ax\rangle &= y[\alpha y + \beta u(\tau)] - \gamma\int_0^\tau u(\eta)\cdot \partial u(\eta)\, d\eta \\ &= \alpha y^2 + \beta y u(\tau) - (\gamma/2)\left[(u(\tau))^2 - (u(0))^2\right] \\ &= (\alpha + \gamma/2)y^2 + \beta y u(\tau) - (\gamma/2)(u(\tau))^2\end{aligned}$$

and

$$\omega\langle x, x\rangle = \omega y^2 + \omega\gamma\int_0^\tau u(\eta)u(\eta)\, d\eta.$$

Hence, for $\omega I - A$ to be monotone we require $\omega \geqslant 0$, $\omega \geqslant \alpha + \gamma/2$, and $4(\omega - \alpha - \gamma/2)\gamma/2 \geqslant \beta^2$; clearly, we can choose

$$\omega = \max\{0, \alpha + \gamma/2 + \beta^2/2\gamma\}.$$

The best result (lowest ω) is obtained if we define $\gamma = |\beta|$; then $\omega I - A$ is accretive for $\omega \equiv \max\{0, \alpha + |\beta|\}$.

By Theorem 3.2 we conclude that A generates a linear dynamical system $\{S(t)\}_{t \geqslant 0}$ on $\mathcal{X}$, and $S(\cdot)x_0\colon \mathcal{R}^+ \to \mathcal{X}$ provides the (unique) strong solution of the abstract evolution equation for every $x_0 \in \mathcal{D}(A)$. Furthermore, if $\gamma \equiv |\beta|$, which implies

$$\|x\|^2 \equiv y^2 + |\beta|\int_0^1 u(\eta)u(\eta)\, d\eta, \qquad x = (y, u) \in \mathcal{X},$$

then $\|S(t)x\| \leqslant e^{\omega t}\|x\|$ for every $x \in \mathcal{X}$, $t \in \mathcal{R}^+$, where $\omega \equiv \max\{0, \alpha + |\beta|\}$.

In the foregoing example we allowed for certain adjustments of the norm, and we were able to select the norm so as to optimize (minimize) the value of ω for this particular class of norms (parameterized by $\gamma > 0$).

We have consistently claimed, but have not yet shown, that all strong solutions of (1) are motions of $\{S(t)\}_{t \geqslant 0}$ when A is the infinitesimal generator of $\{S(t)\}_{t \geqslant 0}$. This claim requires proof.

Theorem 3.4. If $A\colon (\mathcal{D}(A) \subset \mathcal{X}) \to \mathcal{X}$ is the infinitesimal generator of a linear dynamical system $\{S(t)\}_{t \geqslant 0}$ on $\mathcal{X}$, a Banach space, then $\mathcal{D}(A)$ is positive invariant and $S(t)x_0$ is the unique strong solution of (1),

$$\dot{x}(t) = Ax(t), \qquad t \in \mathcal{R}^+,$$

$$x(0) = x_0 \in \mathcal{D}(A) \subset \mathcal{X}, \tag{1}$$

for every $x_0 \in \mathcal{D}(A)$.

Proof. In view of Theorem 2.1, we need show only that (1) has no solutions other than the motions of $\{S(t)\}_{t \geq 0}$, equivalently, that strong solutions of (1) are unique. Let $x(\cdot)$ and $y(\cdot)$ be strong solutions on $\mathcal{R}^+$. Let $\|\cdot\|_e$ be the equivalent norm asserted to exist in Theorem 3.2, and let $\omega \in \mathcal{R}$ be as there.

For simplicity, we first assume that $(\mathcal{X}, \|\cdot\|_e)$ is a Hilbert space with inner product $\langle\ ,\ \rangle$. Then, with $g(t) \equiv x(t) - y(t)$ and by the remark following Proposition II.4.1,

$$\frac{d}{dt}\langle g(t), g(t)\rangle = 2\,\mathrm{Re}\langle g(t), \dot{g}(t)\rangle = 2\,\mathrm{Re}\langle g(t), Ag(t)\rangle$$

$$\leqslant 2\omega\langle g(t), g(t)\rangle, \qquad t \in \mathcal{R}^+,$$

since $\omega I - A$ is $\|\cdot\|_e$-accretive (monotone) by Theorem 3.2. But this implies that

$$\|x(t) - y(t)\|_e^2 \leqslant e^{2\omega t}\|x(0) - y(0)\|_e^2, \qquad t \in \mathcal{R}^+,$$

and, therefore, strong solutions of (1) are unique.

If $\|\cdot\|_e$ does not satisfy the parallelogram law, the proof becomes more difficult. Since $g(\cdot)$ is everywhere differentiable on $\mathcal{R}^+$, it is everywhere Lipschitz continuous. Since

$$|\,\|g(t)\|_e - \|g(\tau)\|_e| \leqslant \|g(t) - g(\tau)\|_e, \qquad t, \tau \in \mathcal{R}^+,$$

it follows that $\|g(\cdot)\|_e$ is everywhere Lipschitz continuous and is therefore absolutely continuous and differentiable almost everywhere on compact subsets of $\mathcal{R}^+$ (Ref. 7). This implies that

$$\|g(t)\|_e - \|g(0)\|_e = \int_0^t \frac{d}{ds}\|g(s)\|_e\,ds, \qquad t \in \mathcal{R}^+,$$

with the integrand replaced by zero wherever $\| g(\cdot)\|_e$ is not differentiable. Using essentially the same argument as in the proof of Proposition II.4.2, we also find that $(d^-/dt)\| g(t)\|_e$ exists for every $t > 0$ such that $\dot{g}^-(t)$ exists, and

$$\frac{d^-}{dt}\| g(t)\|_e = -\lim_{h\searrow 0}\frac{1}{h}(\| g(t) - h\dot{g}^-(t)\|_e - \| g(t)\|_e).$$

Now, $\dot{g}(t) = \dot{x}(t) - \dot{y}(t) = Ax(t) - Ay(t)$ a.e., and $\|\cdot\|_e$-accretiveness of $\omega I - A$ implies that $\|x - y - h(Ax - Ay)\|_e \geqslant (1 - \omega h)\|x - y\|_e$ for all $x, y \in \mathfrak{D}(A)$, $h > 0$, $\omega h < 1$. Hence,

$$\frac{d}{dt}\| g(t)\|_e \leqslant \omega\| g(t)\|_e$$

wherever this derivative exists; therefore,

$$e^{-\omega t}\| g(t)\|_e - e^{\omega 0}\| g(0)\|_e = \int_0^t \frac{d}{ds}(e^{-\omega s}\| g(s)\|_e)\, ds$$

$$= \int_0^t e^{-\omega s}\left(-\omega\| g(s)\|_e + \frac{d}{ds}\| g(s)\|_e\right) ds \leqslant 0,$$

which implies that

$$\|x(t) - y(t)\|_e \leqslant e^{\omega t}\|x(0) - y(0)\|_e, \qquad t \in \mathfrak{R}^+.$$

As this assures uniqueness of solutions of (1), the proof is complete. □

Proof of Theorem 3.2 (Necessity). By Theorem 3.1, there exists $\omega \in \mathfrak{R}$ and an equivalent norm $\|\cdot\|_e$ such that $\|S(t)x\|_e \leqslant e^{\omega t}\|x\|_e$ for all $t \in \mathfrak{R}^+$, $x \in \mathfrak{X}$. Defining $F_\lambda \in \mathfrak{B}(\mathfrak{X}, \mathfrak{X})$ for $\lambda > 0$, $\lambda\omega < 1$, by

$$F_\lambda x \equiv \frac{1}{\lambda}\int_0^\infty e^{-t/\lambda} S(t)x\, dt, \qquad x \in \mathfrak{X},$$

we see that

$$\|F_\lambda x\|_e \leqslant \frac{1}{\lambda}\int_0^\infty e^{(\omega - 1/\lambda)t}\, dt = \|x\|_e/(1 - \lambda\omega)$$

for all $x \in \mathfrak{X}$, $\lambda > 0$, with $\lambda\omega < 1$. Also if $h > 0$, $x \in \mathfrak{X}$,

$$\begin{aligned}\frac{1}{h}[S(h) - I]F_\lambda x &= \frac{1}{\lambda h}\int_0^\infty e^{-t/\lambda}[S(t+h)x - S(t)x]\,dt \\ &= \frac{1}{\lambda h}\int_h^\infty e^{-(\tau-h)/\lambda}S(\tau)x\,d\tau - \frac{1}{\lambda h}\int_0^\infty e^{-t/\lambda}S(t)\,dt \\ &= \frac{1}{h}(e^{h/\lambda} - 1)F_\lambda x - \frac{1}{\lambda h}e^{h/\lambda}\int_0^h e^{-\tau/\lambda}S(\tau)\,d\tau \\ &\to \frac{1}{\lambda}F_\lambda x - \frac{1}{\lambda}x \qquad \text{as } h \searrow 0;\end{aligned}$$

it follows that $F_\lambda x \in \mathfrak{D}(A)$ and $\lambda A F_\lambda x = F_\lambda x - x$ for every $x \in \mathfrak{X}$. This implies that $\mathfrak{R}(I - \lambda A) = \mathfrak{X}$ and $(I - \lambda A)F_\lambda = I$. Using the linearity and closedness of A, as well as the definition of the Riemann integral, we also see that for $x \in \mathfrak{D}(A)$,

$$\begin{aligned}\lambda F_\lambda Ax &= \int_0^\infty e^{-t/\lambda}S(t)Ax\,dt = \int_0^\infty A[e^{-t/\lambda}S(t)x]\,dt \\ &= A\left(\int_0^\infty e^{-t/\lambda}S(t)x\,dt\right) = \lambda A F_\lambda x;\end{aligned}$$

hence, $(I - \lambda A)F_\lambda x = F_\lambda(I - \lambda A)x = x$ for $x \in \mathfrak{D}(A)$, and we conclude that $I - \lambda A$ is invertible with $(I - \lambda A)^{-1} = F_\lambda \in \mathfrak{B}(\mathfrak{X}, \mathfrak{X})$. Moreover, for $\lambda > 0$, $\lambda\omega < 1$, $\mu \equiv \lambda/(1 - \lambda\omega)$, our estimate on $\|F_\lambda x\|_e$ implies that

$$\begin{aligned}\|x + \mu(\omega I - A)x\|_e &= (1 - \lambda\omega)^{-1}\|(1 - \lambda\omega)x + \lambda(\omega I - A)x\|_e \\ &= (1 - \lambda\omega)^{-1}\|(I - \lambda A)x\|_e > \|x\|_e\end{aligned}$$

for all $x \in \mathfrak{D}(A)$ and all sufficiently small $\mu > 0$. Hence, $\omega I - A$ is $\|\cdot\|_e$-accretive, the denseness of $\mathfrak{D}(A)$ follows from Theorem 2.1, and the necessity proof is complete. □

In the foregoing proof, we note that $(I - \lambda A)^{-1}$ is a closed operator, since $\mathfrak{R}(I - \lambda A) = \mathfrak{X}$ and $(I - \lambda A)^{-1} \in \mathfrak{B}(\mathfrak{X}, \mathfrak{X})$ for sufficiently small $\lambda > 0$; hence, A is a closed operator, and there is no disagreement with (i) of Theorem 2.1.

The following sufficiency proof is not the least difficult and is borrowed largely from Crandall and Liggett (Ref. 8). The advantage is that it can be partially extended to the nonlinear case (Ref. 8), as we shall do in Section 5.

Proof of Theorem 3.2 (Sufficiency). Given $\|\cdot\|_e$ such that $\|x + \mu(\omega I - A)x\|_e \geqslant \|x\|_e$ for all $x \in \mathfrak{D}(A)$, $\mu > 0$, and given $\lambda_0 > 0$, $\lambda_0|\omega| < 1$, such that $\mathfrak{R}(I - \lambda A) = \mathfrak{X}$ for all $\lambda \in (0, \lambda_0)$, we note that

$$\left\| x - \frac{\mu}{1 + \mu\omega} Ax \right\|_e \geqslant (1 + \mu\omega)^{-1} \|x\|_e$$

for $x \in \mathfrak{X}$, $\mu > 0$, $\mu\omega > -1$; hence, defining $\lambda \equiv \mu(1 + \mu\omega)^{-1}$, we see that $\|x - \lambda Ax\| \geqslant (1 - \lambda\omega)\|x\|_e$ for all $x \in \mathfrak{D}(A)$, $0 < \lambda < \lambda_0$. It follows that $(I - \lambda A)$ is invertible on its range $\mathfrak{R}(I - \lambda A) = \mathfrak{X}$ for $0 < \lambda < \lambda_0$, and we define $J_\lambda \equiv (I - \lambda A)^{-1} \in \mathfrak{B}(\mathfrak{X}, \mathfrak{X})$ for every $\lambda \in (0, \lambda_0)$; clearly, $\|J_\lambda x\|_e \leqslant (1 - \lambda\omega)^{-1}\|x\|_e$ for $x \in \mathfrak{X}$, and both J_λ and A are closed operators.

For $t \in \mathfrak{R}^+$, we claim that an operator $S(t) \in \mathfrak{B}(\mathfrak{X}, \mathfrak{X})$ is well defined by

$$S(t)x \equiv \lim_{n\to\infty} J_{t/n}^n x, \qquad x \in \mathfrak{X},$$

the limit being taken in $\mathfrak{X}$ (not in $\mathfrak{B}(\mathfrak{X}, \mathfrak{X})$). To verify this, we shall use an estimate, obtained with considerable difficulty in Ref. 8, which is valid under our assumptions on A. For $\lambda_0 > \lambda \geqslant \gamma > 0$, integers $n \geqslant m > 0$, and $x \in \mathfrak{D}(A)$,

$$\begin{aligned} \|J_\gamma^n x - J_\lambda^m x\|_e \leqslant \Big\{ & \big[(n\gamma - \lambda m)^2 + n\gamma(\lambda - \gamma)\big]^{1/2} \exp\{2|\omega|(n\gamma + m\lambda)\} \\ & + \big[m\lambda(\lambda - \gamma) + (m\lambda - n\gamma)^2\big]^{1/2} \exp\{4|\omega| n\gamma\} \Big\} \|Ax\|_e. \end{aligned} \tag{2}$$

Setting $\gamma = t/n$, $\lambda = t/m$ in (2) we find that, for all $n \geqslant m > t/\lambda_0$ and $x \in \mathfrak{D}(A)$,

$$\begin{aligned} \|J_{t/n}^n x - J_{t/m}^m x\|_e &\leqslant 2te^{4|\omega|t}(1/m - 1/n)^{1/2}\|Ax\|_e \\ &\to 0 \qquad \text{as } n, m \to \infty; \end{aligned}$$

hence, $S(t)x$ is well defined for $x \in \mathfrak{D}(A)$. Because $\mathfrak{D}(A)$ is dense, $(1 - t\omega/n)^{-n} \to e^{\omega t}$ as $n \to \infty$, and

$$\|J_{t/n}^n x - J_{t/n}^n y\|_e \leqslant (1 - t\omega/n)^{-n}\|x - y\|_e \qquad \text{for all } x, y \in \mathfrak{X},$$

it follows that $S(t)x$ is well defined for all $x \in \mathfrak{X}$ and $\|S(t)x\|_e \leqslant e^{\omega t}\|x\|_e$; it is apparent that $S(t)\colon \mathfrak{X} \to \mathfrak{X}$ is linear.

To show continuity of $S(\cdot)x\colon \mathcal{R}^+ \to \mathcal{X}$, consider (2) with $m = n$, $\gamma = t/n$, $\lambda = \tau/n$, $\tau > t$, and let $n \to \infty$; we obtain

$$\|S(t)x - S(\tau)x\|_e \leqslant \left[\exp\{2|\omega|(t+\tau)\} + \exp\{4|\omega|t\}\right]|t-\tau|\,\|Ax\|_e$$

for $x \in \mathcal{D}(A)$. Hence, for $x \in \mathcal{D}(A)$, $S(\cdot)x$ is uniformly Lipschitz continuous on bounded subsets of $\mathcal{R}^+$, and it follows from the denseness of $\mathcal{D}(A)$ that $S(\cdot)x$ is continuous on $\mathcal{R}^+$ for every $x \in \mathcal{X}$.

To verify the semigroup property, we note that the uniform Lipschitz equicontinuity of the family $\{J^n_{t/n}\}_{n \geqslant N}$, for sufficiently large N, together with the convergence of $J^n_{t/n}x$ as $n \to \infty$, implies that

$$[S(t)]^m x = \lim_{n\to\infty} [J^n_{t/n}]^m x = \lim_{n\to\infty} [J^m_{t/n}]^n x, \qquad x \in \mathcal{X},$$

where m is a positive integer. Hence,

$$\begin{aligned} S(mt)x &= \lim_{n\to\infty} J^n_{mt/n}x = \lim_{p\to\infty} J^{mp}_{t/p}x = \lim_{p\to\infty}[J^m_{t/p}]^p x \\ &= [S(t)]^m x, \qquad x \in \mathcal{X}. \end{aligned}$$

Now consider rational $t = k/s$, $\tau = p/r$, for positive integers k, s, p, r; then, for every $x \in \mathcal{X}$,

$$\begin{aligned} S(t+\tau)x &= S\left(\frac{kr+ps}{sr}\right)x = \left[S\left(\frac{1}{sr}\right)\right]^{(kr+ps)} x \\ &= \left[S\left(\frac{1}{sr}\right)\right]^{kr}\left[S\left(\frac{1}{sr}\right)\right]^{ps} x = S\left(\frac{k}{s}\right)S\left(\frac{p}{r}\right)x = S(t)S(\tau)x. \end{aligned}$$

As the rationals are dense in compact intervals, $S(\cdot)x\colon \mathcal{R}^+ \to \mathcal{X}$ is continuous, and $S(t)\colon \mathcal{X} \to \mathcal{X}$ is Lipschitz continuous, we see that $S(t+\tau) = S(t)S(\tau)$ for all $t, \tau \in \mathcal{R}^+$.

We also note that for $\lambda > 0$, $\lambda\omega < 1$,

$$\|J^n_\lambda x\|_e \leqslant (1-\lambda\omega)^{-n}\|x\|_e, \qquad n = 1, 2, \ldots,$$

and consequently,

$$\begin{aligned} \|S(t)x\| &= \lim_{n\to\infty} \|J^n_{t/n}x\|_e \\ &\leqslant \|x\|_e \lim_{n\to\infty}\left(1 - \frac{t}{n}\omega\right)^{-n} = e^{\omega t}\|x\|_e, \qquad t \geqslant 0. \end{aligned}$$

It only remains to show that A is the infinitesimal generator of the linear dynamical system $\{S(t)\}_{t>0}$. As A is linear, $AJ^n_{t/n}x = J^n_{t/n}Ax$ for $x \in \mathfrak{D}(A)$, $t \in \mathfrak{R}^+$, $n = 1, 2, \ldots$; as A is closed, this implies that $AS(t)x = S(t)Ax$ for $x \in \mathfrak{D}(A)$, $t \in \mathfrak{R}^+$. We also note that

$$\|J^k_\lambda x - x\|_e = \left\| \sum_{p=1}^{k} J^p_\lambda x - J^{p-1}_\lambda x \right\|_e$$

$$\leq \sum_{p=1}^{k} \|J^p_\lambda x - J^{p-1}_\lambda x\|_e = \sum_{p=1}^{k} \|J^{p-1}_\lambda Ax\|_e$$

$$\leq \lambda \sum_{p=1}^{k} (1 - \lambda\omega)^{-p+1} \|Ax\|_e$$

$$\leq \lambda(1 - \lambda|\omega|)^{-k+1} \|Ax\|_e, \qquad x \in \mathfrak{D}(A).$$

Hence, $\lim_{t\searrow 0} J^k_{t/n}x = x$, converging uniformly in n, for $n \geq k \geq 0$ and $x \in \mathfrak{D}(A)$, and thus also for $x \in \mathfrak{X}$, by the denseness of $\mathfrak{D}(A)$. Therefore, considering

$$\frac{1}{t}\left[S(t)x - x\right] = \frac{1}{t} \lim_{n\to\infty} \left[J^n_{t/n}x - x\right]$$

$$= \lim_{n\to\infty} \frac{1}{t} \sum_{p=1}^{n} J^{p-1}_{t/n}\left[J_{t/n} - I\right]x$$

$$= \lim_{n\to\infty} \frac{1}{n} \sum_{p=1}^{n} J^p_{t/n}Ax, \qquad x \in \mathfrak{D}(A),$$

we see that

$$\lim_{t\searrow 0} \frac{1}{t}\left[S(t)x - x\right] = Ax, \qquad x \in \mathfrak{D}(A).$$

It can be verified that A admits no proper extension satisfying the necessary conditions, and it follows that A is the infinitesimal generator of $\{S(t)\}_{t>0}$. The sufficiency proof is complete. □

The idea behind this sufficiency proof is that of approximating equation (1) by the backward-difference equation

$$\frac{1}{t/n}\left[z(mt/n) - z(mt/n - t/n)\right] = Az(mt/n),$$

$$t \in \mathfrak{R}^+, m = 1, 2, \ldots, n,$$

$$z(0) = x_0 \in \mathfrak{D}(A) \subset \mathfrak{X}. \tag{3}$$

The solution of this equation is apparent:

$$z(t) = J_{t/n}^{n} x_0, \qquad J_\lambda \equiv (I - \lambda A)^{-1},$$

provided that $(I - tA/n)^{-1}$ exists and $z_0 \in \mathfrak{D}(J_{t/n}^{n}) = \mathfrak{R}((I - tA/n)^n)$. Letting $n \to \infty$, this approximation scheme seems to suggest that $J_{t/n}^{n} x_0 \to S(t)x_0$ in some sense, where $S(\cdot)x_0$ is a solution of (1). Our sufficiency proof shows this to be the case. Note that this backward-difference approximation scheme does not seem to require linearity of A; we shall discuss nonlinear dynamical systems in Section 5.

4. Choosing the State Space in Applications

It might now appear that, for linear autonomous systems, Section 3 provides a fairly complete answer to the modeling problem mentioned at the beginning of this chapter. Unfortunately, for the infinite-dimensional case, there still remains the problem of choosing the Banach space $\mathfrak{X}$. As a dynamical system $\{S(t)\}_{t \geq 0}$ is defined such that the mappings $S(t): \mathfrak{X} \to \mathfrak{X}$, $S(\cdot)x: \mathfrak{R}^+ \to \mathfrak{X}$, are continuous for all $t \in \mathfrak{R}^+$, $x \in \mathfrak{X}$, it is obvious that the topology of $\mathfrak{X}$ is very important. If only a formal evolution equation (e.g., a P.D.E. or F.D.E.) is known for a given physical system, the choice of a suitable state space $\mathfrak{X}$ may be considerably less than obvious. On the other hand, by our results in Section 3 for the linear heat conduction problem, there may be more than one suitable state space for a given physical system; this is demonstrated further by the following result.

Theorem 4.1. Let $A: (\mathfrak{D}(A) \subset \mathfrak{X}) \to \mathfrak{X}$, $\mathfrak{X}$ a Banach space, be the infinitesimal generator of a linear dynamical system $\{S(t)\}_{t \geq 0}$ on $\mathfrak{X}$, with $\|S(t)\|_{\mathfrak{B}(\mathfrak{X}, \mathfrak{X})} \leq Me^{\omega t}$ for all $t \in \mathfrak{R}^+$. For some $n = 1, 2, \ldots,$ let $\hat{A}$ denote the restriction of A to $\mathfrak{D}(A^{n+1})$ and let $\hat{\mathfrak{X}}$ be the Banach space $(\mathfrak{D}(A^n), \|\cdot\|_n)$, where either

$$\|x\|_n \equiv \|x\| + \|Ax\| + \cdots + \|A^n x\|$$

or

$$\|x\|_n \equiv \left(\|x\|^2 + \|Ax\|^2 + \cdots + \|A^n x\|^2\right)^{1/2}$$

for $x \in \hat{\mathfrak{X}}$. Then $\hat{A}: (\mathfrak{D}(\hat{A}) \subset \hat{\mathfrak{X}}) \to \hat{\mathfrak{X}}$ is the infinitesimal generator of the linear dynamical system $\{\hat{S}(t)\}_{t \geq 0}$ on $\hat{\mathfrak{X}}$, with $\|\hat{S}(t)\|_{\mathfrak{B}(\hat{\mathfrak{X}}, \hat{\mathfrak{X}})} \leq Me^{\omega t}$ for all

$t \in \mathcal{R}^+$; moreover, $\hat{S}(t)$ is the restriction of $S(t)$ to $\hat{\mathcal{X}} \subset \mathcal{X}$ for each $t \in \mathcal{R}^+$.

Proof. Without loss of generality, we consider only $n = 1$ and define $\hat{S}(t)$ to be the restriction of $S(t)$ to $\mathcal{D}(A)$, $t \geqslant 0$. Using (ii) of Theorem 2.1 and recalling that $J_\lambda \equiv (I - \lambda A)^{-1} \in \mathcal{B}(\mathcal{X}, \mathcal{X})$ for all $\lambda \in (0, \lambda_0)$, some $\lambda_0 > 0$, we see that $\hat{S}(t) = J_\lambda S(t)(I - \lambda A)$, $t \geqslant 0$, $\lambda \in (0, \lambda_0)$. Clearly, $\hat{S}(0) = I$ on $\mathcal{D}(A)$, $\hat{S}(t + \tau) = \hat{S}(t)\hat{S}(\tau)$, and $\mathcal{R}(\hat{S}(t)) \subset \mathcal{R}(J_\lambda) \subset \mathcal{D}(A)$ for all $t, \tau \in \mathcal{R}^+$, $\lambda \in (0, \lambda_0)$. Defining $\hat{\mathcal{X}} = (\mathcal{D}(A), \|\cdot\|_{\hat{\mathcal{X}}})$ with $\|x\|_{\hat{\mathcal{X}}} \equiv \|x\|_{\mathcal{X}} + \|Ax\|_{\mathcal{X}}$ for $x \in \mathcal{D}(A)$, we see that $\hat{\mathcal{X}}$ is a Banach space since A is closed; hence, $\hat{S}(t)$: $\hat{\mathcal{X}} \to \hat{\mathcal{X}}$ for $t \geqslant 0$. We note that the mappings $(I - \lambda A)$: $\hat{\mathcal{X}} \to \mathcal{X}$, $S(t)$: $\mathcal{X} \to \mathcal{X}$, and J_λ: $\mathcal{X} \to \hat{\mathcal{X}}$ are continuous, $t \in \mathcal{R}^+$, $0 < \lambda < \lambda_0$; hence $\hat{S}(t) = J_\lambda S(t)(I - \lambda A) \in \mathcal{B}(\hat{\mathcal{X}}, \hat{\mathcal{X}})$. As $S(\cdot)(I - \lambda A)x$: $\mathcal{R}^+ \to \mathcal{X}$ and J_λ: $\mathcal{X} \to \hat{\mathcal{X}}$ are continuous mappings, it also follows that $\hat{S}(\cdot)x$: $\mathcal{R}^+ \to \mathcal{X}$ is continuous, and we conclude that $\{\hat{S}(t)\}_{t \geqslant 0}$ is a linear dynamical system on $\hat{\mathcal{X}}$. Moreover, we note that

$$\begin{aligned} \|\hat{S}(t)x\|_{\hat{\mathcal{X}}} &= \|S(t)x\|_{\mathcal{X}} + \|AS(t)x\|_{\mathcal{X}} \\ &= \|S(t)x\|_{\mathcal{X}} + \|S(t)Ax\|_{\mathcal{X}} \\ &\leqslant Me^{\omega t}\|x\|_{\mathcal{X}} + Me^{\omega t}\|Ax\|_{\mathcal{X}} = Me^{\omega t}\|x\|_{\hat{\mathcal{X}}} \end{aligned}$$

for the first choice of $\|\cdot\|_{\hat{\mathcal{X}}}$ (and similarly for the second), for all $t \in \mathcal{R}^+$, $x \in \hat{\mathcal{X}} = \mathcal{D}(A)$; hence, $\|\hat{S}(t)\|_{\mathcal{B}(\hat{\mathcal{X}}, \hat{\mathcal{X}})} \leqslant Me^{\omega t}$ for all $t \in \mathcal{R}^+$.

Finally, we recall that the infinitesimal generator of $\{\hat{S}(t)\}_{t \geqslant 0}$ is the linear operator $\hat{A}$: $(\mathcal{D}(\hat{A}) \subset \hat{\mathcal{X}}) \to \hat{\mathcal{X}}$ defined by

$$\hat{A}x \equiv \lim_{t \searrow 0} \frac{1}{t}[\hat{S}(t)x - x],$$

where $\mathcal{D}(\hat{A})$ is the subset of $\hat{\mathcal{X}}$ for which the limit exists (in the topology of $\hat{\mathcal{X}}$). Noting that J_λ: $\mathcal{X} \to \hat{\mathcal{X}}$ is a homeomorphism (Ref. 3) for $\lambda \in (0, \lambda_0)$ and that

$$\hat{S}(t)x - x = J_\lambda[S(t) - I](I - \lambda A)x$$

for $x \in \hat{\mathcal{X}} = \mathcal{D}(A)$, it follows from the definition of the infinitesimal generator A: $(\mathcal{D}(A) \subset \hat{\mathcal{X}}) \to \mathcal{X}$ that $\mathcal{D}(\hat{A}) = \{x \in \hat{\mathcal{X}} | (I - \lambda A)x \in \mathcal{D}(A)\} = \mathcal{D}(A^2)$ and $\hat{A}$ is the restriction of A to $\mathcal{D}(A^2)$. The proof is complete. □

Example 4.1. The heat conduction problem of Example 3.1 was found to lead to a linear dynamical system $\{S(t)\}_{t \geqslant 0}$ on the Banach space

$$\mathcal{X} \equiv \{x \in \mathcal{C}[0, 1] | x(\eta) = 0 \quad \text{at} \quad \eta = 0, 1\}, \qquad \|x\|_{\mathcal{X}} \equiv \max_{\eta} |x(\eta)|,$$

satisfying $\|S(t)x\|_{\mathfrak{X}} \leq \|x\|_{\mathfrak{X}}$ for all $t \in \mathfrak{R}^+$, $x \in \mathfrak{X}$. The infinitesimal generator was shown to be defined by

$$Ax \equiv \partial^2 x, \qquad x \in \mathfrak{D}(A) \equiv \{x \in \mathcal{C}^2[0, 1] | x(\eta) = 0 = \partial^2 x(\eta) \quad \text{at} \quad \eta = 0, 1\}.$$

Choosing some $n = 1, 2, \ldots,$ it follows from Theorem 4.1 that the heat conduction problem also leads to a linear dynamical system $\{\hat{S}(t)\}_{t \geq 0}$ on the Banach space

$$\hat{\mathfrak{X}} \equiv \{x \in \mathcal{C}^{2n}[0, 1] | \partial^{2m} x(\eta) = 0 \text{ at } \eta = 0, 1; m = 0, 1, \ldots, n\},$$

$$\|x\|_{\hat{\mathfrak{X}}} \equiv \sum_{m=0}^{n} \max_{\eta} |\partial^{2m} x(\eta)|,$$

satisfying $\|\hat{S}(t)x\|_{\hat{\mathfrak{X}}} \leq \|x\|_{\hat{\mathfrak{X}}}$ for all $t \in \mathfrak{R}^+$, $x \in \hat{\mathfrak{X}}$. Moreover, the infinitesimal generator $\hat{A}$: $(\mathfrak{D}(\hat{A}) \subset \hat{\mathfrak{X}}) \to \hat{\mathfrak{X}}$ is defined by

$$\hat{A}x \equiv \partial^2 x, \qquad x \in \mathfrak{D}(\hat{A}),$$

$$\mathfrak{D}(\hat{A}) \equiv \{x \in \mathcal{C}^{2(n+1)}[0, 1] | \partial^{2m} x(\eta) = 0 \text{ at } \eta = 0, 1; m = 0, 1, \ldots, n + 1\}.$$

Exercise 4.1. Referring to Example 3.2, show that the heat conduction problem leads to a linear dynamical system $\{\hat{S}(t)\}_{t \geq 0}$ on the Hilbert space

$$\hat{\mathfrak{X}} \equiv \{x \in \mathfrak{W}_2^{2n}(0, 1) | \partial^{2m} x(\eta) = 0 \text{ at } \eta = 0, 1; m = 0, 1, \ldots, n - 1\},$$

$$\langle x, y \rangle_{\hat{\mathfrak{X}}} \equiv \int_0^1 \left[\sum_{m=0}^{n} \partial^{2m} x(\eta) \cdot \partial^{2m} y(\eta) \right] d\eta,$$

for any $n = 1, 2, \ldots,$ and $\|\hat{S}(t)x\|_{\hat{\mathfrak{X}}} \leq \exp\{-\pi^2 t\} \|x\|_{\hat{\mathfrak{X}}}$ for all $t \geq 0$, $x \in \mathfrak{X}$. Define the infinitesimal generator $\hat{A}$: $(\mathfrak{D}(\hat{A}) \subset \hat{\mathfrak{X}}) \to \hat{\mathfrak{X}}$.

Exercise 4.2. Referring to Exercises 3.3 and 3.4, show that the heat conduction problem leads to a linear dynamical system on the Hilbert space

$$\hat{\mathfrak{X}} \equiv \{x \in \mathfrak{W}_2^{2n+1}(0, 1) | \partial^{2m} x(\eta) = 0 \text{ at } \eta = 0, 1; m = 0, 1, \ldots, n\},$$

$$\langle x, y \rangle_{\hat{\mathfrak{X}}} \equiv \int_0^1 \sum_{m=0}^{2n} [\partial^m x(\eta) \cdot \partial^m y(\eta) + \partial^{m+1} x(\eta) \cdot \partial^{m+1} y(\eta)] \, d\eta,$$

or on the Banach space.

$$\tilde{\mathfrak{X}} \equiv \{x \in \mathcal{C}^{2n+1}[0, 1] | \partial^{2m} x(\eta) = 0 \text{ at } \eta = 0, 1; \; m = 0, 1, \ldots, n\},$$

$$\|x\|_{\tilde{\mathfrak{X}}} \equiv \sum_{m=0}^{2n+1} \left[\max_{\eta} |\partial^m x(\eta)| + \max_{\eta} |\partial^{m+1} x(\eta)| \right],$$

for any $n = 1, 2, \ldots$. Find the respective infinitesimal generators.

From Theorem 4.1 we see that if a linear dynamical system has an unbounded generator, then various restrictions of the dynamical system lead to an infinite family of dynamical systems on differing spaces which have differing topologies; a result in Ref. 9 generalizes this result even further. This suggests that a given linear autonomous physical system often leads to dynamical systems on a wide variety of state spaces, and perhaps our search for a "suitable" state space will not be difficult (for the heat conduction problem, it was not). This optimism is somewhat diminished by consideration of the formal wave equation mentioned in Chapter II,

$$\left(\frac{\partial}{\partial t}u(\eta, t), \frac{\partial}{\partial t}v(\eta, t)\right) = \left(v(\eta, t), \frac{\partial^2}{\partial \eta^2}u(\eta, t)\right), \qquad 0 \leqslant \eta \leqslant 1, \quad t \geqslant 0,$$

$$u(0, t) = 0 = u(1, t), \qquad t \geqslant 0,$$

with given initial data

$$(u(\eta, 0), v(\eta, 0)) = (u_0(\eta), v_0(\eta)), \qquad 0 \leqslant \eta \leqslant 1.$$

It is clear that if this wave equation leads to a dynamical system on a Banach space $\mathfrak{X}$, then $x \in \mathfrak{X}$ means $x = (u, v)$ where $u: [0, 1] \to \mathfrak{R}$, $v: [0, 1] \to \mathfrak{R}$ satisfy the boundary conditions at $\eta = 0, 1$. The first difficulty is to choose a suitable topology. It can be shown that neither

$$\mathfrak{X} \equiv \mathfrak{L}_2(0, 1) \times \mathfrak{L}_2(0, 1)$$

nor

$$\mathfrak{X} \equiv \{(u, v) \in \mathcal{C}[0, 1] \times \mathcal{C}[0, 1] | u(\eta) = 0 = v(\eta) \text{ at } \eta = 0, 1\}$$

has a suitable topology, i.e., the wave equation does not lead to a dynamical system on either of these Banach spaces. The same is true for (the appropriate subspace of) $\mathfrak{W}_2^n(0, 1) \times \mathfrak{W}_2^n(0, 1)$ and $\mathcal{C}^n[0, 1] \times \mathcal{C}^n[0, 1]$, for any $n = 1, 2, \ldots$.

One way of finding an appropriate state space for a physical system, in terms of whose norm we must have $\|S(t)x\| \leqslant Me^{\omega t}\|x\|$ for some $M, \omega \in \mathfrak{R}$ and all $x \in \mathfrak{X}$, $t \geqslant 0$, is to use outside information about the physical system. If the wave equation models a vibrating string, for example, physical considerations suggest that the total mechanical energy cannot increase from its initial value,

$$\frac{1}{2}\int_0^1 \left[|\partial u(\eta, t)|^2 + |v(\eta, t)|^2\right] d\eta \leqslant \frac{1}{2}\int_0^1 \left[|\partial u_0(\eta)|^2 + |v_0(\eta)|^2\right] d\eta, \qquad t \geqslant 0,$$

just as in Example I.8.2. Multiplying by two and taking the square root of both sides, we see that we have an "*a priori* estimate" of the form

$$\|(u(t), v(t))\| \leqslant e^{0t}\|(u_0, v_0)\|$$

provided that $\|\cdot\|$ is defined by

$$\|x\| \equiv \left(\int_0^1 \left[|\partial u(\eta)|^2 + |v(\eta)|^2\right] d\eta\right)^{1/2}, \qquad x = (u, v).$$

The following example utilizes this suggested norm.

Example 4.2. Let us set the particular wave equation just described in the form

$$\begin{aligned} \dot{x}(t) &= Ax(t), \qquad t \in \mathcal{R}^+, \\ x(0) &= x_0 \in \mathcal{D}(A) \subset \mathcal{X}, \end{aligned} \tag{1}$$

by defining a Hilbert space

$$\mathcal{X} \equiv \left\{(u, v) \in \mathcal{W}_2^1(0, 1) \times \mathcal{L}_2(0, 1) | u(\eta) = 0 \text{ at } \eta = 0, 1\right\},$$

$$\langle x_1, x_2\rangle \equiv \int_0^1 [\partial u_1(\eta)\cdot \partial u_2(\eta) + v_1(\eta)\cdot v_2(\eta)]\, d\eta, \qquad (u_i, v_i) = x_i \in \mathcal{X}, \qquad i = 1, 2,$$

and a linear operator A: $(\mathcal{D}(A) \subset \mathcal{X}) \to \mathcal{X}$,

$$Ax \equiv (v, \partial^2 u), \qquad x = (u, v) \in \mathcal{D}(A),$$

$$\mathcal{D}(A) \equiv \left\{(u, v) \in \mathcal{W}_2^2(0, 1) \times \mathcal{W}_2^1(0, 1) | u(\eta) = 0 = v(\eta) \text{ at } \eta = 0, 1\right\}.$$

We now note that for $x = (u, v) \in \mathcal{D}(A) \cap (\mathcal{C}^2 \times \mathcal{C}^1)$,

$$\langle x, Ax\rangle = \int_0^1 [\partial u(\eta)\cdot \partial v(\eta) + v(\eta)\cdot \partial^2 u(\eta)]\, d\eta = 0.$$

Using the continuity of the mapping $A: (\mathcal{D}(A), \|\cdot\|_g) \to \mathcal{X}$ and the $\|\cdot\|_g$-denseness of $\mathcal{D}(A) \cap (\mathcal{C}^2 \times \mathcal{C}^1)$ in $\mathcal{D}(A)$, it follows that $\langle x, Ax\rangle = 0$ for all $x \in \mathcal{D}(A)$. Hence, $-A$ is monotone (accretive). As the graph norm is equivalent to $\|\cdot\|_{\mathcal{W}_2^2 \times \mathcal{W}_2^1}$, $(\mathcal{D}(A), \|\cdot\|_g)$ is complete; therefore, A: $(\mathcal{D}(A) \subset \mathcal{X}) \to \mathcal{X}$ is a closed operator.

We now ask whether $\mathcal{R}(I - \lambda A) = \mathcal{X}$ for all sufficiently small $\lambda > 0$. To this end, we define

$$\mathcal{E} = \left\{x \in \mathcal{X} \middle| x(\eta) = \sum_{n=1}^{N} (\alpha_n, \beta_n) \sin n\pi\eta; \;\; (\alpha_1, \beta_1), \ldots, (\alpha_N, \beta_N) \in \mathcal{R}^2; \right.$$

$$\left. N = 1, 2, \ldots \right\},$$

noting that $\mathcal{E} \subset \mathcal{D}(A) \subset \mathcal{X}$; as $\mathcal{E}$ is dense in $\mathcal{X}$, so is $\mathcal{D}(A)$. We see that the equation $(I - \lambda A)x = \hat{x}$, given $\hat{x} \in \mathcal{E} \subset \mathcal{X}$, reduces to

$$\alpha_n - \lambda\beta_n = \hat{\alpha}_n,$$

$$\beta_n + \lambda n^2\pi^2\alpha_n = \hat{\beta}_n, \qquad n = 1, 2, \ldots, N,$$

which has a solution $(\alpha_n, \beta_n) \in \mathcal{R}^2$, $n = 1, 2, \ldots, N$, because

$$\det\begin{vmatrix} 1 & -\lambda \\ \lambda n^2\pi^2 & 1 \end{vmatrix} = 1 + \lambda^2 n^2\pi^2 \neq 0, \qquad n = 1, 2, \ldots, \quad \lambda > 0.$$

Hence, given $\hat{x} \in \mathcal{E} \subset \mathcal{X}$, there exists $x \in \mathcal{E} \subset \mathcal{D}(A)$ such that $(I - \lambda A)x = \hat{x}$, and this implies that $\mathcal{R}(I - \lambda A) \supset \mathcal{E}$. As $-A$ is accretive and closed, $J_\lambda \equiv (I - \lambda A)^{-1}$ is a closed and bounded operator; hence, $\mathcal{D}(J_\lambda) = \mathcal{R}(I - \lambda A)$ must be closed in $\mathcal{X}$. Denseness of $\mathcal{E} \subset \mathcal{R}(I - \lambda A)$ now implies that $\mathcal{R}(I - \lambda A) = \mathcal{X}$.

From Theorem 3.2 we conclude that A is the infinitesimal generator of a linear dynamical system on $\mathcal{X}$; moreover, $\|S(t)x\| \leqslant \|x\|$ for all $x \in \mathcal{X}$, $t \in \mathcal{R}^+$.

Even if we know a "good topology," the choice of the norm is sometimes critical, if we wish to use Theorem 3.2 to verify that A is an infinitesimal generator. Suppose, in Example 4.2, we had equipped $\mathcal{X}$ with the equivalent norm

$$\|(u, v)\|_e \equiv \left(\int_0^1 \left[\alpha|u(\eta)|^2 + \beta|\partial u(\eta)|^2 + 2\gamma u(\eta)\cdot v(\eta) + |v(\eta)|^2\right] d\eta\right)^{1/2},$$

for some $\alpha, \beta, \gamma \in \mathcal{R}$ such that $\beta > 0$ and $\pi^2\beta + \alpha > \gamma^2$. By the result of Example 4.2, A still generates $\{S(t)\}_{t \geqslant 0}$, and $\|S(t)x\|_e \leqslant Me^{\omega t}\|x\|_e$ for the *same* $\omega = 0$ and *some* $M \geqslant 1$; however, $\omega I - A$ is *not* $\|\cdot\|_e$-accretive for *any* $\omega \in \mathcal{R}$, unless $\beta = 1$. This is because, in the preceding estimate, it can be shown that $M > 1$ (strictly) unless $\beta \equiv 1$. Our "energy estimate" was very fortunate; it not only provided a good topology for $\mathcal{X}$, but also a good norm for application of Theorem 3.2.

When such nice outside information is not readily available, the following result suggests a formalized approach to choosing an appropriate state space, equipped with a good norm, assuming that some appropriate state space is a Hilbert space.

Theorem 4.2. Consider a linear operator $\tilde{A}$: $(\mathcal{D}(\tilde{A}) \subset \mathcal{H}) \to \mathcal{H}$, $\mathcal{H}$ a Hilbert space. Let there exist $\omega \in \mathcal{R}$, $\gamma \in \mathcal{R}^+$, and a symmetric linear B: $(\mathcal{D}(\tilde{A}) \subset \mathcal{H}) \to \mathcal{H}$ such that

(i) $\langle By, y\rangle_{\mathcal{H}} \geqslant \gamma\|y\|^2_{\mathcal{H}}$,
(ii) $\mathrm{Re}\langle By, (\omega I - \tilde{A})y\rangle_{\mathcal{H}} \geqslant 0$,

for all $y \in \mathfrak{D}(\tilde{A})$. Define a Hilbert space $\mathfrak{X}$ to be the $\|\cdot\|_{\mathfrak{X}}$-completion of $\mathfrak{D}(\tilde{A})$, where

$$\langle x, y\rangle_{\mathfrak{X}} \equiv \langle Bx, y\rangle_{\mathcal{H}}, \qquad x, y \in \mathfrak{D}(\tilde{A}),$$

and let A: $\mathfrak{D}(A) \to \mathfrak{X}$ be the restriction of $\tilde{A}$ to $\mathfrak{D}(A) \equiv \{x \in \mathfrak{D}(\tilde{A}) | \tilde{A}x \in \mathfrak{X}\}$. If $\mathfrak{R}(I - \lambda A) = \mathfrak{X}$ for all $\lambda \in (0, \lambda_0)$, some $\lambda_0 > 0$, then A is the infinitesimal generator of a linear dynamical system $\{S(t)\}_{t \geq 0}$ on $\mathfrak{X}$, satisfying $\|S(t)x\|_{\mathfrak{X}} \leq e^{\omega t}\|x\|_{\mathfrak{X}}$ for all $x \in \mathfrak{X}$, $t \in \mathfrak{R}^+$.

This result simplifies an approach suggested in Ref. 10; we defer the proof until the end of this section. We demonstrate this approach by application to our wave equation, ignoring the energy estimate.

Example 4.3. For our wave equation, we know that the state space has elements of the form (u, v), u: $[0, 1] \to \mathfrak{R}$, v: $[0, 1] \to \mathfrak{R}$, with $u(\eta) = 0 = v(\eta)$ at $\eta = 0, 1$. Our first step is to choose a Hilbert space $\mathcal{H}$ of this type, having a very simple inner product; for example, we can choose

$$\mathcal{H} \equiv \mathfrak{L}_2(0, 1) \times \mathfrak{L}_2(0, 1),$$

$$\langle y_1, y_2\rangle \equiv \int_0^1 [u_1(\eta)\cdot u_2(\eta) + v_1(\eta)\cdot v_2(\eta)]\, d\eta, \qquad y_i = (u_i, v_i) \in \mathcal{H}, \qquad i = 1, 2.$$

Our second step is to use the formal equations to write an artificial evolution equation of the form

$$\dot{y}(t) = \tilde{A}y(t), \qquad t \geq 0,$$

$$y(0) = y_0 \in \mathfrak{D}(\tilde{A}) \subset \mathcal{H};$$

hence, we define $\tilde{A}$: $(\mathfrak{D}(\tilde{A}) \subset \mathcal{H}) \to \mathcal{H}$ by

$$\tilde{A}x \equiv (v, \partial^2 u), \qquad x = (u, v) \in \mathfrak{D}(\tilde{A}),$$

$$\mathfrak{D}(\tilde{A}) \equiv \hat{\mathfrak{W}}_2^2(0, 1) \times \mathfrak{L}_2(0, 1),$$

where

$$\hat{\mathfrak{W}}_2^n(0, 1) \equiv \{x \in \mathfrak{W}_2^n(0, 1) | \partial^{2m}x(\eta) = 0 \text{ at } \eta = 0, 1; \quad m = 0, \ldots, < n/2\}.$$

Without stopping to ask whether $\tilde{A}$ is an infinitesimal generator, our third step is to seek $\omega \in \mathfrak{R}$ and a symmetric linear B: $(\mathfrak{D}(\tilde{A}) \subset \mathcal{H}) \to \mathcal{H}$, satisfying the conditions of Theorem 4.2. Writing By as

$$By = \begin{bmatrix} B_{11} & B_{12} \\ B_{21} & B_{22} \end{bmatrix} \begin{pmatrix} u \\ v \end{pmatrix}, \qquad y = (u, v) \in \mathfrak{D}(\tilde{A}) = \mathfrak{D}(B),$$

we see that

$$B_{11}\colon \hat{\mathcal{W}}_2^2 \to \mathcal{L}_2, \qquad B_{12}\colon \mathcal{L}_2 \to \mathcal{L}_2,$$

$$B_{21}\colon \hat{\mathcal{W}}_2^2 \to \mathcal{L}_2, \qquad B_{22}\colon \mathcal{L}_2 \to \mathcal{L}_2,$$

and symmetry of B is equivalent to

$$\langle u, B_{11}\hat{u}\rangle_{\mathcal{L}_2} = \langle B_{11}u, \hat{u}\rangle_{\mathcal{L}_2}, \qquad \langle v, B_{22}\hat{v}\rangle_{\mathcal{L}_2} = \langle B_{22}v, \hat{v}\rangle_{\mathcal{L}_2},$$

$$\langle u, B_{12}\hat{v}\rangle_{\mathcal{L}_2} + \langle v, B_{21}\hat{u}\rangle_{\mathcal{L}_2} = \langle B_{21}u, \hat{v}\rangle_{\mathcal{L}_2} + \langle B_{12}v, \hat{u}\rangle_{\mathcal{L}_2},$$

for all $u, \hat{u} \in \hat{\mathcal{W}}_2^2$ and $v, \hat{v} \in \mathcal{L}_2$. To simplify, let us tentatively choose $B_{12} = 0$, $B_{21} = 0$; then condition (i) requires that

$$\langle u, B_{11}u\rangle_{\mathcal{L}_2} \geq 0, \qquad \langle v, B_{22}v\rangle_{\mathcal{L}_2} \geq 0,$$

and (ii) requires $\omega \in \mathcal{R}$ such that

$$\langle B_{11}u, \omega u - v\rangle_{\mathcal{L}_2} + \langle B_{22}v, \omega v - \partial^2 u\rangle_{\mathcal{L}_2} \geq 0,$$

for all $u \in \hat{\mathcal{W}}_2^2$, $v \in \mathcal{L}_2$. The last condition is

$$\int_0^1 \left[v(\eta)\cdot B_{11}u(\eta) + B_{22}v(\eta)\cdot \partial^2 u(\eta)\right] d\eta$$
$$\leq \omega \int_0^1 [u(\eta)\cdot B_{11}u(\eta) + v(\eta)\cdot B_{22}v(\eta)]\, d\eta.$$

Inspecting these conditions and noting that

$$\langle u, -\partial^2 \hat{u}\rangle_{\mathcal{L}_2} = \langle \partial u, \partial \hat{u}\rangle_{\mathcal{L}_2},$$

$$\langle u, -\partial^2 u\rangle_{\mathcal{L}_2} \geq \pi^2 \langle u, u\rangle_{\mathcal{L}_2},$$

for all $u, \hat{u} \in \hat{\mathcal{W}}_2^2$, we see that all conditions can be satisfied (for $\omega = 0$!) if we choose $B_{22} \equiv I$, $B_{11} \equiv -\partial^2$.

Our fourth step is define a Hilbert space $\mathcal{X}$ as the $\|\cdot\|_{\mathcal{X}}$-completion of $\mathcal{D}(\tilde{A}) = \hat{\mathcal{W}}_2^2(0, 1) \times \mathcal{L}_2(0, 1)$, where

$$\langle y_1, y_2\rangle_{\mathcal{X}} \equiv \langle By_1, y_2\rangle = \int_0^1 \left[-\partial^2 u_1(\eta)\cdot u_2(\eta) + v_1(\eta)\cdot v_2(\eta)\right] d\eta$$
$$= \int_0^1 [\partial u_1(\eta)\cdot \partial u_2(\eta) + v_1(\eta)\cdot v_2(\eta)]\, d\eta,$$

$$y_i = (u_i, v_i) \in \mathcal{D}(\tilde{A}),\ i = 1, 2.$$

Clearly, $\mathcal{X} = \hat{\mathcal{W}}_2^1(0, 1) \times \mathcal{L}_2(0, 1)$. We also define $A\colon (\mathcal{D}(A) \subset \mathcal{X}) \to \mathcal{X}$ as the restriction of $\tilde{A}$ to

$$\begin{aligned}\mathfrak{D}(A) &\equiv \{x \in \mathfrak{D}(\tilde{A}) | \tilde{A}x \in \mathfrak{X}\} \\ &= \{(u, v) \in \hat{\mathfrak{W}}_2^2 \times \mathfrak{L}_2 | v \in \hat{\mathfrak{W}}_2^1, \partial^2 u \in \mathfrak{L}_2\} \\ &= \hat{\mathfrak{W}}_2^2(0, 1) \times \hat{\mathfrak{W}}_2^1(0, 1).\end{aligned}$$

We note that $\mathfrak{X}$, $\|\cdot\|_{\mathfrak{X}}$, and $A: (\mathfrak{D}(A) \subset \mathfrak{X}) \to \mathfrak{X}$ are exactly as defined in Example 4.1; there we found that $\mathfrak{R}(I - \lambda A) = \mathfrak{X}$, $\lambda > 0$, and all conclusions of that example now follow from Theorem 4.2.

Exercise 4.3. With $\hat{\mathfrak{W}}_2^n(0, 1)$ as defined in Example 4.3, show that our wave equation leads to a dynamical system on each of the spaces $\mathfrak{X}_n \equiv \hat{\mathfrak{W}}_2^{n+1}(0, 1) \times \hat{\mathfrak{W}}_2^n(0, 1)$, $n = 1, 2, \ldots$, and show that the norm can be defined such that $\|S_n(t)x\|_n \leqslant \|x\|_n$ for all $t \in \mathfrak{R}^+$, $x \in \mathfrak{X}_n$. Then $\{S_n(t)\}_{t \geqslant 0}$ is a *nonexpansive* dynamical system on $\mathfrak{X}_n$. *Hint:* Use Theorem 4.1.

When a "suitable" *a priori* estimate is available, there is no real need to use the approach suggested by Theorem 4.2.

Exercise 4.4. Again consider the wave equation, but with the boundary conditions changed to $u(0, t) = 0 = (\partial/\partial\eta)u(1, t)$, $t > 0$. Set up an associated abstract evolution equation (1) on (the appropriate subspace $\mathfrak{X}$ of) $\mathfrak{W}_2^1(0, 1) \times \mathfrak{L}_2(0, 1)$. Choose a norm, equivalent to $\|\cdot\|_{\mathfrak{W}_2^1 \times \mathfrak{L}_2}$ on $\mathfrak{X}$, such that A generates a nonexpansive dynamical system. *Hint*: The expression for total mechanical energy is unchanged, and there is still no physical reason to believe that it can increase with increasing $t \in \mathfrak{R}^+$.

Exercise 4.5. Consider bending deflection of a slender uniform bar, formally described by the Bernoulli–Euler beam equation

$$\rho \frac{\partial^2}{\partial t^2} u(\eta, t) + \alpha \frac{\partial^4}{\partial \eta^4} u(\eta, t) + \beta \frac{\partial^2}{\partial \eta^2} u(\eta, t) = 0, \qquad 0 \leqslant \eta \leqslant 1, \quad t \geqslant 0,$$

where ρ denotes mass/(unit length) ($\rho > 0$), α is the bending stiffness ($\alpha > 0$), and β is a constant axial load (assume $\beta < \alpha\pi^2$). Assume pinned ends,

$$u(\eta, t) = 0 = \frac{\partial^2}{\partial \eta^2} u(\eta, t) \qquad \text{at} \quad \eta = 0, 1 \qquad \text{for} \quad t \geqslant 0,$$

and initial data

$$u(\eta, 0) = u_0(\eta), \qquad \frac{\partial}{\partial t} u(\eta, 0) = v_0(\eta), \qquad 0 \leqslant \eta \leqslant 1.$$

Describe this problem in the form (1), A an infinitesimal generator of a linear dynamical system on a suitable Hilbert space $\mathcal{X}$. *Hint:* For any $\beta \in \mathcal{R}$, the total mechanical energy is of the form

$$\frac{1}{2}\int_0^1 [\alpha|\partial^2 u(\eta, t)|^2 - \beta|\partial u(\eta, t)|^2 + \rho|v(\eta, t)|^2]\, d\eta,$$

and it is nonnegative for $\beta \leqslant \alpha\pi^2$. Under our assumption that $\beta < \alpha\pi^2$, show that a norm can be used for which $\{S(t)\}_{t \geqslant 0}$ is nonexpansive.

Even when some sort of *a priori* estimate is available, it may not suggest a suitable norm; then the approach suggested by Theorem 4.2 can be used, but the *a priori* estimate may still help to reduce the guesswork (in choosing $\mathcal{H}$, $\tilde{A}$, and B) that is inherent with this approach.

Exercise 4.6. Reconsider the pin-ended beam, of the previous exercise, for axial load $\beta > \alpha\pi^2$. Obtain a related dynamical system, its infinitesimal generator, and an estimate of the form $\|S(t)x\| \leqslant e^{\omega t}\|x\|$ for all $t \geqslant 0$, $x \in \mathcal{X}$, and some $\omega \in \mathcal{R}$. Note that the total mechanical energy is not nonnegative. *Hint:* Use the method suggested by Theorem 4.2 with $\mathcal{H} = \mathcal{L}_2(0, 1) \times \mathcal{L}_2(0, 1)$, choosing B (independent of β) so that $\mathcal{X}$ has the same topology as in the previous exercise.

We see that *a priori* estimates, when available, can be quite helpful in arriving at a suitable state space (and even in suggesting a norm useful in Theorem 3.2); however, we also see that such estimates usually arise from the underlying physics of the system, and not from the formal equation itself. For some formal equations, whether or not they arise from physical systems, no such *a priori* estimates are readily available; then the method suggested by Theorem 4.2 becomes even more useful, assuming that some appropriate state space happens to be a Hilbert space.

Exercise 4.7. Given $\alpha, \beta, \tau \in \mathcal{R}$, where $0 < |\beta| < 1$ and $\tau > 0$, consider a scalar differential–difference equation of neutral type,

$$\dot{v}(t) = \alpha v(t) + \beta \frac{d}{dt} v(t - \tau), \qquad t \geqslant 0,$$

with initial data

$$v(-\eta) = u_0(\eta), \qquad 0 \leqslant \eta \leqslant \tau,$$

where $v: [-\tau, \infty) \to \mathcal{R}$. Defining $w(t) \equiv v(t) - \beta v(t - \tau)$, $u(\eta, t) \equiv v(t - \eta)$ for $t \in \mathcal{R}^+$, $\eta \in [0, \tau]$, we restate the formal equation as

$$\dot{w}(t) = \alpha w(t) + \alpha\beta u(\tau, t), \qquad t \geqslant 0,$$

$$\frac{\partial}{\partial t} u(\eta, t) = -\frac{\partial}{\partial \eta} u(\eta, t), \qquad 0 \leqslant \eta \leqslant \tau, \quad u(0, t) = w(t) + \beta u(\tau, t),$$

with initial data $w(0) = u_0(0) - \beta u_0(\tau)$, $u(\eta, 0) = u_0(\eta)$. Consider $(w,u) = y \in \mathcal{H} \equiv \mathcal{R} \times \mathcal{L}_2(0, \tau)$, and an artificial operator $\tilde{A}: (\mathcal{D}(\tilde{A}) \subset \mathcal{H}) \to \mathcal{H}$ given by

$$\tilde{A}y = (\alpha w + \alpha\beta u(\tau), -\partial u), \qquad y = (w, u) \in \mathcal{D}(\tilde{A}),$$

$$\mathcal{D}(\tilde{A}) \equiv \{(w, u) \in \mathcal{R} \times \mathcal{W}_2^1(0, \tau) | w = u(0) - \beta u(\tau)\}.$$

Use the approach suggested by Theorem 4.2 to obtain a suitable state space $\mathcal{X}$ and corresponding infinitesimal generator $A: (\mathcal{D}(A) \subset \mathcal{X}) \to \mathcal{X}$; try to obtain a low value for ω in the estimate $\|S(t)\|_{\mathcal{B}(\mathcal{X}, \mathcal{X})} \leqslant Me^{\omega t}$, $t \geqslant 0$.

Notice should be taken of γ in condition (i) of Theorem 4.2. We insist only that $\gamma \in \mathcal{R}^+$ and might choose $\gamma = 0$; however, the choice of γ involves a subtlety. If $\gamma > 0$, then $\|\cdot\|_{\mathcal{X}}$ is necessarily a norm for the linear space $\mathcal{D}(\tilde{A})$, and $\mathcal{X} \subset \mathcal{H}$. If $\gamma = 0$, but $\langle By, y\rangle_{\mathcal{H}} > 0$ for all nonzero $y \in \mathcal{D}(\tilde{A})$, then $\|\cdot\|_{\mathcal{X}}$ is still a norm for the linear space $\mathcal{D}(\tilde{A})$, but we may not have $\mathcal{X} \subset \mathcal{H}$. If $\gamma = 0$ and $\langle By, y\rangle_{\mathcal{H}} > 0$ for some $y \in \mathcal{D}(\tilde{A})$, then $\|\cdot\|_{\mathcal{X}}$ may be only a *seminorm* (Ref. 3) for $\mathcal{D}(\tilde{A})$, and $(\mathcal{D}(\tilde{A}), \|\cdot\|_{\mathcal{X}})$ may have to be viewed as a normed linear space of equivalence classes of elements of $\mathcal{D}(\tilde{A})$. In fact, Theorem 4.2 obviously holds for $B \equiv 0$ (implying $\gamma = 0$), but then $\mathcal{X}$ has zero dimension ($\mathcal{X} = \{0\}$), and all elements of $\mathcal{D}(\tilde{A})$ belong to the same equivalence class, which is $0 \in \mathcal{X}$; this is a worthless result. Even assuming $B \neq 0$, we may still "lose part of the problem" if $\gamma = 0$ in Theorem 4.2; consequently, care should be exercised when choosing B, so that nothing "important" is lost.

On several occasions we have implied that an appropriate state space $\mathcal{X}$ can be found by use of the approach suggested in Theorem 4.2, provided that some suitable state space happens to be a Hilbert space. This implication is not obvious, even using the result of Theorem 3.1. Suppose that $A: (\mathcal{D}(A) \subset \mathcal{X}) \to \mathcal{X}$ is an infinitesimal generator of a dynamical system $\{S(t)\}_{t \geqslant 0}$ on $\mathcal{X}$, a Hilbert space, with $\|S(t)\|_{\mathcal{B}(\mathcal{X}, \mathcal{X})} \leqslant Me^{\omega t}$ for all $t \in \mathcal{R}^+$, some $M, \omega \in \mathcal{R}$; then Theorem 3.1 asserts only the existence of an equivalent norm $\|\cdot\|_e$ such that $\|S(t)x\|_e \leqslant e^{\omega t}\|x\|_e$ for all $t \geqslant 0$, $x \in \mathcal{X}$. However, Theorem 3.1 gives no assurance that $(\mathcal{X}, \|\cdot\|_e)$ will also be a Hilbert space. The following result shows that our implication is nevertheless justified (Refs. 10, 11).

Theorem 4.3. Let $\{S(t)\}_{t \geqslant 0}$ be a linear dynamical system on a Hilbert space $\mathcal{X}$, with $\|S(t)\|_{\mathcal{B}(\mathcal{X}, \mathcal{X})} \leqslant Me^{\omega t}$ for all $t \geqslant 0$. Then, for any $\hat{\omega} > \omega$, there exists a Hilbert space $\hat{\mathcal{X}}$, $c > 0$, and a linear dynamical system $\{\hat{S}(t)\}_{t \geqslant 0}$ on $\hat{\mathcal{X}}$ such that $\hat{\mathcal{X}} \supset \mathcal{X}$, $\|x\|_{\hat{\mathcal{X}}} \leqslant c\|x\|_{\mathcal{X}}$ for all $x \in \mathcal{X}$, $\|\hat{S}(t)\|_{\mathcal{B}(\hat{\mathcal{X}}, \hat{\mathcal{X}})} \leqslant e^{\hat{\omega} t}$ for all $t \geqslant 0$, and $S(t)$ is the restriction of $\hat{S}(t)$ to $\mathcal{X}$ for $t \geqslant 0$.

Proof. Choosing any $\hat{\omega} > \omega$, define

$$\langle x, y\rangle_{\hat{\mathfrak{X}}} \equiv \int_0^\infty e^{-2\hat{\omega}\tau}\langle S(\tau)x, S(\tau)y\rangle_{\mathfrak{X}}\, d\tau, \qquad x, y \in \mathfrak{X},$$

$$\|x\|_{\hat{\mathfrak{X}}} \equiv \langle x, x\rangle_{\hat{\mathfrak{X}}}^{1/2}, \qquad x \in \mathfrak{X},$$

noting that $\|x\|_{\hat{\mathfrak{X}}} \leqslant M\|x\|_{\mathfrak{X}}$ for all $x \in \mathfrak{X}$. We see that $(\mathfrak{X}, \|\cdot\|_{\hat{\mathfrak{X}}}, \langle\, ,\, \rangle_{\hat{\mathfrak{X}}})$ is a pre-Hilbert space, and we denote its $\|\cdot\|_{\hat{\mathfrak{X}}}$-completion by $\hat{\mathfrak{X}}$. Hence, $\hat{\mathfrak{X}}$ is a Hilbert space and $\mathfrak{X} \subset \hat{\mathfrak{X}}$.

At each $t \in \mathfrak{R}^+$, $x \in \mathfrak{X}$, we see that

$$\begin{aligned}\|S(t)x\|_{\hat{\mathfrak{X}}}^2 &= \int_0^\infty e^{-2\hat{\omega}\tau}\langle S(t+\tau)x, S(t+\tau)x\rangle_{\mathfrak{X}}\, d\tau\\ &= e^{2\hat{\omega}t}\int_0^\infty e^{-2\hat{\omega}\eta}\langle S(\eta)x, S(\eta)x\rangle_{\mathfrak{X}}\, d\eta \leqslant e^{2\hat{\omega}t}\|x\|_{\hat{\mathfrak{X}}}^2,\end{aligned}$$

and we denote by $\hat{S}(t)$ the (unique) continuous linear extension of $S(t)$ to all of $\hat{\mathfrak{X}}$.

To verify that $\{\hat{S}(t)\}_{t\geqslant 0}$ is a dynamical system on $\hat{\mathfrak{X}}$, consider $t, \tau \in \mathfrak{R}^+$, $x \in \hat{\mathfrak{X}}$, and sequences $\{t_n\} \subset \mathfrak{R}^+$, $\{x_n\} \subset \mathfrak{X}$, such that $t_n \to t$ and $x_n \stackrel{\hat{\mathfrak{X}}}{\to} x$ as $n \to \infty$; then

$$\begin{aligned}\|\hat{S}(t+\tau)x - \hat{S}(t)\hat{S}(\tau)x\|_{\hat{\mathfrak{X}}} &\leqslant \|\hat{S}(t+\tau)x - \hat{S}(t+\tau)x_n\|_{\hat{\mathfrak{X}}}\\ &\quad + \|\hat{S}(t+\tau)x_n - \hat{S}(t)\hat{S}(\tau)x_n\|_{\hat{\mathfrak{X}}}\\ &\quad + \|\hat{S}(t)\hat{S}(\tau)x_n - \hat{S}(t)\hat{S}(\tau)x\|_{\hat{\mathfrak{X}}}\\ &\leqslant 2e^{\hat{\omega}(t+\tau)}\|x - x_n\|_{\hat{\mathfrak{X}}} + M\|S(t+\tau)x_n - S(t)S(\tau)x_n\|_{\mathfrak{X}}\\ &= 2e^{\hat{\omega}(t+\tau)}\|x - x_n\|_{\hat{\mathfrak{X}}} \to 0 \qquad \text{as} \quad n \to \infty,\end{aligned}$$

and it follows that $\hat{S}(t+\tau)x = \hat{S}(t)\hat{S}(\tau)x$. Also, making use of the remark following Definition 1.1, we see that

$$\begin{aligned}\|\hat{S}(t_n)x - \hat{S}(t)x\|_{\hat{\mathfrak{X}}} &\leqslant \|\hat{S}(t_n)x - \hat{S}(t_n)x_n\|_{\hat{\mathfrak{X}}} + \|\hat{S}(t_n)x_n - \hat{S}(t)x_n\|_{\hat{\mathfrak{X}}}\\ &\quad + \|\hat{S}(t)x_n - \hat{S}(t)x\|_{\hat{\mathfrak{X}}}\\ &\leqslant (e^{\hat{\omega}t_n} + e^{\hat{\omega}t})\|x_n - x\|_{\hat{\mathfrak{X}}} + M\|S(t_n)x_n - S(t)x_n\|_{\mathfrak{X}}\\ &\to 0 \qquad \text{as} \quad n \to \infty.\end{aligned}$$

Consequently, $\hat{S}(\cdot)x\colon \mathcal{R}^+ \to \hat{\mathcal{X}}$ is continuous, and the proof is complete. □

This result permits a certain amount of confidence in the method suggested by Theorem 4.2, assuming that there exists some appropriate state space that happens to be a Hilbert space. If both A and $-A$ are infinitesimal generators (that is, if A generates a group), it has been shown (Ref. 12) that the conclusion of Theorem 4.3 holds with $\hat{\mathcal{X}} = \mathcal{X}$ and $\|\cdot\|_{\hat{\mathcal{X}}}$ equivalent to $\|\cdot\|_{\mathcal{X}}$. If $-A$ is not a generator, there is a partially open question here.

Exercise 4.8. Consider A as defined in Example 4.2 and show that $-A$ is an infinitesimal generator. Can the same be said for A as defined in Example 3.2?

These remarks were made because in Theorem 4.2 both $\mathcal{H}$ (by assumption) and $\mathcal{X}$ (by conclusion) were Hilbert spaces. Although a version of Theorem 4.2 can be given in terms of Banach spaces, we have not done so because considerable complexity would be involved. Many simplifications occur for Hilbert spaces; for example, the following result simplifies the proof of Theorem 4.2.

Lemma 4.1. Let $A\colon (\mathcal{D}(A) \subset \mathcal{X}) \to \mathcal{X}$ be linear, $\mathcal{X}$ a Hilbert space. If $\omega I - A$ is accretive (monotone) for some $\omega \in \mathcal{R}$, and if $\mathcal{R}(I - \lambda A) = \mathcal{X}$ for all sufficiently small $\lambda > 0$, then $\mathcal{D}(A)$ is dense.

Proof. Suppose $\mathcal{D}(A)$ is not dense; then there exists nonzero $x_0 \in \mathcal{X}$ such that $\langle x_0, x\rangle = 0$ for all $x \in \mathcal{D}(A)$, since $\mathcal{D}(A)$ is a linear manifold (Ref. 3). As $\mathcal{R}(I - \lambda A) = \mathcal{X}$ for some $\lambda > 0$, $\lambda\omega < 1$, it follows that $x_0 = (I - \lambda A)y_0$ for some nonzero $y_0 \in \mathcal{D}(A)$; therefore,

$$0 = \langle y_0 - \lambda A y_0, y_0\rangle = (1 - \lambda\omega)\|y_0\|^2 + \lambda\langle(\omega I - A)y_0, y_0\rangle$$

$$> \lambda\langle(\omega I - A)y_0, y_0\rangle,$$

which contradicts the monotonicity of $\omega I - A$. Consequently, $\mathcal{D}(A)$ is dense and the proof is complete. □

Proof of Theorem 4.2. By definition, $\mathcal{D}(A) \subset \mathcal{D}(\tilde{A}) \subset \mathrm{Cl}_{\mathcal{X}}\, \mathcal{D}(\tilde{A}) = \mathcal{X}$; hence, $A\colon (\mathcal{D}(A) \subset \mathcal{X}) \to \mathcal{X}$. By (ii), $\mathrm{Re}\langle x, (\omega I - A)x\rangle_{\mathcal{X}} \geq 0$ for all $x \in \mathcal{D}(A) \subset \mathcal{D}(\tilde{A})$, and $\omega I - A$ is $\|\cdot\|_{\mathcal{X}}$-accretive. By assumption, $\mathcal{R}(I - \lambda A) = \mathcal{X}$ for all $\lambda \in (0, \lambda_0)$, some $\lambda_0 > 0$, and Lemma 4.1 shows that $\mathcal{D}(A)$ is dense in $\mathcal{X}$. Applying Theorem 3.2 completes the proof. □

5. Generation of Nonlinear Dynamical Systems

By Theorem 2.1 we saw that every linear dynamical system $\{S(t)\}_{t\geq 0}$, on a Banach space $\mathfrak{B}$, is associated with some autonomous evolution equation

$$\begin{aligned} \dot{x}(t) &= Ax(t), \qquad t \in \mathfrak{R}^+, \\ x(0) &= x_0 \in \mathfrak{D}(A) \subset \mathfrak{B}, \end{aligned} \tag{4}$$

when $A : (\mathfrak{D}(A) \subset \mathfrak{B}) \to \mathfrak{B}$, with $\mathfrak{D}(A)$ dense, is the infinitesimal generator of $\{S(t)\}_{t\geq 0}$. Unfortunately, severe difficulties arise in attempts to state an equally simple relationship between nonlinear dynamical systems and nonlinear autonomous equations of evolution. Avoiding this problem for the moment, we recall that there is another sense in which a linear dynamical system can be said to be generated. By the proof of Theorem 3.2, there exists a family $\{J_\lambda\}_{\lambda \in [0, \lambda_0)}$ of continuous operators, $J_\lambda : \mathfrak{B} \to \mathfrak{B}$, such that $J_0 \equiv I$ and $J^n_{t/n}x \to S(t)x$ as $n \to \infty$, converging uniformly on compact t-intervals, for each $x \in \mathfrak{B}$. This idea motivates the following definition of a "generated" dynamical system, linear or not, on a general metric space $\mathfrak{X}$ (Ref. 13)

Definition 5.1. Let $\{S(t)\}_{t\geq 0}$ be a dynamical system on a metric space $\mathfrak{X}$. Let there exist a family $\{J_\lambda\}_{\lambda \in [0, \lambda_0)}$, $\lambda_0 > 0$, of continuous operators $J_\lambda : \mathfrak{X} \to \mathfrak{X}$, $J_0 \equiv I$, such that

(i) $J^n_\lambda x \to x$ as $\lambda \searrow 0$, $n = 1, 2, \ldots,$ for every $x \in \mathfrak{X}$,

(ii) $\lim_{n\to\infty} J^n_{t/n}x = S(t)x$ exists for all $x \in \mathfrak{X}$, $t \in \mathfrak{R}^+$, converging uniformly on compact t-intervals.

Then $\{S(t)\}_{t\geq 0}$ is said to be *generated* by a product formula.

At present an open question seems to be whether all (nonlinear) dynamical systems are generated in this sense, but it does appear that every dynamical system "closely associated" with a differential evolution equation is of this type. Before discussing such possible associations in any detail, let us consider whether (4) is actually the evolution equation that we wish to discuss, assuming now that A is a nonlinear operator.

Suppose that $\mathfrak{B} = \mathfrak{R}^1$ and that $A : \mathfrak{B} \to \mathfrak{B}$ is defined by

$$\begin{aligned} Ax &\equiv -1 \qquad \text{if } x > 0, \\ Ax &\equiv \alpha \qquad \text{if } x = 0, \\ Ax &\equiv 0 \qquad \text{if } x < 0, \end{aligned}$$

for some fixed $\alpha \in \mathfrak{R}$. We see that (4) has a solution on $\mathfrak{R}^+$ for $x_0 < 0$; but if $\alpha \neq 0$, there is no solution at all for $x_0 = 0$, and for $x_0 > 0$ a solution

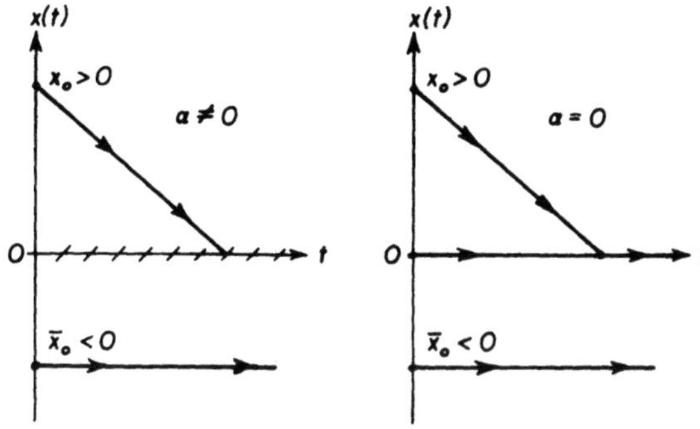

Fig. 4. Some solutions.

exists only on the time interval [0, 1). By defining $\alpha = 0$ we can obtain a solution for $x_0 = 0$, as well as $x_0 < 0$, but a problem remains for $x_0 > 0$; although the solution on $[0, x_0)$ can be "continuously joined" with another solution on $[x_0, \infty)$, the resulting "motion" is not differentiable at $t = x_0$. See Figure 4.

It appears that we may wish to weaken (4), considering instead

$$\begin{aligned} \dot{x}(t) &= Ax(t), \qquad \text{a.e. } t \in \mathfrak{R}^+, \\ x(0) &= x_0 \in \mathrm{Cl}_{\mathfrak{B}}\mathfrak{D}(A), \end{aligned} \tag{5}$$

insisting that $x(\cdot) : \mathfrak{R}^+ \to \mathfrak{B}$ be continuous (and have several other properties), but not requiring differentiability for all $t \in \mathfrak{R}^+$. We see that our simple example can be nicely fitted into the form (5) with $\mathfrak{D}(A) \equiv \mathfrak{R}^1 = \mathfrak{B}$, provided that we have the insight to choose $\alpha = 0$ before doing so.

As an example in which considerably more insight seems to be required (in order to make good use of the description (5)), consider a simple spring-mass system with Coulomb friction (dry friction), defined by

$$(\dot{u}(t), \dot{v}(t)) = (v(t), -u(t) - f(v(t))) \in \mathfrak{R}^2, \qquad \text{a.e. } t \in \mathfrak{R}^+,$$

where $|f(\eta)| \leqslant \mu$ for all $\eta \in \mathfrak{R}$, and $f(\eta) = \mu\eta/|\eta|$ for $\eta \neq 0$, for given $\mu > 0$. How should we define $f(0)$? If we want to put this problem in the form (5) and have solutions for all $x_0 = (u_0, v_0) \in \mathfrak{R}^2 = \mathfrak{D}(A) = \mathfrak{B}$, it can be shown that $f(0)$ must be defined by

$$\begin{aligned} f(0) &\equiv \mu && \text{if } u < -\mu, \\ f(0) &\equiv -u && \text{if } |u| \leqslant \mu, \\ f(0) &\equiv -\mu && \text{if } u > \mu, \end{aligned}$$

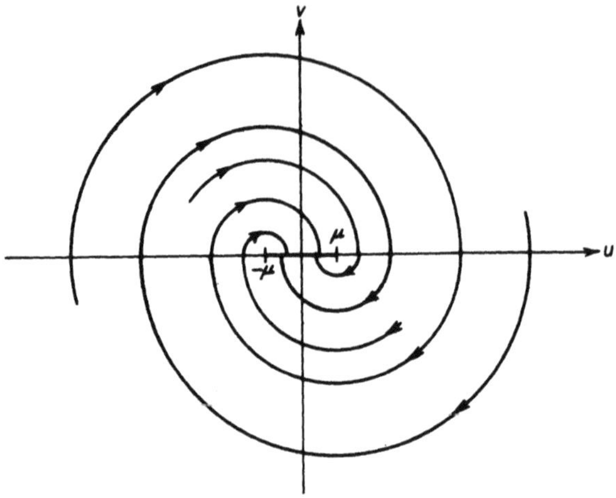

Fig. 5. Some positive orbits.

assuming $\pm\mu$ to be extremal values of the dry friction force. The motions, when projected on $\mathcal{X} = \mathcal{R}^2$, have the form sketched in Figure 5.

Our "suitable" definition of $f(0)$ can be arrived at by a detailed analysis of the physical system. Unfortunately, sometimes physical reasoning is not enough, as in the Navier–Stokes equations for an incompressible fluid; there, the pressure gradient is a similarly mysterious quantity. In fact, even our definition of $f(0)$ is rather mysterious, since its "value" is the *set* $[-\mu, \mu] \subset \mathcal{R}$, rather than a *point* in $\mathcal{R}$. Let $2^{\mathcal{R}}$ denote the set of all subsets of $\mathcal{R}$ (including $\mathcal{R}$ and the empty set $\varnothing$); then we see that we have defined a function $f : \mathcal{R} \to 2^{\mathcal{R}}$, *not* a function from $\mathcal{R}$ into $\mathcal{R}$. Actually, it will be more convenient to call f a multivalued operator from $\mathcal{R}$ into $\mathcal{R}$, when what we really mean is a function from $\mathcal{R}$ into $2^{\mathcal{R}}$; note that every function $F : \mathcal{R} \to \mathcal{R}$ is also a "multivalued operator" from $\mathcal{R}$ into $\mathcal{R}$ (i.e. $F : \mathcal{R} \to 2^{\mathcal{R}}$, since $Fx \in \mathcal{R}$ implies $Fx \in 2^{\mathcal{R}}$), but *not* conversely.

Definition 5.2. For $\mathcal{B}$ a Banach space, a *multivalued operator* $F : (\mathcal{D}(F) \subset \mathcal{B}) \to \mathcal{B}$ is a function $F : \mathcal{B} \to 2^{\mathcal{B}}$ where $2^{\mathcal{B}}$ is the set of all subsets of $\mathcal{B}$; the *domain* is

$$\mathcal{D}(F) \equiv \{x \in \mathcal{B} \mid Fx \neq \varnothing\},$$

and the *range* is

$$\mathcal{R}(F) \equiv \bigcup_{x \in \mathcal{D}(F)} Fx \subset \mathcal{X}.$$

We also define

$$|Fx| \equiv \inf_{y \in Fx} \|y\|, \qquad x \in \mathfrak{D}(F),$$

and the *canonical restriction* of F is a multivalued operator $F^0 : (\mathfrak{D}(F^0) \subset \mathfrak{B}) \to \mathfrak{B}$ defined by $F^0 x \equiv \{y \in Fx \mid \|y\| = |F|\}$; clearly, $\mathfrak{D}(F^0) \equiv \{x \in \mathfrak{D}(F) \mid F^0 x \neq \varnothing\} \subset \mathfrak{D}(F)$, and $\mathfrak{R}(F^0) \subset \mathfrak{R}(F)$. Given another multivalued operator $G : (\mathfrak{D}(G) \subset \mathfrak{B}) \to \mathfrak{B}$, we define $(F + G)$ by $(F + G)x \equiv \{y + z \mid y \in Fx, z \in Gx\}$ for $x \in \mathfrak{D}(F + G) \equiv \mathfrak{D}(F) \cap \mathfrak{D}(G)$. F is *closed* if its *graph* $\{(x, y) \mid y \in Fx, x \in \mathfrak{D}(F)\}$ is a closed set in $\mathfrak{B} \times \mathfrak{B}$ equipped with the norm $\|(x, y)\|_{\mathfrak{B} \times \mathfrak{B}} \equiv \|x\| + \|y\|$. F is *accretive* if, for every real $\lambda > 0$, $\|x - \hat{x} + \lambda(y - \hat{y})\| \geqslant \|x - \hat{x}\|$ for all $x, \hat{x} \in \mathfrak{D}(F), y \in Fx, \hat{y} \in F\hat{x}$. If $\mathfrak{B}$ is a Hilbert space, F is *monotone* if $\mathrm{Re}\langle x - \hat{x}, y - \hat{y}\rangle \geqslant 0$ for all $x, \hat{x} \in \mathfrak{D}(F), y \in Fx, \hat{y} \in Fx$. For $x_0 \in \mathfrak{B}$ and $\alpha \in \mathfrak{F}$, the field of $\mathfrak{B}$, the set $\{\alpha y + x_0 \mid y \in Fx\}$ is denoted by $\alpha Fx + x_0$ for each $x \in \mathfrak{D}(F)$.

Exercise 5.1. Consider a multivalued operator $F : (\mathfrak{D}(F) \subset \mathfrak{B}) \to \mathfrak{B}$, $\mathfrak{B}$ a Hilbert space. Show that F is monotone if and only if F is accretive. *Hint:* See the proof of Proposition II.2.1.

The point of all this discussion is that rather that (4) or (5), we feel that we should consider nonlinear autonomous evolution equations of the form

$$\dot{x}(t) \in Ax(t), \qquad \text{a.e. } t \in \mathfrak{R}^+,$$
$$x(0) = x_0 \in \mathrm{Cl}_{\mathfrak{B}}\mathfrak{D}(A), \tag{6}$$

for a multivalued operator $A : (\mathfrak{D}(A) \subset \mathfrak{B}) \to \mathfrak{B}$, with $\mathfrak{B}$ a Banach space, and $\mathfrak{D}(A)$ not necessarily dense. The theory would not be simplified, and the results would be less useful, if we were to consider only single-valued A.

Let $\mathfrak{L}^1_{\mathrm{loc}}(\mathfrak{R}^+; \mathfrak{B})$ denote the space of functions $g : \mathfrak{R}^+ \to \mathfrak{B}$ that are (Bochner) integrable on compact subsets of $\mathfrak{R}^+$, equipped with the topology of $\mathfrak{L}^1$-convergence on compact subsets of $\mathfrak{R}^+$. Let $\mathcal{C}(\mathfrak{R}^+; \mathfrak{B})$ denote the space of continuous functions $g : \mathfrak{R}^+ \to \mathfrak{B}$, equipped with the topology of uniform convergence on compact subsets of $\mathfrak{R}^+$. We say that u is a *strong solution* of (6) if $u(0) = x_0$, $u \in \mathcal{C}(\mathfrak{R}^+; \mathfrak{B}) \cap \mathfrak{W}^{1,1}_{\mathrm{loc}}(\mathfrak{R}^+; \mathfrak{B})$, $u(t) \in \mathfrak{D}(A)$ a.e. $t \in \mathfrak{R}^+$, and $\dot{u}(t) \in Au(t)$ a.e. $t \in \mathfrak{R}^+$. The condition, $u \in \mathcal{C}(\mathfrak{R}^+; \mathfrak{B}) \cap \mathfrak{W}^{1,1}_{\mathrm{loc}}(\mathfrak{R}^+; \mathfrak{B})$ means that there is a $v \in \mathfrak{L}^1_{\mathrm{loc}}(\mathfrak{R}^+; \mathfrak{B})$ such that

$$u(t) - u(\tau) = \int_\tau^t v(s)\,ds \qquad \text{for all } t > \tau \in \mathfrak{R}^+;$$

in this case u is differentiable a.e. and $\dot{u}(t) = v(t)$ a.e. $t \in \mathcal{R}^+$. Notice that if u is a solution of (6), then $u(t)$ must belong to $\mathrm{Cl}_{\mathcal{B}}\mathcal{D}(A)$ for every $t \in \mathcal{R}^+$.

The conditions that we shall place on A are sufficient to ensure that strong solutions are Lipschitz continuous; we remark that a Lipschitz-continuous $g : \mathcal{R}^+ \to \mathcal{B}$ belongs to $\mathcal{C}(\mathcal{R}^+; \mathcal{B}) \cap \mathcal{W}^{1,1}_{\mathrm{loc}}(\mathcal{R}^+; \mathcal{B})$ if and only if it is differentiable a.e. (Ref. 3). We also mention that Lipschitz continuity implies differentiability a.e. when $\mathcal{B}$ is reflexive (e.g., a Hilbert space)(Ref. 14).

Our approach to (6) is to first consider an "approximating" backward-difference equation similar to (3) in Section 3.

$$\frac{1}{t/n}\left[z(mt/n) - z(mt/n - t/n)\right] \in Az(mt/n),$$

$$t \in \mathcal{R}^+, m = 1, 2, \ldots, n, \qquad z(0) = x_0 \in \mathrm{Cl}_{\mathcal{B}}\mathcal{D}(A), \tag{7}$$

under the assumptions that $\omega I - A$ is accretive for some $\omega \in \mathcal{R}$ and that $\mathcal{R}(I - \lambda A) \supset \mathrm{Cl}_{\mathcal{B}}\mathcal{D}(A)$ for all sufficiently small $\lambda > 0$. By accretiveness of $\omega I - A$,

$$\|(1 + \mu\omega)(x - \hat{x}) - \mu(y - \hat{y})\| \geqslant \|x - \hat{x}\|$$

for all $x, \hat{x} \in \mathcal{D}(A)$, $\mu > 0, y \in Ax, \hat{y} \in A\hat{x}$; defining $\lambda \equiv \mu/(1 + \mu\omega)$ for $\mu > 0$, $\mu\omega < 1$, we see that

$$\|x - \hat{x} - \lambda(y - \hat{y})\| \geqslant (1 - \lambda\omega)\|x - \hat{x}\|$$

for all $x, \hat{x} \in \mathcal{D}(A), y \in Ax, \hat{y} \in A\hat{x}, \lambda > 0, \lambda\omega < 1$. It follows that the sets $(I - \lambda A)x$ and $(I - \lambda A)\hat{x}$ are disjoint for $x \neq \hat{x}$; therefore, there exists a single-valued inverse $J_\lambda \equiv (I - \lambda A)^{-1}$, $J_\lambda : (\mathcal{R}(I - \lambda A) \subset \mathcal{B}) \to \mathcal{B}$ for all $\lambda \in (0, \lambda_0)$, $\lambda_0 > 0$. In particular, we note that $\mathcal{R}(J_\lambda) \subset \mathcal{D}(A)$ and

$$\|J_\lambda y - J_\lambda \hat{y}\| \leqslant (1 - \lambda\omega)^{-1}\|y - \hat{y}\|$$

for all $y, \hat{y} \in \mathcal{R}(I - \lambda A)$, $\lambda \in (0, \lambda_0)$, and $\mathcal{R}(I - \lambda A) \supset \mathrm{Cl}_{\mathcal{B}}\mathcal{D}(A)$ by assumption. Hence, (7) has a solution $z(t) \equiv J^n_{t/n}x_0$ for every $x_0 \in \mathrm{Cl}_{\mathcal{B}}\mathcal{D}(A)$, $t \in \mathcal{R}^+$.

By this result, we suspect that a strong solution of (6) might be given as the limit of $J^n_{t/n}x_0$ as $n \to \infty$. With this in mind, define $\mathcal{X}$ to be the closure in $\mathcal{B}$ of $\mathcal{D}(A)$ ($\mathcal{X} \equiv \mathrm{Cl}_{\mathcal{B}}\mathcal{D}(A)$), and equip $\mathcal{X}$ with the metric induced by the norm; then $\mathcal{X}$ is a complete metric space with $\mathcal{D}(A)$ dense in $\mathcal{X}$. (Compare these ideas with Definition 5.1.)

Theorem 5.1. Let a multivalued $A : (\mathfrak{D}(A) \subset \mathfrak{B}) \to \mathfrak{B}$, $\mathfrak{B}$ a Banach space, be such that $\omega I - A$ is accretive for some $\omega \in \mathfrak{R}$, and $\mathfrak{R}(I - \lambda A) \supset \mathfrak{X} \equiv \mathrm{Cl}_{\mathfrak{B}}\mathfrak{D}(A)$ for all $\lambda \in (0, \lambda_0)$, some $\lambda_0 > 0$. Then a dynamical system $\{S(t)\}_{t \geqslant 0}$ is generated on $\mathfrak{X}$ by the product formula

$$S(t)x \equiv \lim_{n\to\infty} J^n_{t/n}x, \qquad x \in \mathfrak{X}, \quad t \in \mathfrak{R}^+,$$

where $J_\lambda \equiv (I - \lambda A)^{-1}$, $\lambda \in (0, \lambda_0)$, and $J_0 \equiv I$. Moreover,

$$\|S(t)x - S(t)\hat{x}\| \leqslant e^{\omega t}\|x - \hat{x}\|$$

for all $x, \hat{x} \in \mathfrak{X}$, $t \in \mathrm{R}^+$, and $S(\cdot)x : \mathfrak{R}^+ \to \mathfrak{X}$ is Lipschitz continuous for every $x \in \mathfrak{D}(A)$.

Sketch of Proof. Under these assumptions, Crandall and Liggett (Ref. 8) have shown that the estimate (2), used in the sufficiency proof of Theorem 3.2, remains valid with $\|Ax\|$ replaced by $|Ax|$; hence, all current conclusions follow from essentially the same arguments used in that proof. □

When A and $\{S(t)\}_{t \geqslant 0}$ are as described in Theorem 5.1, we shall say that A *generates* $\{S(t)\}_{t \geqslant 0}$.

Example 5.1. Consider the dry friction problem described earlier, assuming $\mu > 0$ and defining $\mathfrak{D}(A) \equiv \mathfrak{R}^2 \equiv \mathfrak{B}$; hence, $\mathfrak{X} = \mathfrak{R}^2$. We define a multivalued operator $A : \mathfrak{R}^2 \to \mathfrak{R}^2$ by

$$Ax \equiv \left\{ (v, -u - w) \in \mathfrak{R}^2 \,\middle|\, \begin{array}{l} w \equiv \mu \text{ if } v > 0,\ w \equiv -\mu \text{ if } v < 0 \\ w \in [-\mu, \mu] \text{ if } v = 0 \end{array} \right\}, \qquad x = (u, v) \in \mathfrak{R}^2.$$

It is easy to verify that $\mathfrak{R}(I - \lambda A) = \mathfrak{R}^2$ for all $\lambda > 0$, and we note that for $x = (u, v) \in \mathfrak{R}^2$, $y = (v, -u - w) \in Ax$, we have

$$\langle x - \hat{x}, y - \hat{y} \rangle = (v - \hat{v})(-w + \hat{w}) = \left\{ \begin{array}{ll} = 0 & \text{if } v\hat{v} > 0 \\ \leqslant 0 & \text{if } v\hat{v} = 0 \\ = -2\mu|v - \hat{v}| & \text{if } v\hat{v} < 0 \end{array} \right\} \leqslant 0;$$

hence, $-A$ is monotone (accretive). By Theorem 5.1 a dynamical system is generated by the product formula $S(t)x \equiv \lim_{n\to\infty} J^n_{t/n}x$, $x \in \mathfrak{R}^2$, $t \geqslant 0$, where $J_0 \equiv I$ and $J_\lambda \equiv (I - \lambda A)^{-1}$. Moreover,

$$\|S(t)x - S(t)y\| \leqslant \|x - y\| \quad \text{for all } x, y \in \mathfrak{R}^2,\ t \geqslant 0;$$

hence, $\{S(t)\}_{t \geqslant 0}$ is nonexpansive.

Notice that the canonical restriction of A is defined by

$$A^0x \equiv (v, -u - \mu v/|v|) \qquad \text{if } v \neq 0,$$

$$A^0x \equiv (0, 0) \qquad \text{if } v = 0, |u| < \mu,$$

$$A^0x \equiv (0, -u + \mu u/|u|) \qquad \text{if } v = 0, |u| > \mu,$$

for $x = (u, v) \in \mathcal{R}^2 = \mathcal{D}(A^0)$; $-A^0$ is accretive (since $-A$ is) and also single valued, but $\mathcal{R}(I - \lambda A^0) \neq \mathcal{X}$ for any $\lambda > 0$.

Exercise 5.2. Consider the other problem described earlier, defining $\mathcal{X} \equiv \mathcal{R}^1$ and a multivalued operator $A : \mathcal{R}^1 \to \mathcal{R}^1$ by

$$Ax = \left\{ w \in \mathcal{R}^1 \,\middle|\, \begin{matrix} w \equiv -1 \text{ for } x > 0, w \equiv 0 \text{ for } x < 0 \\ w \in [-1, 0] \text{ for } x = 0 \end{matrix} \right\}, \qquad x \in \mathcal{R}^1.$$

Show that $-A$ is accretive and that $\mathcal{R}(I - \lambda A) = \mathcal{R}^1$ for every $\lambda > 0$. Apply Theorem 5.1. Find $\mathcal{D}(A^0)$, A^0, and $\mathcal{R}(I - \lambda A^0)$ for $\lambda > 0$.

If A is a linear function and satisfies the conditions of Theorem 5.1, then, as we saw in Section 3, the motion $S(\cdot)x_0 : \mathcal{R}^+ \to \mathcal{X}$ is a strong solution of (1) (hence, of (6)) for every $x_0 \in \mathcal{D}(A)$. If A is not linear this may not be true, as $S(\cdot)x_0$ may not be differentiable a.e., or even at $t = 0$; this difficulty may arise even when A is single-valued but nonlinear.

Theorem 5.2. Under the assumptions of Theorem 5.1, and with $\{S(t)\}_{t \geq 0}$ as defined there on $\mathcal{X} \equiv \mathrm{Cl}_{\mathcal{B}}\mathcal{D}(A)$, $\mathcal{B}$ a Banach space,

(i) strong solutions of (6) are unique and are motions of $\{S(t)\}_{t \geq 0}$;

(ii) if $x(\cdot)$ is a strong solution of (6), then $x(t) \in \mathcal{D}(A^0)$ and $\dot{x}(t) \in A^0x(t)$ a.e. $t \in \mathcal{R}^+$;

(iii) if A is closed, $x_0 \in \mathcal{X}$, and the motion $S(\cdot)x_0 : \mathcal{R}^+ \to \mathcal{X}$ is differentiable a.e., then (6) has a (unique) strong solution, given by $x(\cdot) = S(\cdot)x_0$, and $(d/dt)S(t)x_0 \in A^0S(t)x_0$ at every $t \in \mathcal{R}^+$ such that $(d/dt)S(t)x_0$ exists;

(iv) if A is closed and $\mathcal{B}$ is a Hilbert space, then $\mathcal{D}(A^0) = \mathcal{D}(A)$ and A^0 is single-valued; moreover, for every $x_0 \in \mathcal{D}(A)$, $S(\cdot)x_0$ is a (unique) strong solution of (6), $S(t)x_0 \in \mathcal{D}(A^0)$ for all $t \in \mathcal{R}^+$, $(d^+/dt)S(t)x_0$ exists for all $t \in \mathcal{R}^+$, and

$$\frac{d^+}{dt}S(t)x_0 = A^0S(t)x_0 \qquad \text{for all } t \in \mathcal{R}^+.$$

A proof of Theorem 5.2. will be sketched at the end of this section. Notice how nicely conclusion (iv) applies to Example 5.1.

For nonlinear problems, the major difficulty lies in verifying the range condition of Theorem 5.1, $\mathcal{R}(I - \lambda A) \supset \mathcal{X} \equiv \mathrm{Cl}_{\mathcal{B}}\mathcal{D}(A)$ for all $\lambda \in (0, \lambda_0)$, some $\lambda_0 > 0$. This has led to many "perturbation results" (Refs. 7, 15–17), of which the following are good examples.

Proposition 5.1. Let a multivalued operator $A : (\mathcal{D}(A) \subset \mathcal{B}) \to \mathcal{B}$, $\mathcal{B}$ a Banach space, be such that $\omega I - A$ is accretive for some $\omega \in \mathcal{R}$, and $\mathcal{R}(I - \lambda A) = \mathcal{B}$ for all $\lambda \in (0, \lambda_0)$, some $\lambda_0 > 0$. Let $F : \mathcal{B} \to \mathcal{B}$ be continuous and such that $\mu I - F$ is accretive for some $\mu \in \mathcal{R}$. Then $A + F$, defined on $\mathcal{D}(A + F) = \mathcal{D}(A)$, is such that $(\mu + \omega)I - (A + F)$ is accretive, and $\mathcal{R}(I - \lambda(A + F)) = \mathcal{B}$ for all $\lambda \in (0, \lambda_1)$, some $\lambda_1 > 0$.

Proposition 5.2. For $\mathcal{B}$ a Hilbert space, consider multivalued operators $A : (\mathcal{D}(A) \subset \mathcal{B}) \to \mathcal{B}$, $F : (\mathcal{D}(F) \subset \mathcal{B}) \to \mathcal{B}$, $\mathcal{D}(F) \supset \mathcal{D}(A)$, such that $\mathcal{R}(I - \lambda A) = \mathcal{B} = \mathcal{R}(I - \lambda F)$ for all $\lambda \in (0, \lambda_0)$, some $\lambda_0 > 0$; moreover, let both $\omega I - A$ and $\mu I - F$ be monotone (accretive) for some $\omega, \mu \in \mathcal{R}$. If, for each $r > 0$, there exist numbers $c_1(r)$ and $c_2(r) < 1$ such that $\|F^0 x\| \leq c_1(r) + c_2(r)\|A^0 x\|$ for all $x \in \mathcal{D}(A)$ with $\|x\| \leq r$, then $(\mu + \omega)I - (A + F)$ is accretive and $\mathcal{R}(I - \lambda(A + F)) = \mathcal{B}$ for all $\lambda \in (0, \lambda_1)$, some $\lambda_1 > 0$, where $\mathcal{D}(A + F) = \mathcal{D}(A)$.

A proof of Proposition 5.1 is sketched in Ref. 16 for the case $\omega = 0 = \mu$ (see Section 4 of Ref. 16), while Proposition 5.2 is proved in Ref. 17 for the case $\omega = 0 = \mu$ (see Theorem 4.3 of Ref. 17). Note that $(A + F)^0 \neq A^0 + F^0$, in general.

Exercise 5.3. Assuming that Proposition 5.1 and Proposition 5.2 are true for $\omega = 0 = \mu$, show that they must also be true for general $\omega, \mu \in \mathcal{R}$.

Example 5.2. Consider the following nonlinear modification of the heat conduction problem,

$$\frac{\partial}{\partial t}\theta(\eta, t) = \frac{\partial^2}{\partial \eta^2}\theta(\eta, t) + f(\theta(\eta, t)), \qquad 0 < \eta < 1, \quad t > 0,$$

$$\theta(0, t) = 0 = \theta(1, t), \qquad t > 0,$$

with initial data $\theta(\eta, 0) = x_0(\eta)$, $0 < \eta < 1$. We assume that $f : \mathcal{R} \to \mathcal{R}$ is continuous, $f(0) = 0$, and $\mu I - f$ is monotone for some $\mu \in \mathcal{R}$, i.e.,

$$\mu|\xi - \hat{\xi}|^2 \geq (\xi - \hat{\xi})[f(\xi) - f(\hat{\xi})] \qquad \text{for all } \xi, \hat{\xi} \in \mathcal{R}.$$

Let us define a Banach space

$$\mathcal{B} \equiv \{x \in \mathcal{C}[0, 1] | x(0) = 0 = x(1)\}, \qquad \|x\| \equiv \max_{\eta} |x(\eta)|,$$

and operators $A : (\mathfrak{D}(A) \subset \mathfrak{B}) \to \mathfrak{B}$, $F : \mathfrak{B} \to \mathfrak{B}$, by

$$Ax \equiv \partial^2 x, \qquad x \in \mathfrak{D}(A) \equiv \{x \in \mathcal{C}^2[0,1] | x(\eta) = 0 = \partial^2 x(\eta) \text{ at } \eta = 0, 1\},$$

$$(Fx)(\eta) \equiv f(x(\eta)), \qquad 0 \leq \eta \leq 1, \quad x \in \mathfrak{B}.$$

By Example II.2.4 we see that $\mathfrak{D}(A)$ is dense, $-A$ is accretive, and $\mathfrak{R}(I - \lambda A) = \mathfrak{B}$ for all $\lambda > 0$.

We claim that $F : \mathfrak{B} \to \mathfrak{B}$ is continuous; to see this, fix $\hat{x} \in \mathfrak{B}$ and consider a sequence $\{x_n\}$, $x_n \to \hat{x}$ as $n \to \infty$. Note that $\|x_n\| < \|\hat{x}\| + 1$ for all sufficiently large n, and also note that $f : \mathfrak{R} \to \mathfrak{R}$ is uniformly continuous on the compact set $[-1 - \|\hat{x}\|, 1 + \|\hat{x}\|] \subset \mathfrak{R}$; since $\max_\eta |x_n(\eta) - \hat{x}(\eta)| \to 0$ for $n \to \infty$, it now follows that

$$\|Fx_n - F\hat{x}\| = \max_\eta |f(x_n(\eta)) - f(\hat{x}(\eta))| \to 0 \qquad \text{as } n \to \infty,$$

and $F : \mathfrak{B} \to \mathfrak{B}$ is shown to be continuous at every $\hat{x} \in \mathfrak{B}$.

We also claim that $\mu I - F$ is accretive; to see this, recall that monotonicity of $(\mu I - f) : \mathfrak{R} \to \mathfrak{R}$ implies accretiveness,

$$|(1 + \lambda\mu)(\xi - \hat{\xi}) - \lambda[f(\xi) - f(\hat{\xi})]| \geq |\xi - \hat{\xi}|$$

for all $\lambda > 0$ and $\xi, \hat{\xi} \in \mathfrak{R}$. If $\eta_0 \in [0, 1]$ is such that

$$\|x - \hat{x}\| = |x(\eta_0) - \hat{x}(\eta_0)|,$$

it follows that

$$|(1 + \lambda\mu)[x(\eta_0) - \hat{x}(\eta_0)] - \lambda[Fx(\eta_0) - F\hat{x}(\eta_0)]| \geq |x(\eta_0) - \hat{x}(\eta_0)| = \|x - \hat{x}\|, \lambda > 0;$$

hence,

$$\|(1 + \lambda\mu)(x - \hat{x}) - \lambda(Fx - F\hat{x})\| \geq \|x - \hat{x}\|, \qquad \lambda > 0,$$

and we see that $\mu I - F$ is accretive.

As a result of Proposition 5.1 we see that $\mu I - A - F$ is accretive and $\mathfrak{R}(I - \lambda(A + F)) = \mathfrak{B}$ for all $\lambda \in (0, \lambda_1)$, some $\lambda_1 > 0$; hence, by Theorem 5.1, $A + F$ generates a dynamical system $\{S(t)\}_{t \geq 0}$ on $\mathrm{Cl}_{\mathfrak{B}}\mathfrak{D}(A) = \mathfrak{B}$,

$$\|S(t)x - S(t)y\| \leq e^{\mu t}\|x - y\|$$

for $x, y \in \mathfrak{B}$, $t \in \mathfrak{R}^+$, and the motion $S(\cdot)x_0 : \mathfrak{R}^+ \to \mathfrak{B}$ is Lipschitz continuous for $x_0 \in \mathfrak{D}(A)$. By Theorem 5.2 we see that $S(\cdot)x_0$ is the (unique) strong solution of

$$\dot{x}(t) = (A + F)x(t), \qquad \text{a.e. } t \in \mathfrak{R}^+,$$

$$x(0) = x_0 \in \mathrm{Cl}_{\mathfrak{B}}\mathfrak{D}(A) = \mathfrak{B},$$

for every x_0 such that this equation has a strong solution (note that $A + F$ is single valued). As $\mathcal{B}$ is not a Hilbert space, conclusion (iv) of Theorem 5.2 does not apply; hence, existence of strong solutions is not assured, even for $x_0 \in \mathcal{D}(A)$.

Exercise 5.4. Set up the heat conduction problem of Example 5.2 in the space $\mathcal{B} \equiv \mathcal{L}_2(0, 1)$, under the additional assumption that $(\mu I + f) : \mathcal{R} \to \mathcal{R}$ is monotone; that is, assume $f : \mathcal{R} \to \mathcal{R}$ to be uniformly Lipschitz continuous. Referring to Example II.2.5 and Proposition 5.1, as well as to Theorems 5.1 and 5.2, what can you conclude regarding generation of a dynamical system, exponential estimates, and existence of strong solutions?

Exercise 5.5. Consider the following scalar functional differential equation,

$$\dot{y}(t) = \alpha y(t) + \int_0^\tau g(y(t-\eta))d\eta, \qquad t > 0,$$

for some fixed $\alpha \in \mathcal{R}$, $\tau > 0$, with initial data $y(-\eta) = u_0(\eta)$, $0 < \eta < 1$. Assume that $g : \mathcal{R} \to \mathcal{R}$ is uniformly Lipschitz continuous, with Lipschitz constant μ. Referring to Example 3.3, Proposition 5.1, Theorem 5.1, and Theorem 5.2, what can you conclude regarding generation of a dynamical system on $\mathcal{R} \times \mathcal{L}_2(0, \tau)$, exponential estimates, and existence of strong solutions?

Exercise 5.6. The following nonlinear equation is called the sine-Gordon equation (Ref. 18):

$$\frac{\partial^2}{\partial t^2}u(\eta, t) + 2\alpha\frac{\partial}{\partial t}u(\eta, t) - \frac{\partial^2}{\partial \eta^2}u(\eta, t) - \beta\sin(u(\eta, t)) = 0, \qquad 0 < \eta < 1, t > 0,$$

$$u(0, t) = 0 = u(1, t), \qquad t > 0.$$

Here α and β are given real numbers, and the initial data is $u(\eta, 0) = u_0(\eta)$, $(\partial/\partial t)u(\eta, 0) = v_0(\eta)$, for $0 < \eta < 1$. Show that this formal equation leads to a dynamical system on $\mathcal{B} \equiv \mathcal{W}_2^1(0, 1) \times \mathcal{L}_2(0, 1)$, where $\mathcal{W}_2^1(0, 1) \equiv \{u \in \mathcal{W}_2^1(0, 1) | u(0) = 0 = u(1)\}$, generated by $A + F$ with $\mathcal{D}(F) = \mathcal{B}$ and $\mathcal{D}(A)$ dense. Obtain an exponential estimate and discuss strong solutions.

Exercise 5.7. Consider a simply supported beam, surrounded by a viscous fluid of negligible density, and formally described by

$$\rho\frac{\partial^2}{\partial t^2}u(\eta, t) + f\left(\frac{\partial}{\partial t}u(\eta, t)\right) + \alpha\frac{\partial^4}{\partial \eta^4}u(\eta, t) = 0, \qquad 0 < \eta < 1, \quad t > 0,$$

$$u(\eta, t) = 0 = \frac{\partial^2}{\partial \eta^2}u(\eta, t) \qquad \text{at } \eta = 0, 1, \quad t > 0,$$

with initial data $u(\eta, 0) = u_0(\eta)$, $(\partial/\partial t)u(\eta, 0) = v_0(\eta)$, for $0 < \eta < 1$. Here, α and ρ are positive real numbers and the continuous function $f : \mathcal{R} \to \mathcal{R}$ satisfies

$$(\xi - \hat{\xi})[f(\xi) - f(\hat{\xi})] > 0, \qquad |f(\xi)| < \beta|\xi|,$$

for all $\xi, \hat{\xi} \in \mathfrak{R}$, some $\beta \in \mathfrak{R}^+$. Set up this problem so as to generate a nonexpansive dynamical system on some Hilbert space, and discuss strong solutions. *Hint*: See Exercise 4.5 and use Proposition 5.2.

Beginning with Theorem 5.1, we have assumed throughout this section that a "suitable" norm (as well as a "suitable" space) was reasonably apparent; our examples and exercises have involved "perturbations" of problems for which a suitable state space and norm were known on the basis of prior work. However, in general, the problem of finding a suitable space and a suitable norm represents an even more significant difficulty for nonlinear systems than it does for linear systems. If the problem is not a perturbation of some known problem, but some *a priori* estimate is available, then a suitable space and norm may be suggested as in Section 4 for linear systems. If not, then the following result suggests a more formal approach, if some suitable $\mathfrak{B}$ happens to be a Hilbert space. The proof is an obvious generalization of the proof of Theorem 4.2.

Theorem 5.3. Consider a multivalued operator $\tilde{A} : (\mathfrak{D}(\tilde{A}) \subset \mathcal{H}) \to \mathcal{H}$, $\mathcal{H}$ a Hilbert space. Let there exist $\omega \in \mathfrak{R}$, $\gamma \in \mathfrak{R}^+$, and a symmetric linear $B : (\mathfrak{D}(B) \subset \mathcal{H}) \to \mathcal{H}$, $\mathfrak{D}(B) \supset \mathfrak{D}(\tilde{A})$, such that

(i) $\langle By, y\rangle_{\mathcal{H}} \geq \gamma\|y\|^2_{\mathcal{H}}$ for all $y \in \mathfrak{D}(B)$,

(ii) $\mathrm{Re}\langle By - B\hat{y}, z - \hat{z}\rangle_{\mathcal{H}} \leq \omega\langle By - B\hat{y}, y - \hat{y}\rangle_{\mathcal{H}}$ for all $y, \hat{y} \in \mathfrak{D}(\tilde{A})$, $z \in \tilde{A}y$, $\hat{z} \in \tilde{A}\hat{y}$.

Define a Hilbert space $\mathfrak{B}$ to be the $\|\cdot\|_{\mathfrak{B}}$-completion of $\mathfrak{D}(B)$ where

$$\langle x, y\rangle_{\mathfrak{B}} \equiv \langle Bx, y\rangle_{\mathcal{H}}, \qquad x, y \in \mathfrak{D}(B),$$

and let $A : (\mathfrak{D}(A) \subset \mathfrak{B}) \to \mathfrak{B}$ be a restriction of $\tilde{A}$ to the set $\mathfrak{D}(A) \equiv \{x \in \mathfrak{D}(\tilde{A}) | \tilde{A}x \in \mathfrak{B}\}$. If $\mathfrak{R}(I - \lambda A) \supset \mathfrak{X} \equiv \mathrm{Cl}_{\mathfrak{B}}\mathfrak{D}(A)$ for all $\lambda \in (0, \lambda_0)$, some $\lambda_0 > 0$, then A generates a dynamical system $\{S(t)\}_{t \geq 0}$ on $\mathfrak{X}$ (equipped with the metric induced by $\|\cdot\|_{\mathfrak{B}}$), satisfying

$$\|S(t)x - S(t)z\|_{\mathfrak{B}} \leq e^{\omega t}\|x - z\|_{\mathfrak{B}}$$

for all $x, z \in \mathfrak{X}$, $t \in \mathfrak{R}^+$.

Proof. Just as in the proof of Theorem 4.2, we see that the multivalued operator $\omega I - A : (\mathfrak{D}(A) \subset \mathfrak{B}) \to \mathfrak{B}$ is accretive. Our conclusion now follows from Theorem 5.1. □

Sketch of Proof, Theorem 5.2. First we will show that every strong solution $u : \mathfrak{R}^+ \to \mathfrak{B}$ of (6) satisfies

$$e^{-\omega t}\|u(t) - x\| - e^{-\omega t}\|u(\tau) - x\|$$

$$\leqslant \int_\tau^t e^{-\omega s} \lim_{\lambda \searrow 0} \frac{1}{\lambda}\big[\|u(s) - x + \lambda y\| - \|u(s) - x\|\big] ds$$

$$\text{for all } t > \tau \geqslant 0,\ x \in \mathfrak{D}(A), y \in Ax, \tag{8}$$

assuming that $\omega I - A$ is accretive. To this end, note that if a mapping $u : \mathfrak{R}^+ \to \mathfrak{B}$ is left-differentiable at some $t > 0$, then the argument used in the proof of Proposition II.4.2 shows that $(d^-/dt)\|u(t)\|$ exists and

$$\frac{d^-}{dt}\|u(t)\| = \lim_{\lambda \searrow 0} \frac{1}{\lambda}\big[\|u(t)\| - \|u(t) - \lambda \dot{u}^-(t)\|\big].$$

Now consider $f, g \in \mathfrak{L}^1_{\text{loc}}(\mathfrak{R}^+; \mathfrak{B})$ and $u, v \in \mathcal{C}(\mathfrak{R}^+; \mathfrak{B}) \cap \mathfrak{W}^{1,1}_{\text{loc}}(\mathfrak{R}^+; \mathfrak{B})$ such that

$$\dot{u}(t) \in Au(t) + f(t) \qquad \text{a.e. } t \in \mathfrak{R}^+,$$

$$\dot{v}(t) \in Av(t) + g(t) \qquad \text{a.e. } t \in \mathfrak{R}^+.$$

As $\omega I - A$ is accretive, we see that for $\lambda > 0$

$$\|(1 + \lambda\omega)[u(t) - v(t)] - \lambda[\dot{u}(t) - \dot{v}(t)]\|$$

$$= \|2(u(t) - v(t)) + \lambda[\omega\{u(t) - v(t)\} - \{(\dot{u}(t) - f(t)) - (\dot{v}(t) - g(t))\}]$$

$$-[(u(t) - v(t)) + \lambda(f(t) - g(t))]\|$$

$$\geqslant 2\|u(t) - v(t)\| - \|(u(t) - v(t)) + \lambda(f(t) - g(t))\|.$$

Hence, for almost every $t \in \mathfrak{R}^+$,

$$\frac{d}{dt}\big[e^{-\omega t}\|u(t) - v(t)\|\big] = e^{-\omega t} \lim_{\lambda \searrow 0} \frac{1}{\lambda}\big[\|u(t) - v(t)\| - \|(u(t) - v(t))$$

$$-\lambda\{(\dot{u}(t) - \dot{v}(t)) - \omega(u(t) - v(t))\}\|\big]$$

$$\leqslant e^{-\omega t} \lim_{\lambda \searrow 0} \frac{1}{\lambda}\big[\|u(t) - v(t) + \lambda(f(t) - g(t))\| - \|u(t) - v(t)\|\big],$$

and it follows after integration that for all $t > \tau \geqslant 0$,

$$e^{-\omega t}\|u(t) - v(t)\| - e^{-\omega\tau}\|u(\tau) - v(\tau)\|$$

$$\leqslant \int_\tau^t e^{-\omega s} \lim_{\lambda \searrow 0} \frac{1}{\lambda} [\|(u(s) - v(s)) + \lambda(f(s) - g(s))\| - \|u(s) - v(s)\|] ds \tag{9}$$

(we refer the reader to Refs. 16, 19 for certain details). Assuming u to be a strong solution ($\dot{u}(t) \in Au(t)$ a.e.) and choosing $v(t) \equiv x \in \mathfrak{D}(A)$, $-g(t) \equiv y \in Ax$, we see that (9) implies that u satisfies (8) for all $x \in \mathfrak{D}(A), y \in Ax$.

Following Benilan (Ref. 19), any $u \in \mathcal{C}(\mathfrak{R}^+; \mathfrak{B})$ with $u(0) \in \mathfrak{X} \equiv \mathrm{Cl}_{\mathfrak{B}}\mathfrak{D}(A)$, is called an *integral solution* of (6) if it satisfies (8). Assuming that $\mathfrak{R}(I - \lambda A) \supset \mathfrak{X} \equiv \mathrm{Cl}_{\mathfrak{B}}\mathfrak{D}(A)$ for all $\lambda \in (0, \lambda_0)$, some $\lambda_0 > 0$, Benilan (Ref. 19) has shown that every motion $S(\cdot)x_0$, $x_0 \in \mathfrak{X}$, is an integral solution, where $\{S(t)\}_{t \geqslant 0}$ is generated by A; furthermore, if u is an integral solution, it also satisfies (9) with $v(t) \equiv S(t)x_0$, $f(t) \equiv 0$, $g(t) \equiv 0$, for every $x_0 \in \mathfrak{X}$. Hence, by (9) for $t > \tau \geqslant 0$,

$$e^{-\omega t}\|u(t) - S(t)x_0\| \leqslant e^{-\omega\tau}\|u(\tau) - S(\tau)x_0\|$$

for every integral solution u and all $x \in \mathfrak{X}$. Choosing $\tau = 0$, $x_0 = u(0) \in \mathfrak{X}$, it follows that every integral solution is a motion of $\{S(t)\}_{t \geqslant 0}$; hence, integral solutions are unique. As strong solutions are integral solutions, we have obtained (i).

If u is a strong solution of (6) and τ is such that $u(\tau) \in \mathfrak{D}(A)$, we find by setting $x \equiv u(\tau)$ in (8) that

$$\|u(t) - u(\tau)\| \leqslant \int_\tau^t e^{\omega(t-s)} ds \Big(\inf_{y \in Au(\tau)} \|y\| \Big)$$

$$= \int_\tau^t e^{\omega(t-s)} ds |Au(\tau)|, \qquad t > \tau.$$

If $\dot{u}(\tau)$ exists, it now follows that $\|\dot{u}(\tau)\| \leqslant |Au(\tau)|$; as $\dot{u}(\tau) \in Au(\tau)$, this implies that $u(\tau) \in \mathfrak{D}(A^0)$, $\dot{u}(\tau) \in A^0 u(\tau)$, and we have obtained (ii).

To prove (iii) it is sufficient to show that if $(d/dt)S(t)x_0$ exists, then $S(t)x_0 \in \mathfrak{D}(A)$ and $(d/dt)S(t)x_0 \in AS(t)x_0$; assuming that A is closed, this follows from a result of Crandall and Liggett (Ref. 8), as extended by Miyadera (Ref. 20). [It could also be shown by using (8); see Refs. 16, 19.] Since $S(\cdot)z$ is Lipschitz continuous for every $z \in \mathfrak{D}(A)$, and since we have assumed $S(\cdot)x_0$ differentiable a.e., implying $S(t)x_0 \in \mathfrak{D}(A)$ a.e. $t \in \mathfrak{R}^+$, it follows that $S(\cdot)x_0 \in \mathcal{C}(\mathfrak{R}^+; \mathfrak{B}) \cap \mathfrak{W}^{1,1}_{\text{loc}}(\mathfrak{R}^+; \mathfrak{B})$ and (iii) is obtained.

To prove (iv), we note that a Hilbert space is reflexive (Ref. 3), implying that a Lipschitz-continuous motion $S(\cdot)x_0$ must be differentiable a.e. (Ref. 14); hence, result (iii) applies for $x_0 \in \mathfrak{D}(A)$, using Theorem 5.1. As A is closed, $\omega I - A$ is accretive, and a Hilbert space is uniformly convex, the set Ax is closed and convex for given $x \in \mathfrak{D}(A)$, containing a unique element A^0x of minimal norm (Refs. 7, 20); hence, $\mathfrak{D}(A^0) = \mathfrak{D}(A)$, A^0 is single-valued, and

$$\frac{d}{dt} S(t)x_0 = A^0 S(t)x_0 \qquad \text{a.e. } t \in \mathfrak{R}^+,$$

for every $x_0 \in \mathfrak{D}(A)$. The remaining conclusion follows as in Ref. 20, where it is shown that, for $x_0 \in \mathfrak{D}(A)$, $S(\cdot)x_0$ is everywhere right-differentiable and

$$\frac{d^+}{dt} S(t)x_0 = A^0 S(t)x_0 \qquad \text{for all } t \in \mathfrak{R}^+.$$

The sketch of the proof is complete. □

6. Comments and Extensions

Many of the results of Section 5 can be extended to processes; a process on $\mathfrak{X} \times \mathfrak{R}_T$, $\mathfrak{X}$ a metric space and $\mathfrak{R}_T \equiv [T, \infty)$, being as in Definition I.4.1 with $\mathfrak{X}$ replacing $\mathfrak{R}^n$ and $\mathfrak{R}_T$ replacing $\mathfrak{R}$. The results of Ref. 7, 19, 21, 22 provide partial extensions of many of the ideas of Section 5 to a process on $\mathfrak{X} \times \mathfrak{R}_T$ that is "generated" by a family $\{A(t)\}_{t \geq T}$, $A(t) : (\mathfrak{D}(A(T)) \subset \mathfrak{B}) \to \mathfrak{B}$, $\mathfrak{B}$ a Banach space and $\mathfrak{X} \equiv \text{Cl}_{\mathfrak{B}} \mathfrak{D}(A(0))$. Then (6) is replaced by a nonautonomous equation, and there are several new difficulties. The results are similar to, but more complicated than, those of Section 5, and we shall not pursue the subject further; the interested reader is referred to Refs. 7, 19, 21, 22.

The results of Sections 2–4 provide a complete description of a linear dynamical system on a Banach space $\mathfrak{X}$, generalizing the finite-dimensional ideas of Section I.5. The results of Section 5 refer to nonlinear dynamical systems and their relationship with nonlinear autonomous evolution equations; it is reasonable to ask whether the sufficient conditions of Section 5 also completely generalize the finite-dimensional results of Chapter I. Exercise I.12.2 seems to suggest that the answer is negative; however, this judgment may be too hasty, as the following example shows.

Example 6.1. Consider the two-dimensional problem

$$\dot{z}(t) = -y(t)\left[(z(t))^2 + (y(t))^2\right],$$

$$\dot{y}(t) = z(t)\left[(z(t))^2 + (y(t))^2\right], \qquad t > 0,$$

with $(z(0), y(0)) = (z_0, y_0) \in \mathfrak{R}^2$. By the results of Chapter I, this leads to a dynamical system $\{S(t)\}_{t \geq 0}$ on $\mathfrak{R}^2$. Defining $A : \mathfrak{R}^2 \to \mathfrak{R}^2$ by

$$Ax = (-yz^2 - y^3, z^3 + zy^2), \qquad x = (z, y) \in \mathfrak{R}^2,$$

it can be shown that $\omega I - A$ is not accretive for any $\omega \in \mathfrak{R}$, in terms of any norm for $\mathfrak{R}^2$. Hence, Theorem 5.1 does not apply to $A : \mathfrak{R}^2 \to \mathfrak{R}^2$. The basic reason for this difficulty is that there exists no (finite) ω such that

$$\|S(t)x - S(t)\hat{x}\|_e \leq e^{\omega t}\|x - \hat{x}\|_e \qquad \text{for all } x, \hat{x} \in \mathfrak{R}^2, t \in \mathfrak{R}^+,$$

in terms of any norm $\|\cdot\|_e$ for $\mathfrak{R}^2$.

Let us try another approach, defining $V : \mathfrak{R}^2 \to \mathfrak{R}$ by $V(x) = z^2 + y^2$, $x = (z, y) \in \mathfrak{R}^2$, and letting $A_\alpha : (\mathfrak{D}(A_\alpha) \subset \mathfrak{R}^2) \to \mathfrak{R}^2$ be the restriction of A to $\mathfrak{D}(A_\alpha) \equiv \{x \in \mathfrak{R}^2 | V(x) \leq \alpha\}$, some (finite) $\alpha > 0$. It is easily verified that $\mathfrak{R}(I - \lambda A_\alpha) \supset \mathfrak{D}(A_\alpha)$ for all $\lambda > 0$; moreover, using the Euclidean inner product,

$$\begin{aligned}\langle x - \hat{x}, A_\alpha x - A_\alpha \hat{x}\rangle &= (z - \hat{z})\left[-y(z^2 + y^2) + \hat{y}(\hat{z}^2 + \hat{y}^2)\right] \\ &\quad + (y - \hat{y})\left[z(z^2 + y^2) - \hat{z}(\hat{z}^2 + \hat{y}^2)\right] \leq \omega_\alpha\|x - \hat{x}\|^2\end{aligned}$$

for some $\omega_\alpha \in \mathfrak{R}$, all $x, \hat{x} \in \mathfrak{D}(A_\alpha)$. It follows from Theorem 5.1 that A_α generates a dynamical system $\{S_\alpha(t)\}_{t \geq 0}$ on $\mathfrak{X}_\alpha \equiv \mathfrak{D}(A_\alpha)$, satisfying

$$\|S_\alpha(t) - S_\alpha(t)\hat{x}\| \leq \exp\{\omega_\alpha t\}\|x - \hat{x}\| \qquad \text{for all } x, \hat{x} \in \mathfrak{X}_\alpha, t \geq 0.$$

Noting that $S_\beta(t)$ is the restriction of $S_\alpha(t)$ to $\mathfrak{X}_\beta \subset \mathfrak{X}_\alpha$ for $\beta < \alpha$, and that $\mathfrak{R}^2 = \bigcup_{\alpha \in \mathfrak{R}} \mathfrak{X}_\alpha$, it follows from Definition 5.1 that a dynamical system $\{S(t)\}_{t \geq 0}$ is generated (by a product formula) on all of $\mathfrak{R}^2$. This conclusion does not follow directly from Theorem 5.1 because $\omega_\alpha \to \infty$ as $\alpha \to \infty$.

Exercise 6.1. Consider the two-dimensional problem

$$\dot{u}(t) = v(t),$$

$$\dot{v}(t) = -(u(t))^3, \qquad t \geq 0,$$

with $(u(0), v(0)) = (u_0, v_0) \in \mathfrak{R}^2$. Using only the ideas of this chapter, show that this problem leads to a dynamical system $\{S(t)\}_{t \geq 0}$ on $\mathfrak{R}^2$. *Hint*: Define $V : \mathfrak{R}^2 \to \mathfrak{R}$ by $V(x) = 2v^2 + u^4$, $x = (u, v)$, and proceed as in Example 6.1.

Although $\omega I - A$ of the foregoing example was not accretive for any $\omega \in \mathfrak{R}$, our approach was to form a family of restrictions A_α of A (by the use of a "well-chosen" function V), such that $\omega_\alpha I - A_\alpha$ was accretive for some $\omega_\alpha \in \mathfrak{R}$, and $\mathfrak{R}(I - \lambda A_\alpha) \supset \mathrm{Cl}\, \mathfrak{D}(A_\alpha)$ for small $\lambda > 0$. This approach has also been employed for infinite-dimensional systems in Refs. 23, 24 and has been formalized in Ref. 13. Recalling the material of Chapter I, we notice that the well-chosen function V is a Liapunov function for the resulting dynamical system on $\mathfrak{R}^2$.

This deficiency of Theorem 5.1 is the primary motivation for our general definition of a generated dynamical system (Definition 5.1). There we have in mind some multivalued operator $A : (\mathfrak{D}(A) \subset \mathfrak{B}) \to \mathfrak{B}$, $\mathfrak{B}$ a Banach space with $\mathfrak{X} \equiv \mathrm{Cl}_{\mathfrak{B}} \mathfrak{D}(A)$ and $d(x, y) \equiv \|x - y\|$, such that $I - \lambda A$ has a continuous inverse J_λ defined on all of $\mathfrak{X}$ for all $\lambda \in (0, \lambda_0)$, some $\lambda_0 > 0$. Then the principal difficulty lies in showing that the product formula $S(t)x \equiv \lim_{n\to\infty} J^n_{t/n} x$ actually leads to a dynamical system $\{S(t)\}_{t\geq 0}$ on $\mathfrak{X}$. Theorem 5.1 resolves this difficulty under the additional assumption that $\omega I - A$ is accretive for some $\omega \in \mathfrak{R}$; however, Example 6.1 shows that this assumption is not strictly necessary, if we do not insist that $\{S(t)\}_{t\geq 0}$ satisfy the exponential estimate of Theorem 5.1

The technique used in Example 6.1 is generalized by the following theorem, which is a restriction of a result obtained in Ref. 13.

Theorem 6.1. For $\mathfrak{B}$ a Banach space, consider a multivalued operator $A : (\mathfrak{D}(A) \subset \mathfrak{B}) \to \mathfrak{B}$, a continuous $V : \mathfrak{B} \to \mathfrak{R}$, and $\alpha \in \mathfrak{R}$ such that $\mathfrak{X}_\alpha \subset \mathrm{Cl}_{\mathfrak{B}} \mathfrak{D}(A)$, where $\mathfrak{X}_\alpha$ is a disjoint component of the (closed) set $\{x \in \mathfrak{B} \mid V(x) \leq \alpha\}$. Let there exist an equivalent norm $\|\cdot\|_\alpha$ and real numbers ω_α, $\lambda_\alpha > 0$ such that, for all $\lambda \in (0, \lambda_\alpha)$,

(i) $V(x - \lambda y) \geq V(x)$ for each $y \in Ax$, $x \in \mathfrak{D}(A)$, such that $x - \lambda y \in \mathfrak{X}_\alpha$;

(ii) $\mathfrak{R}(I - \lambda A) \supset \mathfrak{X}_\alpha$;

(iii) $\omega_\alpha I - A_\alpha$ is $\|\cdot\|_\alpha$-accretive, where A_α is the maximal restriction of A to $\mathfrak{D}(A_\alpha) \equiv \mathfrak{D}(A) \cap \mathfrak{X}_\alpha$.

Then a dynamical system $\{S_\alpha(t)\}_{t\geq 0}$ is generated, on the (complete) metric space $\mathfrak{X}_\alpha$, by the product formula

$$S_\alpha(t)x \equiv \lim_{n\to\infty} J^n_{t/n} x, \qquad x \in \mathfrak{X}, t \in \mathfrak{R}^+,$$

where $J_\lambda \equiv (I - \lambda A_\alpha)^{-1}$, $\lambda \in (0, \lambda_\alpha)$, and $J_0 \equiv I$. Moreover,

$$\|S_\alpha(t)x - S_\alpha(t)\hat{x}\|_\alpha \leq \exp\{\omega_\alpha t\} \|x - \hat{x}\|_\alpha$$

for all $x, \hat{x} \in \mathfrak{X}$, $t \in \mathfrak{R}^+$, and $S_\alpha(\cdot) : \mathfrak{R}^+ \to \mathfrak{X}$ is Lipschitz continuous for every $x \in \mathfrak{D}(A_\alpha)$.

Proof. Using condition (iii) and noting that the continuity of V implies $\mathcal{X}_\alpha = \mathrm{Cl}_{\mathcal{B}}\mathcal{D}(A_\alpha)$, we see by Theorem 5.1 that we need only show that $\mathcal{R}(I - \lambda A_\alpha) \supset \mathcal{X}_\alpha$ for $\lambda \in (0, \lambda_0)$. To this end, note that if $x \in \mathcal{D}(A), y \in Ax$, and $x - \lambda y \in \mathcal{X}_\alpha$, then condition (i) implies that $x \in \mathcal{X}_\alpha \cap \mathcal{D}(A) \equiv \mathcal{D}(A_\alpha)$. Using condition (ii), it now follows that $\mathcal{R}(I - \lambda A_\alpha) \supset \mathcal{X}_\alpha$. The proof is complete. □

Example 6.2. We consider a nonlinear wave equation,

$$\frac{\partial^2}{\partial t^2}u(\eta, t) + 2\beta\frac{\partial}{\partial t}u(\eta, t) - \frac{\partial^2}{\partial \eta^2}u(\eta, t) + \gamma[u(\eta, t)]^3 = 0, \quad 0 \leqslant \eta \leqslant 1, \quad t \geqslant 0,$$

$$u(0, t) = 0 = u(1, t), \qquad t \geqslant 0.$$

The numbers $\beta \geqslant 0, \gamma > 0$, are given, and the initial data is $u(\eta, 0) = u_0(\eta)$, $(\partial/\partial t)u(\eta, 0) = v_0(\eta)$, for $0 \leqslant \eta \leqslant 1$. We wish to show that this formal equation leads to a dynamical system on $\mathcal{B} \equiv \hat{\mathcal{W}}_2^1(0, 1) \times \mathcal{L}_2(0, 1)$ where $\hat{\mathcal{W}}_2^1(0, 1) \equiv \{u \in \mathcal{W}_2^1 | u(0) = 0 = u(1)\}$. We define $A : (\mathcal{D}(A) \subset \mathcal{B}) \to \mathcal{B}$ as $Ax \equiv (v, -2\beta v + \partial^2 u - \gamma Fu)$, $x = (u, v) \in \mathcal{D}(A)$, $\mathcal{D}(A) \equiv \{(u, v) \in \mathcal{B} | u \in \mathcal{W}_2^2(0, 1), v \in \hat{\mathcal{W}}_2^1(0, 1)\}$, $Fu(\eta) \equiv (u(\eta))^3$ a.e. $\eta \in (0, 1)$, $u \in \mathcal{D}(F) \equiv \hat{\mathcal{W}}_2^1 \cap \mathcal{W}_2^2$. We note that $Fu \in \mathcal{L}_2(0, 1)$ for all $u \in \mathcal{D}(F)$, because $\mathcal{C}[0, 1] \supset \mathcal{D}(F)$, and A is well defined. Unfortunately, $\omega I - A$ is not accretive for any $\omega \in \mathcal{R}$, in terms of any norm for $\mathcal{B}$ equivalent to $\|\cdot\|_{\hat{\mathcal{W}}_2^1 \times \mathcal{L}_2}$.

Choosing some finite $\alpha > 0$, we define

$$\|x\| \equiv \left(\int_0^1 [|\partial u(\eta)|^2 + |v(\eta)|^2] d\eta\right)^{1/2}, \qquad x = (u, v) \in \mathcal{B},$$

$$V(x) \equiv 2\|x\|^2 + \gamma\int_0^1 |u(\eta)|^4 d\eta, \qquad x = (u, v) \in \mathcal{B},$$

$$\mathcal{X}_\alpha \equiv \{x \in \mathcal{B} | V(x) \leqslant \alpha\}.$$

We see that $V : \mathcal{B} \to \mathcal{R}^+$ is continuous, because $\|\cdot\|^2 : \mathcal{B} \to \mathcal{R}^+$ is continuous and

$$\|x + \hat{x}\|^2\|x - \hat{x}\|^2 \geqslant \left(\int_0^1 |\partial u(\eta) + \partial\hat{u}(\eta)| d\eta\right)^2\left(\int_0^1 |\partial u(\eta) - \partial\hat{u}(\eta)| d\eta\right)^2$$

$$\geqslant \left(\max_\eta |u(\eta) + \hat{u}(\eta)|\right)^2\left(\max_\eta |u(\eta) - \hat{u}(\eta)|\right)^2$$

$$\geqslant \max_\eta ([u(\eta) + \hat{u}(\eta)][u(\eta) - \hat{u}(\eta)])^2 \geqslant \int_0^1 [|u(\eta)|^4 - |\hat{u}(\eta)|^4] d\eta$$

for all $x, \hat{x} \in \mathfrak{B}$. Furthermore, for $x = (u, v) \in \mathfrak{D}(A)$,

$$V(x - \lambda Ax) = 2\int_0^1 |\partial u(\eta) - \lambda \partial v(\eta)|^2 d\eta$$

$$+ 2\int_0^1 |v(\eta) - \lambda[-2\beta v(\eta) + \partial^2 u(\eta) - \gamma(u(\eta))^3]|^2 d\eta + \gamma\int_0^1 |u(\eta) - \lambda v(\eta)|^4 d\eta$$

$$> V(x) + 8\beta\lambda\int_0^1 |v(\eta)|^2 d\eta + \gamma\lambda^2\int_0^1 |v(\eta)|^2[6|u(\eta)|^2 - 4\lambda u(\eta)\cdot v(\eta) + \lambda^2|v(\eta)|^2]d\eta$$

$$> V(x) + 8\beta\lambda\int_0^1 |v(\eta)|^2 d\eta > V(x).$$

Hence, condition (i) of Theorem 6.1 is met. It is difficult but possible to show also that $\mathfrak{R}(I - \lambda A) \supset \mathfrak{X}_\alpha$ for all $\lambda \in (0, \lambda_\alpha)$, some $\lambda_\alpha > 0$; we omit the demonstration here. Finally, with considerable algebraic manipulation, it can also be shown that there exists $\omega_\alpha \in \mathfrak{R}$ such that $\omega_\alpha I - A_\alpha$ is accretive, where $A_\alpha x \equiv Ax$, $x \in \mathfrak{D}(A_\alpha) \equiv \mathfrak{D}(A) \cap \mathfrak{X}_\alpha$.

Applying Theorem 6.1, we see that A_α generates a dynamical system on $\mathfrak{X}_\alpha$ in the sense of Theorem 5.1, and

$$\|S_\alpha(t)x - S_\alpha(t)y\| < \exp\{\omega_\alpha t\}\|x - y\|$$

for all $t \in \mathfrak{R}$, $x, y \in \mathfrak{X}_\alpha$. Noting that α was arbitrary and $\mathfrak{B} = \cup_{\alpha \in \mathfrak{R}} \mathfrak{X}_\alpha$, we obtain a dynamical system $\{S(t)\}_{t>0}$ defined on all of $\mathfrak{B}$, where $S_\alpha(t)$ is the restriction to $\mathfrak{X}_\alpha$ of $S(t) : \mathfrak{B} \to \mathfrak{B}$ for all $t \in \mathfrak{R}^+$. Although $\{S(t)\}_{t>0}$ admits no uniform exponential estimate, we do have

$$\|S(t)x - S(t)y\| < \exp\{\omega_\alpha t\}\|x - y\|$$

for all $t \in \mathfrak{R}^+$, $x, y \in \mathfrak{X}_\alpha$, where $\omega_\alpha \to \infty$ as $\alpha \to \infty$.

In the following chapter we shall find that the function V used in Theorem 6.1 is necessarily a Liapunov function for the dynamical system $\{S_\alpha(t)\}_{t>0}$ generated by A_α on $\mathfrak{X}_\alpha$; we shall also find that, for any $\beta < \alpha$, the set $\mathcal{G}_\beta \equiv \{x \in \mathfrak{X}_\alpha | V(x) < \beta\}$ is positive invariant under $\{S_\alpha(t)\}_{t>0}$.

The following chapter is concerned with the behavior of motions of a dynamical system $\{S(t)\}_{t>0}$ on a metric space $\mathfrak{X}$; the study of this behavior is usually simplified if one or more equilibria exist. We note that every linear dynamical system ($\mathfrak{X}$ a Banach space) has at least one equilibrium (at $x = 0$), and the set of all equilibria is a linear manifold. If $\{S(t)\}_{t>0}$ is not linear, there may or may not be equilibria. If $\mathfrak{X} \subset \mathfrak{B}$, $\mathfrak{B}$ a Banach space, and $\{S(t)\}_{t>0}$ satisfies the estimate

$$\|S(t)x - S(t)y\| < e^{\omega t}\|x - y\|$$

for some $\omega < 0$, we see that all motions are bounded if there exists an equilibrium x_e, for then

$$\|S(t)x - x_e\| \leq e^{\omega t}\|x - x_e\|;$$

if $\omega < 0$, we see that $S(t)x \to x_e$ as $t \to \infty$, and there can be no other equilibrium. Hence, given such an exponential estimate with $\omega \leq 0$, it is important to know whether or not equilibria exist.

Exercise 6.2. Let a closed multivalued operator $A : (\mathfrak{D}(A) \subset \mathfrak{B}) \to \mathfrak{B}$, $\mathfrak{B}$ a Banach space, satisfy the conditions of Theorem 5.1. Let $\{S(t)\}_{t \geq 0}$ be the dynamical system generated by A on $\mathfrak{X} \equiv \mathrm{Cl}_{\mathfrak{B}}\mathfrak{D}(A)$, and suppose that there exists an equilibrium $x_e \in \mathfrak{X}$. Show that $x_e \in \mathfrak{D}(A^0) \subset \mathfrak{D}(A)$, and that $0 \in A^0 x_e \subset Ax_e$. *Hint*: Use (iii) of Theorem 5.2.

Unfortunately, there may be no equilibria, even if $\omega = 0$ in the exponential estimate of Theorem 5.1.

Example 6.3. Consider $A : \mathfrak{R} \to \mathfrak{R}$ defined by

$$Ax \equiv e^{-x}, \qquad x \in \mathfrak{R},$$

and note that

$$\langle x - \hat{x}, Ax - A\hat{x} \rangle_{\mathfrak{R}} = (x - \hat{x})(e^{-x} - e^{-\hat{x}}) \leq 0, \qquad x, \hat{x} \in \mathfrak{R}.$$

We see that $-A$ is monotone (accretive) and that $\mathfrak{R}(I - \lambda A) = \mathfrak{R}$ for all $\lambda > 0$. From Theorem 5.1 it follows that A generates a dynamical system $\{S(t)\}_{t \geq 0}$ on $\mathfrak{R}$ such that

$$\|S(t)x - S(t)y\| \leq \|x - y\|$$

for all $x, y \in \mathfrak{R}$, $t \in \mathfrak{R}^+$. The motion $S(\cdot)x_0$ is seen to be $S(t)x_0 = \ln(t + e^{x_0})$, $t \geq 0$, for every $x_0 \in \mathfrak{R}$. Hence, there exists no equilibrium and every motion is unbounded.

However, there are sufficient conditions for the existence of equilibria.

Exercise 6.3. Let a dynamical system $\{S(t)\}_{t \geq 0}$ be generated by a product formula (see Definition 5.1) on a metric space $\mathfrak{X}$, and suppose that x_0 is a fixed point of $J_\lambda : \mathfrak{X} \to \mathfrak{X}$ for all $\lambda \in (0, \lambda_0)$, some $\lambda_0 > 0$ (i.e., $J_\lambda x_0 = x_0$). Show that x_0 is an equilibrium of $\{S(t)\}_{t \geq 0}$.

Exercise 6.4. Let a multivalued operator $A : (\mathfrak{D}(A) \subset \mathfrak{B}) \to \mathfrak{B}$ generate a dynamical system $\{S(t)\}_{t \geq 0}$ on $\mathfrak{X} \equiv \mathrm{Cl}_{\mathfrak{B}}\mathfrak{D}(A)$, in the sense of Theorem 5.1, and let $0 \in Ax_0$ for some $x_0 \in \mathfrak{D}(A)$. Show that x_0 is an equilibrium. *Hint*: Make use of (i) of Theorem 5.2.

Exercise 6.5. Let a multivalued operator $A : (\mathfrak{D}(A) \subset \mathfrak{B}) \to \mathfrak{B}$ generate a dynamical system $\{S(t)\}_{t \geq 0}$ on $\mathfrak{X} \equiv \mathrm{Cl}_{\mathfrak{B}} \mathfrak{D}(A)$, in the sense of Theorem 5.1, and assume that $\omega I - A$ is accretive for some $\omega < 0$. Show that there exists an equilibrium x_e and that it is unique. *Hint*: Show that J_λ is a (strict) contraction for all $\lambda > 0$, and apply Theorem II.2.2.

References

1. Zubov, V. I., *Methods of A. M. Liapunov and Their Application*, P. Noordhoff, Groningen, 1964.
2. Hille, E., and Phillips, R. S., *Functional Analysis and Semi-groups*, American Mathematical Society Colloquim Publications, Vol. 31, Providence, Rhode Island, 1957.
3. Yosida, K., *Functional Analysis*, Springer-Verlag, New York, 1971.
4. Chernoff, P., and Marsden, J., On continuity and smoothness of group actions, *Bulletin of the American Mathematical Society*, Vol. 76, pp. 1044–1049, 1970.
5. Friedman, A., *Partial Differential Equations*, Holt, Rinehart and Winston, New York, 1969.
6. Kato, T., *Perturbation Theory for Linear Operators*, Springer-Verlag, New York, 1966.
7. Brezis, H., *Opérateurs Maximaux Monotones (et Semi-groupes des Contractions dans les Espaces de Hilbert)*, North Holland–American Elsevier, New York, 1973.
8. Crandall, M. G., and Liggett, T. M., Generation of semi-groups of nonlinear transformations on general Banach spaces, *American Journal of Mathematics*, Vol. 93, pp. 265–298, 1971.
9. Walker, J. A., and Infante, E. F., Some results on the precompactness of orbits of dynamical systems, *Journal of Mathematical Analysis and Applications*, Vol. 51, pp. 56–66, 1975.
10. Walker, J. A., On the application of Liapunov's Direct Method to linear dynamical systems, *Journal of Mathematical Analysis and Applications*, Vol. 53, pp. 187–220, 1976.
11. Slemrod, M., An application of maximal dissipative sets in control theory, *Journal of Mathematical Analysis and Applications*, Vol. 46, pp. 369–387, 1974.
12. Pazy, A., On the applicability of Lyapunov's theorem in Hilbert space, *SIAM Journal of Mathematical Analysis*, Vol. 3, pp. 291–294, 1972.
13. Walker, J. A., Some results on Liapunov functions and generated dynamical systems, *Journal of Differential Equations*, Vol. 30, pp. 424–430, 1978.
14. Komura, Y., Nonlinear semigroups in Hilbert space, *Journal of the Mathematical Society of Japan*, Vol. 19, pp. 493–507, 1967.
15. Browder, F. E., Nonlinear equations of evolution and nonlinear accretive operators in Banach spaces, *Bulletin of the American Mathematical Society*, Vol. 73, pp. 867–874, 1967.
16. Crandall, M. G., An introduction to evolution governed by accretive operators, *Dynamical Systems: An International Symposium*, Vol. 1, Edited by L. Cesari, J. K. Hale, and J. P. LaSalle, Academic Press, New York, 1976.
17. Crandall, M. G., and Pazy, A., Semigroups of nonlinear contractions and dissipative sets, *Journal of Functional Analysis*, Vol. 3, pp. 376–418, 1969.
18. Callegari, A. J., and Reiss, E. L., Nonlinear stability problems for the sine-Gordon equation, *Journal of Mathematical Physics*, Vol. 14, pp. 267–276, 1973.
19. Benilan, P., *Equations d'Evolution dans un Espace de Banach Quelconque et Applications*, University of Paris XI, Orsay, Thesis, 1972.

20. Miyadera, I., Some remarks on semigroups of nonlinear operators, *Tohoku Mathematical Journal*, Vol. 23, pp. 245–258, 1971.
21. Crandall, M. G., and Pazy, A., Nonlinear evolution equations in Banach spaces, *Israel Journal of Mathematics*, Vol. 11, pp. 57–94, 1972.
22. Kato, T., Accretive operators and nonlinear evolution equations in Banach spaces, *Proceedings of Symposia in Pure Mathematics*, Vol. 18, Part I, American Mathematical Society, Providence, Rhode Island, pp. 138–161, 1970.
23. Walker, J. A., and Infante, E. F., On the stability of an operator equation modeling nuclear reactors with delayed neutrons, *Quarterly of Applied Mathematics*, Vol. 34, pp. 421–427, 1977.
24. Infante, E. F., and Walker, J. A., On the stability properties of an equation arising in reactor dynamics, *Journal of Mathematical Analysis and Applications*, Vol. 55, pp. 112–114, 1976.

IV

Some Topological Dynamics

We continue to use semigroup notation for a dynamical system $\{S(t)\}_{t\geq 0}$ on a metric space $\mathfrak{X}$ (see Definitions III.1.1, III.1.2), recalling that the mapping $S(\cdot)x : \mathfrak{R}^+ \to \mathfrak{X}$ is the motion corresponding to the initial state $x \in \mathfrak{X}$, $x_e \in \mathfrak{X}$ is an equilibrium if $S(t)x_e = x_e$ for all $t \in \mathfrak{R}^+$, and a set $\mathbb{S} \subset \mathfrak{X}$ is positive invariant if $x \in \mathbb{S}$ implies that $S(t)x \in \mathbb{S}$ for all $t \in \mathfrak{R}^+$. We define the set $\gamma(x) \equiv \cup_{t\geq 0} S(t)x$ to be the *positive orbit* corresponding to the initial state x. We see that $\mathbb{S} \subset \mathfrak{X}$ is positive invariant if and only if $\gamma(x) \subset \mathbb{S}$ for every $x \in \mathbb{S}$, and $x_e \in \mathfrak{X}$ is an equilibrium if and only if $\gamma(x_e) = \{x_e\}$.

Here we shall use certain ideas of topological dynamics to obtain information about the behavior of motions; we are interested in boundedness of positive orbits, stability of motions (particularly, stability of equilibria), existence of periodic motions, and asymptotic behavior of motions (as $t \to \infty$).

We notice that the exponential estimates of Chapter III may provide quite a bit of this type of information. If $\{S(t)\}_{t\geq 0}$ is a dynamical system on a metric space $\mathfrak{X}$, and if there exists $\omega < 0$ such that $d(S(t)x, S(t)y) \leq e^{\omega t}d(x,y)$ for all $x, y \in \mathfrak{X}$, $t \in \mathfrak{R}^+$, we can immediately conclude that nonconstant periodic motions do not exist; moreover, applying Theorem II.2.2, we see that there exists a unique equilibrium x_e and $d(S(t)x, x_e) \leq e^{\omega t}d(x, x_e)$ for all $x \in \mathfrak{X}$, $t \in \mathfrak{R}^+$, implying that $\gamma(x)$ is bounded and $S(t)x \to x^e$ as $t \to \infty$ for every $x \in \mathfrak{X}$. If a closed multivalued operator $A : (\mathfrak{D}(A) \subset \mathfrak{B}) \to \mathfrak{B}$, $\mathfrak{B}$ a Banach space, generates $\{S(t)\}_{t\geq 0}$ on $\mathfrak{X} \equiv \mathrm{Cl}_{\mathfrak{B}}\mathfrak{D}(A)$ in the sense of Theorem III.5.1 with $\omega < 0$, then we saw by exercise III.6.2 that $x_e \in \mathfrak{D}(A^0) \subset \mathfrak{D}(A)$, and x_e can be found by solving $0 \in A^0 x_e$. It follows that we have a fairly complete description of the behavior of all motions when we have an exponential estimate with $\omega < 0$ (strictly).

However, if we have only an exponential estimate with $\omega \geqslant 0$, then considerably less can be said; there may or may not exist equilibria and/or nontrivial periodic motions, and positive orbits may or may not be bounded. Even if $\omega = 0$, the estimate implies boundedness of all positive orbits only if there exists at least one equilibrium (always true for linear dynamical systems); there may or may not exist periodic motions, and the asymptotic behavior of motions (as $t \to \infty$) remains an open question. To compound our difficulties, we noticed in Section III.6 that it is possible to use the results of Section III.5 to set up a generated nonlinear dynamical system that satisfies no (uniform) exponential estimate at all.

To answer some of these unanswered questions, we shall now discuss certain ideas of topological dynamics. In any application, our most important tool is a Liapunov function. Our principal interest is in dynamical systems that are generated by product formulas, as in Definition III.5.1.

1. Liapunov Functions and Postitive Invariance

Let $\overline{\mathfrak{R}}$ and $\overline{\mathfrak{R}}^+$ denote the extended real line $[-\infty, \infty]$ and extended nonnegative real line $[0, \infty]$, respectively, with $\pm\infty$ considered points, $-\infty < \alpha < \infty$ for every $\alpha \in \mathfrak{R}$. For $\mathfrak{X}$ a metric space, a function $U : \mathfrak{X} \to \overline{\mathfrak{R}}$ is *lower semicontinuous* if the set $\{x \in \mathfrak{X} \mid U(x) \leqslant \alpha\}$ is closed for every $\alpha \in \mathfrak{R}$. Equivalently, $U : \mathfrak{X} \to \overline{\mathfrak{R}}$ is lower semicontinuous if

$$U(x) \leqslant \liminf_{n\to\infty} U(x_n)$$

for every sequence $\{x_n\} \subset \mathfrak{X}$ converging to x, for every $x \in \mathfrak{X}$. It is apparent that every continuous function $U : \mathfrak{X} \to \mathfrak{R}$ is lower semicontinuous (and finite valued). However, even a finite-valued lower semicontinuous function may not be continuous; $U : \mathfrak{X} \to \mathfrak{R}$ is continuous if and only if both U and $-U$ are lower semicontinuous.

Definition 1.1. Let $\{S(t)\}_{t\geqslant 0}$ be a dynamical system on a metric space $\mathfrak{X}$, and let $V : \mathfrak{X} \to \overline{\mathfrak{R}}$ be lower semicontinuous. V is a *Liapunov function* for $\{S(t)\}_{t\geqslant 0}$ on a set $\mathcal{G} \subset \mathfrak{X}$ if $\dot{V}(x) \leqslant 0$ for every $x \in \mathcal{G}$, where $\dot{V} : \mathfrak{X} \to \overline{\mathfrak{R}}$ is the function defined by

$$\dot{V}(x) \equiv \liminf_{t\searrow 0} \frac{1}{t}\left[V(S(t)x) - V(x)\right] \qquad \text{if } |V(x)| < \infty,$$

$$\dot{V}(x) \equiv 0 \text{ if } V(x) = +\infty, \qquad \dot{V}(x) \equiv 1 \text{ if } V(x) = -\infty.$$

The notation $\dot{V}$ is very misleading; unfortunately, it is standard in the literature. We notice that if $V(S(\cdot)x) : \mathfrak{R}^+ \to \overline{\mathfrak{R}}$ happens to be right-differentiable at $t = 0$, then $\dot{V}(x) = (d^+/dt)V(S(0)x)$. We shall say that a Liapunov function V on $\mathcal{G}$ is a *continuous Liapunov function* on $\mathcal{G}$ if V is $\mathfrak{R}$-valued and continuous on $\mathrm{Cl}\,\mathcal{G}$. In Chapter I, for simplicity, we considered only continuous Liapunov functions (see Definition I.8.4).

In the past, Liapunov functions have usually been defined to be continuous $\mathfrak{R}$-valued functions, and most of our applications here involve such Liapunov functions. Although it is not immediately obvious, there are some significant advantages to our more general definition (Refs. 1–3). For example, choosing any $y \in \mathfrak{X}$ and defining $V(x) \equiv \sup_{\tau \geq 0} d(S(\tau)x, y)$, it is not difficult to show that $V : \mathfrak{X} \to \overline{\mathfrak{R}}$ is a Liapunov function on all of $\mathfrak{X}$, although it might not be continuous, or even finite valued, depending on $\{S(t)\}_{t\geq 0}$ (compare this unrestricted result with Theorem I.8.3.). If instead we define

$$V(x) \equiv \sup_{\tau \geq 0}\left[d(S(\tau)x, y)/(1 + d(S(\tau)x, y))\right],$$

we find that $V : \mathfrak{X} \to \mathfrak{R}$ is a Liapunov function on all of $\mathfrak{X}$, $0 \leq V(x) \leq 1$ for all $x \in \mathfrak{X}$, but V might not be continuous. For certain results, such as Liapunov's Direct Method, continuity is needed; in such cases we shall explicity mention continuity.

The useful property of any Liapunov function V is that, under certain conditions, its value can be shown to be nonincreasing along motions of the dynamical system; this property leads to many interesting conclusions. Notice that if *any* lower semicontinuous function $V : \mathfrak{X} \to \overline{\mathfrak{R}}$ has this property along some particular motion $S(\cdot)x : \mathfrak{R}^+ \to \mathfrak{X}$, and $V(x) > -\infty$, the V is automatically a Liapunov function on $\gamma(x)$ by Definition 1.1; the trouble with this idea is that we usually do not have *a priori* knowledge of $S(\cdot)x$, and consequently we usually do not know whether $V(S(\cdot)x)$ is nonincreasing or not. The Liapunov approach is based on the idea of showing $V(S(\cdot)x)$ to be nonincreasing by means that do not require explicit *a priori* knowledge of the motion $S(\cdot)x$. Later we shall provide means of obtaining $\dot{V}$ of Definition 1.1 in the absence of explicit *a priori* knowledge of the motion; accepting this for the moment, our present objective is to obtain similarly indirect conditions that are sufficient to ensure that $V(S(\cdot)x)$ is nonincreasing, when V is a Liapunov function on a set $\mathbb{S}$ and $x \in \mathbb{S}$. We need the following simple lemma.

Lemma 1.1. Let $f : ([0, \beta) \subset \mathfrak{R}) \to \overline{\mathfrak{R}}$ be defined on $[0, \beta)$, $0 < \beta \leq \infty$, with $f(0) < \infty$ and $f(t) > -\infty$ for every $t \in [0, \beta)$, and assume that

(i) f is left lower semicontinuous on $[0, \beta)$, i.e.,

$$\liminf_{h \searrow 0} f(t-h) \geqslant f(t) \qquad \text{for all } t \in (0; \beta);$$

(ii) the lower right derivative is nonpositive on $[0, \beta)$, i.e.,

$$D_+ f(t) \equiv \liminf_{h \searrow 0} \frac{l}{h}\left[f(t+h) - f(t)\right] \leqslant 0 \qquad \text{for all } t \in \left[0\ \beta\right).$$

Then f is nonincreasing and differentiable a.e.; moreover,

$$f(t) \leqslant f(0) + \int_0^t D_+ f(s)\, ds \qquad \text{for all } t \in \left[0, \beta\right).$$

Proof. Choosing some $\varepsilon > 0$, define $f_\varepsilon(t) = f(t) - \varepsilon t$ for $t \in [0, \beta)$. Then f_ε is left lower semicontinuous with $f_\varepsilon(0) = f(0)$ and $f_\varepsilon(t) > -\infty$, $D_+ f_\varepsilon(t) \leqslant -\varepsilon$, for every $t \in [0, \beta)$. If there exists $t_0 \in (0, \beta)$ such that $f_\varepsilon(t_0) > f(0)$, then the left lower semicontinuity of f_ε implies the existence of $t_1 \in [0, t_0)$ such that $f_\varepsilon(t_1) \leqslant f(0)$ and $f_\varepsilon(t) > f(0)$ for all $t \in (t_1, t_0]$; however, this leads to the contradiction $D_+ f_\varepsilon(t_1) \geqslant 0$. We conclude that $f_\varepsilon(t) \leqslant f(0)$ for every $t \in [0, \beta)$ and, as $\varepsilon > 0$ was arbitrary, the same is true for f. Replacing $t = 0$ with $t = \gamma \in (0, \beta)$ and repeating this argument, we find that $f(t) \leqslant f(\gamma)$ for all $t \in [\gamma, \beta)$ and all $\gamma \in [0, \beta)$; hence, f is nonincreasing and finite valued on $[0, \beta)$. By a standard result of integration theory (see Section 34.2 of Ref. 4) It follows that f is a.e. differentiable on compact subsets of $[0, \beta)$, with derivative equal a.e. to $D_+ f(t)$, and that

$$f(t) \leqslant f(0) + \int_0^t D_+ f(s)\, ds$$

for every $t \in [0, \beta)$. The proof is complete. □

The following result now can be obtained quite easily.

Theorem 1.1. Let $\{S(t)\}_{t \geqslant 0}$ be a dynamical system on a metric space $\mathfrak{X}$, and let $V : \mathfrak{X} \to \mathfrak{R}$ be a Liapunov function for $\{S(t)\}_{t \geqslant 0}$ on $\mathfrak{S} \subset \mathfrak{X}$. If $x \in \mathfrak{S}$, $V(x) < \infty$, and $S(t)x \in \mathfrak{S}$ for all $t \in [0, T)$, some $T \in \overline{\mathfrak{R}}^+$, then $V(S(\cdot)x)$ is nonincreasing and differentiable a.e. on $[0, T)$, with

$$V(S(t)x) \leqslant V(x) + \int_0^t \dot{V}(S(\tau)x)\, d\tau \qquad \text{for all } t \in \left[0, T\right).$$

If, in addition, $\dot{V}(y) \leqslant -\alpha V(y)$ for all $y \in \mathfrak{S}$, some $\alpha > 0$, then

$$V(S(t)x) \leqslant e^{-\alpha t} V(x) \qquad \text{for all } t \in \left[0, T\right).$$

Proof. As $S(\cdot)x : \mathcal{R}^+ \to \mathcal{X}$ is continuous and $V : \mathcal{X} \to \overline{\mathcal{R}}$ is lower semicontinuous, we define $f(t) \equiv V(S(t)x)$, $t \in \mathcal{R}^+$, and apply Lemma 1.1 to obtain all but the last conclusion.

Now assuming that $\dot{V}(y) \leqslant -\alpha V(y)$ for all $y \in \mathcal{S}$, we see that

$$V(x) - V(S(t)x) \geqslant \alpha \int_0^t V(S(\tau)x)\, d\tau$$

$$\geqslant \alpha t V(S(t)x), \qquad t \in [0, T),$$

since $\alpha > 0$ and $V(S(\cdot)x)$ is nonincreasing on $[0, T)$. Hence, for $t \in [0, T)$ and $n = 2, 3, \ldots,$

$$V(x) \geqslant (1 + \alpha t/n) V(S(t/n)x) \geqslant (1 + \alpha t/n)^2 V(S(2t/n)x)$$

$$\geqslant (1 + \alpha t/n)^n V(S(t)x) \to e^{\alpha t} V(S(t)x) \qquad \text{as } n \to \infty,$$

and the proof is complete. □

From Theorem 1.1 it is apparent that $V(S(\cdot)x)$ is nonincreasing on $\mathcal{R}^+$, provided that $\gamma(x)$ is contained within a set $\mathcal{S} \subset \mathcal{X}$ such that V is a Liapunov function on $\mathcal{S}$. If V is not a Liapunov function on all of $\mathcal{X}$, the problem now is to ensure that $\gamma(x)$ is actually contained in some set $\mathcal{S}$ on which V is a Liapunov function; this is directly related to the problem of determining positive invariant sets.

Exercise 1.1. Let $\{S(t)\}_{t \geqslant 0}$ be a dynamical system on a metric space $\mathcal{X}$. Show that

(i) the closure of a positive invariant set is positive invariant;
(ii) any union of positive invariant sets is positive invariant;
(iii) any intersection of positive invariant sets is positive invariant, defining the empty set to be positive invariant;
(iv) if a positive invariant set is the union of a family of closed disjoint components, then each component is positive invariant.

The following result provides a partial answer to the problem of showing how a "good" Liapunov function can often be used to determine some positive invariant sets; in fact, this is one of the most important uses for Liapunov functions. We notice that if some bounded (or precompact) set $\mathcal{S}$ can be shown to be positive invariant, then $\gamma(x)$ is bounded (respectively, precompact) for each $x \in \mathcal{S}$, and we have learned something about the behavior of every motion $S(\cdot)x : \mathcal{R}^+ \to \mathcal{X}$ for $x \in \mathcal{S}$.

Theorem 1.2. Let $\{S(t)\}_{t \geqslant 0}$ be a dynamical system on a metric space $\mathcal{X}$, and let $V : \mathcal{X} \to \mathcal{R}$ be a Liapunov function for $\{S(t)\}_{t \geqslant 0}$ on the (closed) set $\overline{\mathcal{G}}_\alpha \equiv \{x \in \mathcal{X} \mid V(x) \leqslant \alpha\}$ for some $\alpha < \infty$. Then for each

$\beta < \alpha$, the sets $\bar{\mathcal{G}}_\beta \equiv \{x \in \bar{\mathcal{G}}_\alpha | V(x) \leqslant \beta\}$, $\mathcal{G}_\beta \equiv \{x \in \bar{\mathcal{G}}_\alpha | V(x) < \beta\}$, and $\mathcal{G}_\alpha \equiv \{x \in \bar{\mathcal{G}}_\alpha | V(x) < \alpha\}$ are positive invariant. Moreover, for every $x \in \mathcal{G}_\alpha$, $V(S(\cdot)x)$ is nonincreasing and differentiable a.e. with

$$V(S(t)x) \leqslant V(x) + \int_0^t \dot{V}(S(\tau)x)\, d\tau, \qquad t \in \mathcal{R}^+.$$

Proof. In view of Theorem 1.1, it only remains to show that $\mathcal{G}_\alpha$, $\mathcal{G}_\beta$, and $\bar{\mathcal{G}}_\beta$ are positive invariant for every $\beta < \alpha$. Since $\mathcal{G}_\alpha = \cup_{\beta<\alpha} \mathcal{G}_\beta$ and $\mathcal{G}_\beta = \cup_{\mathcal{Y}<\beta} \bar{\mathcal{G}}_{\mathcal{Y}}$, the foregoing exercise implies that we need only show $\bar{\mathcal{G}}_\beta$ to be positive invariant for $\beta < \alpha$. To this end, fix $x \in \bar{\mathcal{G}}_\beta$ and note that $V(x) \leqslant \beta < \alpha$ and, by thc continuity of $S(\cdot)x : \mathcal{R}^+ \to \mathcal{X}$, the map $V(S(\cdot)x)$ is lower semicontinuous on $\mathcal{R}^+$. Hence, if there does not exist $T > 0$ such that $s(t)x \in \mathcal{G}_\alpha$ for all $t \in [0, T]$, then either (a) there exists $\delta > 0$ such that $V(S(t)x) > \alpha$ for all $t \in (0, \delta)$ or (b) there exist sequences $\{t_n\}_{n=1,2,\ldots}$, $\{\tau_n\}_{n=1,2,\ldots}$, such that $0 < \tau_{n+1} \leqslant t_n < \tau_n$, $t_n \searrow 0$ as $n \to \infty$, and $V(S(t)x) > \alpha$ for $t \in (t_n, \tau_n)$, $V(S(t)x) \leqslant \alpha$ for $t \in [\tau_{n+1}, t_n]$. Alternative (a) leads to the contradiction $0 \geqslant \dot{V}(x) = +\infty$.

Considering alternative (b), we first note that $\dot{V}(S(t_n)x) \leqslant 0$ implies $V(S(t_n)x) = \alpha$. Then employing Theorem 1.1, we see that $V(S(\cdot)x)$ is nonincreasing on $[\tau_{n+1}, t_n)$; hence, by lower semicontinuity, we must have $V(S(t)x) = \alpha$ for all $t \in [\tau_{n+1}, t_n]$. It follows that $V(S(t)x) \geqslant \alpha$ for all $t \in (0, \tau_1)$, which again leads to the contradiction $0 \geqslant \dot{V}(x) = +\infty$.

We have shown that $S(t)x \in \mathcal{G}_\alpha$ for all $t \in [0, T]$, some $T > 0$; applying Theorem 1.1, we find that $V(S(t)x) \leqslant V(x) \leqslant \beta$ for all $t \in [0, T]$. Repeating this argument, we find that either $S(t)x$ remains in $\bar{\mathcal{G}}_\beta$ for all $t \in \mathcal{R}^+$ or there exists positive $\tau < \infty$ such that $V(S(t)x) \leqslant \beta$ for every $t \in [0, \tau)$ and $V(S(\tau)x) > \beta$. By the lower semicontinuity of $V(s(\cdot)x)$, the latter case is impossible, and we have shown $\bar{\mathcal{G}}_\beta$ to be positive invariant. The proof is complete. □

In most of our applications we will be using continuous Liapunov functions; then the following corollary applies.

Corollary 1.1. Let $\{S(t)\}_{t\geqslant 0}$ be a dynamical system on a metric space $\mathcal{X}$, and let $V : \mathcal{X} \to \mathcal{R}$ be a continuous Liapunov function for $\{S(t)\}_{t\geqslant 0}$ on a disjoint component $\mathcal{G}_\alpha$ of the (open) set $\{x \in \mathcal{X} | V(x) < \alpha\}$ for some $\alpha \leqslant \infty$. Then, for each $\beta \leqslant \alpha$, the sets $\mathcal{G}_\beta \equiv \{x \in \mathcal{G}_\alpha | V(x) < \beta\}$ and $\bar{\mathcal{G}}_\beta \equiv \{x \bar{\mathcal{G}}_\alpha | V(x) \leqslant \beta\}$ are positive invariant, where $\bar{\mathcal{G}}_\alpha \equiv \mathrm{Cl}\, \mathcal{G}_\alpha$. Moreover, for every $x \in \bar{\mathcal{G}}_\alpha$, $V(S(\cdot)x)$ is nonincreasing and differentiable a.e., with

$$V(S(t)x) \leqslant V(x) + \int_0^t \dot{V}(S(\tau)x)\, d\tau, \qquad t \in \mathcal{R}^+, x \in \mathcal{G}_\alpha.$$

Using the following result, Corollary 1.1 is an obvious consequence of Theorem 1.2.

Theorem 1.3. Let V be a continuous Liapunov function on a positive invariant set $\mathbb{S}$ for a dynamical system $\{S(t)\}_{t\geq 0}$ on a metric space space $\mathfrak{X}$, such that $\dot{V}(x) \leq -W(x)$ for all $x \in \mathbb{S}$ and some lower semicontinuous $W : (\mathrm{Cl}\,\mathbb{S} \subset \mathfrak{X}) \to \overline{\mathfrak{R}}^{+}$. Then V is a Liapunov function on $\mathrm{Cl}\,\mathbb{S}$ with $\dot{V}(x) \leq -W(x)$ for all $x \in \mathrm{Cl}\,\mathbb{S}$.

Proof. Consider a sequence $\{x_n\} \subset \mathbb{S}_n$, $x_n \to x \in \mathrm{Cl}\,\mathbb{S}$ as $n \to \infty$; then, by applying Theorem 1.1,

$$\begin{aligned} V(S(t)x_n) &\leq V(x_n) + \int_0^t \dot{V}(S(\tau)x_n)\,d\tau \\ &\leq V(x_n) - \int_0^t W(S(\tau)x_n)\,d\tau \qquad \text{for all } t \in \mathfrak{R}^{+}. \end{aligned}$$

By continuity of V, we see that $V(x_n) \to V(x)$ and $V(S(t)x_n) \to V(S(t)x)$ as $n \to \infty$; hence,

$$\begin{aligned} V(x) - V(S(t)x) &\geq \limsup_{n\to\infty} \int_0^t W(S(\tau)x_n)\,d\tau \\ &\geq \liminf_{n\to\infty} \int_0^t W(S(\tau)x_n)\,d\tau \\ &\geq \int_0^t \Big[\liminf_{n\to\infty} W(S(\tau)x_n)\Big]\,d\tau \\ &\geq \int_0^t W(S(\tau)x)\,d\tau \geq 0, \qquad t \in \mathfrak{R}^{+}, \end{aligned}$$

where we have employed Fatou's lemma (Ref. 4), noting that $W : (\mathrm{Cl}\,\mathbb{S} \subset \mathfrak{X}) \to \overline{\mathfrak{R}}^{+}$ is lower semicontinuous and nonnegative. Therefore,

$$\begin{aligned} -\dot{V}(x) &\equiv -\liminf_{t\searrow 0} \frac{1}{t}\big[V(S(t)x) - V(x)\big] \\ &= \limsup_{t\searrow 0} \frac{1}{t}\big[V(x) - V(S(t)x)\big] \\ &\geq \limsup_{t\searrow 0} \frac{1}{t}\int_0^t W(S(\tau)x)\,d\tau \\ &\geq \limsup_{t\searrow 0} \Big[\inf_{0\leq\tau<t} W(S(\tau)x)\Big] \geq W(x), \end{aligned}$$

where we have again used the lower semicontinuity of W. The proof is complete. □

Theorem 1.3 shows that the assumptions of Corollary 1.1 imply that V is a continuous Liapunov function on $\bar{\mathcal{G}}_\alpha \equiv \text{Cl}\,\mathcal{G}_\alpha$; hence, Corollary 1.1 follows directly from Theorem 1.2.

We have shown how a Liapunov function can be useful in finding positive invariant sets; other uses will be found in Sections 3 and 4. Whether or not a Liapunov function V is "useful" depends entirely on our ability to show that V is nonincreasing along at least some motions of the dynamical system. The foregoing results provide the basic tools for showing that a given Liapunov function is useful; Liapunov functions are easy to find, but useful ones often are not.

Exercise 1.2. Consider a given dynamical system $\{S(t)\}_{t\geqslant 0}$ on a metric space, and any continuous function $V : \mathcal{X} \to \mathcal{R}$. Show that either V is a Liapunov function on some nonempty set or else $-V$ is a Liapunov function on all of $\mathcal{X}$.

In some cases, the results of Chapter III provide a useful Liapunov function on all of $\mathcal{X}$.

Proposition 1.1. For $\mathcal{B}$ a Banach space, let a multivalued operator $A : (\mathcal{D}(A) \subset \mathcal{B}) \to \mathcal{B}$ generate a dynamical system $\{S(t)\}_{t\geqslant 0}$ on $\mathcal{X} \equiv \text{Cl}_{\mathcal{B}}\, \mathcal{D}(A)$, in the sense of Theorem III.5.1 for some $\omega \leqslant 0$; moreover, let there exist $x_e \in \mathcal{D}(A)$ such that $0 \in Ax_e$. Then x_e is an equilibrium and the function $V : \mathcal{X} \to \mathcal{R}^+$ defined by

$$V(x) \equiv \|x - x_e\|^p, \qquad x \in \mathcal{X}, \text{ some } p > 0,$$

is a continuous Liapunov function on $\mathcal{X}$, with $\dot{V}(x) \leqslant p\omega V(x)$ and $V(S(t)x) \leqslant e^{p\omega t}V(x)$ for all $x \in \mathcal{X}$, $t \in \mathcal{R}^+$.

Exercise 1.3. Prove Proposition 1.1 *Hint*: To estimate $\dot{V}$ in Definition 1.1, use the estimate of Theorem III.5.1.

Under the assumptions of Proposition 1.1, Theorem 1.1 leads to the conclusion that $V(S(t)x) \leqslant e^{p\omega t}V(x)$ for all $t \in \mathcal{R}^+$, $x \in \mathcal{X}$; this is equivalent to the statement that $\|S(t)x - x_e\| \leqslant e^{\omega t}\|x - x_e\|$, which also follows directly from the exponential estimate of Theorem III.5.1. Although we seem to have achieved nothing new by the use of Proposition 1.1, this result serves to emphasize the usefulness of selecting a norm such that $\omega I - A$ is accretive for a low value of ω ($\omega \leqslant 0$, it is hoped). In fact, "reselection of norm" can be a very useful way of generating Liapunov functions for a dynamical system that is generated in the sense of Theorem III.5.1. This approach may or may not work, depending on whether there exists an equivalent norm $\|\cdot\|_e$ such that the dynamical system is $\|\cdot\|_e$-nonexpansive. If it works for some $\omega < 0$, we see by the introductory remarks of this chapter that there is little more to be asked about the behavior of any motion; however, if it works, but only for $\omega = 0$, there may remain some unanswered questions.

Example 1.1. Reconsider the linear heat conduction problem of Example III.3.1, where

$$\mathfrak{X} \equiv \{x \in \mathcal{C}[0, 1] | x(0) = 0 = x(1)\}, \qquad \|x\| \equiv \max_{\eta} |x(\eta)|,$$

$$Ax \equiv \partial^2 x, \qquad x \in \mathfrak{D}(A) \equiv \{x \in \mathcal{C}^2[0, 1] | x(\eta) = 0 = \partial^2 x(\eta) \text{ at } \eta = 0, 1\}.$$

There we found that A was the infinitesimal generator of a linear dynamical system $\{S(t)\}_{t \geq 0}$ on $\mathfrak{X}$, and $\|S(t)x\| \leq \|x\|$ for all $t \in \mathfrak{R}^+$, $x \in \mathfrak{X}$.

Defining $V : \mathfrak{X} \to \mathfrak{R}$ by $V(x) \equiv \|x\|$, we see that V is a Liapunov function on all of $\mathfrak{X}$ with $\dot{V}(x) \leq 0$, $x \in \mathfrak{X}$. Aside from noting that the sets $\bar{\mathcal{G}}_\alpha \equiv \{x \in \mathfrak{X} | \|x\| \leq \alpha\}$ and $\mathcal{G}_\alpha \equiv \{x \in \mathfrak{X} | \|x\| < \alpha\}$ are positive invariant for every $\alpha \in \mathfrak{R}$, there is little more that we can say by the use of this Liapunov function. Moreover, there seems to be no equivalent norm under which $\omega I - A$ is accretive for some $\omega < 0$, rather than $\omega = 0$.

Let us try another approach, suggested by the results of Example III.3.2. We define another function $\tilde{V} : \mathfrak{X} \to \mathfrak{R}$ as

$$\tilde{V}(x) \equiv \int_0^1 |x(\eta)|^2 \, d\eta, \qquad x \in \mathfrak{X},$$

noting that $\tilde{V}$ is Fréchet differentiable, and recalling that $S(\cdot)x : \mathfrak{R}^+ \to \mathfrak{X}$ is right-differentiable for every $x \in \mathfrak{D}(A)$. By Proposition II.4.1, we see that $\tilde{V}(S(\cdot)x) : \mathfrak{R}^+ \to \mathfrak{R}$ is right-differentiable and

$$\dot{\tilde{V}}(x) \equiv \liminf_{t \searrow 0} \frac{1}{t}[\tilde{V}(S(t)x) - \tilde{V}(x)] = \tilde{V}_x' Ax$$

$$= 2\int_0^1 x(\eta) \cdot \partial^2 x(\eta) \, d\eta = -2\int_0^1 |\partial x(\eta)|^2 \, d\eta \leq -2\pi^2 \tilde{V}(x)$$

for every $x \in \mathfrak{D}(A)$. It follows that $\tilde{V}$ is a Liapunov function on $\mathfrak{D}(A)$, which is positive invariant by Theorem III.2.1. Defining $W : \mathfrak{X} \to \mathfrak{R}$ by $W(x) \equiv 2\pi^2 \tilde{V}(x)$, $x \in \mathfrak{X}$, we see by Theorem 1.3 that $\dot{\tilde{V}}(x) \leq -2\pi^2 \tilde{V}(x)$ for all $x \in \mathfrak{X}$, and $\tilde{V}$ is a Liapunov function on all of $\mathfrak{X}$; hence, by Theorem 1.1, we see that $\tilde{V}(S(t)x) \leq \exp\{-2\pi^2 t\} \tilde{V}(x)$ for all $t \in \mathfrak{R}^+$, $x \in \mathfrak{X}$, implying that $S(t)x \xrightarrow{\mathcal{L}_2} 0$ as $t \to \infty$, for every $x \in \mathfrak{X}$. Notice that we cannot (yet) conclude that $S(t)x \xrightarrow{\mathcal{C}} 0$, which would be a stronger conclusion. In Section 5 we shall return to this example and obtain this stronger conclusion.

The construction of a useful Liapunov function may be far from simple in a particular application. Even for a linear dynamical system, there is considerable difference between the explicit construction of a useful Liapunov function and the mere knowledge that one exists. For example, Theorem III.3.2 shows $\|\cdot\|_e$ to be a useful Liapunov function if $\omega < 0$ in the estimate $\|S(t)x\|_{\mathfrak{X}} \leq Me^{\omega t}\|x\|_{\mathfrak{X}}$, $x \in \mathfrak{X}$, $t \in \mathfrak{R}^+$, but it provides no means of constructing the "best" $\|\cdot\|_e$ in order to *prove* that $\omega < 0$.

For a linear dynamical system on $\mathfrak{X} = \mathfrak{R}^n$, we recall from Section I.9 that we have an almost fully automatic construction procedure, based on Liapunov's equation (10) of Chapter I. However, if $\mathfrak{X}$ is an infinite-dimensional Banach space, many technical difficulties arise in attempts to extend this construction procedure (Refs. 5–8). Although some partial extensions are available (Refs. 5–8), they are not very helpful in applications; the practical problem of explicitly solving an abstract operator equation having the form of (10) in Chapter I for $B \in \mathfrak{B}(\mathfrak{X}, \mathfrak{X}^*)$, may be insurmountable when $\mathfrak{X}$ has infinite dimension, even when a solution is known to exist. Hence, we shall not go into such extensions of Liapunov's equation.

Sometimes physical considerations may suggest a useful Liapunov function, as we saw in Chapter I. If not, our best alternative usually is to rely on some form of organized guesswork, such as that suggested by Theorem III.4.2 or III.5.3; there, if a suitable $B : (\mathfrak{D}(\tilde{A}) \subset \mathfrak{H}) \to \mathfrak{H}$ exists for some $\omega \leq 0$, then Proposition 1.1 shows that $V(x) \equiv \|x\|_{\mathfrak{X}}^k$, $x \in \mathfrak{X}$, some $k > 0$, will be a continuous Liapunov function on all of $\mathfrak{X}$, with $\dot{V}(x) \leq k\omega V(x)$ for all $x \in \mathfrak{X}$. The problem of constructing useful Liapunov functions provided the original motivation (Ref. 9) for Theorem III.4.2; some extensions of this idea are mentioned in Ref. 9.

Example 1.1 demonstrates the usefulness of Liapunov functions that are unlike equivalent norms, even for nonexpansive linear dynamical systems. In this example we encountered and overcame the problem of computing $\dot{V}$ without explicit knowledge of the motions. Before going further into the uses of Liapunov functions, we shall consider some general approaches to this computational problem.

2. Computation of $\dot{V}$

In order to determine if some lower semicontinuous $V : \mathfrak{X} \to \overline{\mathfrak{R}}$ is a Liapunov function on some set $\mathcal{G}$, it is necessary to evaluate, or at least estimate, the function $\dot{V} : \mathfrak{X} \to \overline{\mathfrak{R}}$ on $\mathcal{G} \subset \mathfrak{X}$ (see Definition 1.1). Often there are severe difficulties involved in computing $\dot{V}$ when, as is usually true in applications, the family $\{S(t)\}_{t \geq 0}$ is not explicitly known (Refs. 3, 9, 10); this computational problem is of considerable research interest. When $V : \mathfrak{X} \to \mathfrak{R}$ is Fréchet differentiable, the following theorem and its corollary generalize the method of computation used in Example 1.1.

Theorem 2.1. For $\mathfrak{B}$ a Hilbert space, let a closed multivalued operator $A : (\mathfrak{D}(A) \subset \mathfrak{B}) \to \mathfrak{B}$ generate a dynamical system $\{S(t)\}_{t \geq 0}$ on $\mathfrak{X} \equiv \mathrm{Cl}_{\mathfrak{B}}\mathfrak{D}(A)$, in the sense of Theorem III.5.1. If $V : \mathfrak{X} \to \mathfrak{R}$ is Fréchet differentiable, then $\mathfrak{D}(A) = \mathfrak{D}(A^0)$ is positive invariant,

$$\dot{V}(x) \equiv \liminf_{t \searrow 0} \frac{1}{t}\left[V(S(t)x) - V(x) \right] = V_x'(A^0 x)$$

for all $x \in \mathfrak{D}(A^0) = \mathfrak{D}(A)$, and V is a continuous Liapunov function on $\{x \in \mathfrak{D}(A) | V_x' A^0 x \leq 0\}$.

If, in addition, $V_x' A^0 x \leq -W(x)$ for all $x \in \mathcal{G}_\alpha \cap \mathfrak{D}(A)$, where $W : (\mathrm{Cl}_{\mathfrak{B}}\mathcal{G}_\alpha \subset \mathfrak{X}) \to \overline{\mathfrak{R}}^+$ is lower semicontinuous and $\mathcal{G}_\alpha \equiv \{x \in \mathfrak{X} | V(x) < \alpha\}$, then V is a continuous Liapunov function on $\overline{\mathcal{G}}_\alpha \equiv \mathrm{Cl}_{\mathfrak{B}}\mathcal{G}_\alpha$, both $\mathcal{G}_\alpha$ and $\overline{\mathcal{G}}_\alpha$ are positive invariant, and $\dot{V}(x) \leq -W(x) \leq 0$ for all $x \in \overline{\mathcal{G}}_\alpha$.

Proof. As $\mathfrak{B}$ is a Hilbert space and A is closed, it follows from Theorem III.5.2 that $\mathfrak{D}(A) = \mathfrak{D}(A^0)$, A^0 is a single-valued operator, $\mathfrak{D}(A)$ is positive invariant, and the motion $S(\cdot)x$ is right-differentiable on $\mathfrak{R}^+$ with $(d^+/dt)S(t)x = A^0 S(t)x$, $t \in \mathfrak{R}^+$, for every $x \in \mathfrak{D}(A)$. As V is Fréchet differentiable, it follows from Proposition II.4.1 that $V(S(\cdot)x)$ is right-differentiable and $(d^+/dt)V(S(t)x) = V'_{S(t)x} A^0 S(t)x$ for all $x \in \mathfrak{D}(A)$, $t \in \mathfrak{R}^+$; hence, $\dot{V}(x) = V_x A^0 x$ for all $x \in \mathfrak{D}(A) = \mathfrak{D}(A^0)$. As $\mathfrak{D}(A)$ is positive invariant, the remaining conclusions now follow from Theorem 1.3 and Corollary 1.1. The proof is complete. □

Corollary 2.1. Let a linear operator $A : (\mathfrak{D}(A) \subset \mathfrak{B}) \to \mathfrak{B}$, $\mathfrak{B}$ a Banach space, be the infinitesimal generator of a linear dynamical system $\{S(t)\}_{t \geq 0}$ on $\mathfrak{B}$. Then all conclusions of Theorem 2.1 hold true.

Proof. We have dropped the assumption of Theorem 2.1 that $\mathfrak{B}$ is a Hilbert space, requiring instead that A be a (linear) infinitesimal generator. In the proof of Theorem 2.1, this assumption was used only to ensure that $\mathfrak{D}(A)$ was positive invariant and that $S(\cdot)x$ was right-differentiable, with right derivative $A^0 x$ at $t = 0$, for $x \in \mathfrak{D}(A)$. However, under our new assumption, these properties are assured by Theorem III.2.1, and all conclusions follow as in the proof of Theorem 2.1. □

Example 2.1. Consider the sine-Gordon equation (Ref. 11)

$$\frac{\partial^2}{\partial t^2} u(\eta, t) + 2\alpha \frac{\partial}{\partial t} u(\eta, t) - \frac{\partial^2}{\partial \eta^2} u(\eta, t) - \beta \sin(u(\eta, t)) = 0, \quad 0 \leq \eta \leq 1, \quad t \geq 0,$$

$$u(0, t) = 0 = u(1, t), \qquad t \geq 0,$$

for some $\alpha > 0$, $\beta \neq 0$. Defining

$$\mathfrak{X} \equiv \hat{\mathfrak{W}}_2^1(0, 1) \times \mathfrak{L}_2(0, 1), \qquad \hat{\mathfrak{W}}_2^1(0, 1) \equiv \{u \in \mathfrak{W}_2^1(0, 1) | u(0) = 0 = u(1)\},$$

$$\|x\| \equiv \left(\int_0^1 [|\partial u(\eta)|^2 + |v(\eta)|^2] \, d\eta\right)^{1/2}, \qquad x = (u, v) \in \mathfrak{X},$$

$$Ax(\eta) \equiv (v(\eta), -2\alpha v(\eta) + \partial^2 u(\eta) + \beta \sin[u(\eta)]) \text{ a.e. } \eta \in [0, 1],$$

$$x = (u, v) \in \mathfrak{D}(A),$$

$$\mathfrak{D}(A) \equiv \{(u, v) \in \mathfrak{X} | \partial^2 u \in \mathfrak{L}_2(0, 1), v \in \hat{\mathfrak{W}}_2^1(0, 1)\},$$

we see as in Exercise III.5.6 that $(|\beta|/\pi)I - A$ is accretive and that $\mathfrak{R}(I - \lambda A) = \mathfrak{X}$ for all sufficiently small $\lambda > 0$. Hence, A generates a dynamical system $\{S(t)\}_{t \geq 0}$ on the Hilbert space $\mathfrak{X}$, in the sense of Theorem III.5.1, and $\|S(t)x - S(t)y\| \leq e^{|\beta| t/\pi} \|x - y\|$ for all $x, y \in \mathfrak{X}$, $t \in \mathfrak{R}^+$.

Seeking a Liapunov function, we consider the total mechanical energy and define

$$V(x) \equiv \int_0^1 [|\partial u(\eta)|^2 + 2\beta \cos(u(\eta)) + |v(\eta)|^2] \, d\eta, \qquad x = (u, v) \in \mathfrak{X},$$

noting that $V : \mathfrak{X} \to \mathfrak{R}$ is Fréchet differentiable with $V_x' \in \mathfrak{B}(\mathfrak{X}, \mathfrak{X})$ defined by

$$V_x' \hat{x} \equiv 2 \int_0^1 [\partial u(\eta) \cdot \partial \hat{u}(\eta) - \beta \hat{u}(\eta) \sin(u(\eta)) + v(\eta) \cdot \hat{v}(\eta)] \, d\eta, \qquad \hat{x} = (\hat{u}, \hat{v}) \in \mathfrak{X},$$

at each $x = (u, v) \in \mathfrak{X}$. As A is closed, Theorem 2.1 implies that

$$\dot{V}(x) = V_x' A x = -4\alpha \int_0^1 |v(\eta)|^2 \, d\eta, \qquad x \in (u, v) \in \mathfrak{D}(A),$$

and V is a Liapunov function on $\mathfrak{D}(A)$ since $\alpha \geq 0$. Also defining a (continuous) function $W : \mathfrak{X} \to \mathfrak{R}^+$ as

$$W(x) \equiv 4\alpha \int_0^1 |v(\eta)|^2 \, d\eta, \qquad x = (u, v) \in \mathfrak{X},$$

we find by Theorem 1.3 that V is a Liapunov function on all of $\mathfrak{X}$, with $\dot{V}(x) \leq -W(x)$ for all $x \in \mathfrak{X}$.

Defining the sets $\bar{\mathcal{G}}_\alpha \equiv \{x \in \mathfrak{X} | V(x) \leq \alpha\}$, $\mathcal{G}_\alpha \equiv \{x \in \mathfrak{X} | V(x) < \alpha\}$, we see by Theorem 1.4 that both $\bar{\mathcal{G}}_\alpha$ and $\mathcal{G}_\alpha$ are positive invariant for each $\alpha \in \mathfrak{R}$. Because result (iv) of Theorem III.5.2 shows $\mathfrak{D}(A)$ to be positive invariant, it also follows that $\bar{\mathcal{G}}_\alpha \cap \mathfrak{D}(A)$ and $\mathcal{G}_\alpha \cap \mathfrak{D}(A)$ are positive invariant for each $\alpha \in \mathfrak{R}$. Noting that $V(x) \to \infty$ as $\|x\| \to \infty$, we see that each $\bar{\mathcal{G}}_\alpha$ is bounded and, as $\mathfrak{X} = \bigcup_{\alpha \in \mathfrak{R}} \bar{\mathcal{G}}_\alpha$, we may now conclude that all positive orbits are bounded.

Theorem III.5.2 implies that $x_e \in \mathfrak{X}$ is an equilibrium if and only if $x_e \in \mathfrak{D}(A)$ and $Ax_e = 0$, since A is closed and single-valued. But we see that $Ax_e = 0$ if and only if $x_e \in \mathfrak{D}(A)$ and $V'_{x_e} = 0 \in \mathfrak{B}(\mathfrak{X}, \mathfrak{X})$. It follows that each disjoint and nonempty component of $\overline{\mathcal{G}}_\alpha \cap \mathfrak{D}(A)$ contains at least one equilibrium; hence, the same is true of $\overline{\mathcal{G}}_\alpha$ and $\mathcal{G}_\alpha$. Notice that $Ax_e = 0$ if and only if $x_e = (u_e, 0) \in \mathfrak{D}(A)$ and $\partial^2 u_e(\eta) = -\beta \sin(u_e(\eta))$ a.e. $\eta \in [0, 1]$; it follows that $0 \in \mathfrak{X}$ is an equilibrium, and it is unique if $\pi^2 > |\beta|$. The study of how the number (and location) of equilibria depends on β is known as a *bifurcation problem* (Refs. 11–13); we see that this problem is directly related to a *minimization problem* for the function V.

Exercise 2.1. In Example 2.1, show that $0 \in \mathfrak{X}$ is an equilibrium and that there are no others if $\pi^2 > |\beta|$.

Exercise 2.2. Consider the nonlinear wave equation of Example III.6.2, as well as the dynamical system $\{S(t)\}_{t \geq 0}$ set up there on the Hilbert space $\mathfrak{B} \equiv \hat{\mathfrak{W}}_2^1(0, 1) \times \mathfrak{L}_2(0, 1)$. Show that the Fréchet-differentiable function $V : \mathfrak{B} \to \mathfrak{R}$ defined by

$$V(x) \equiv \int_0^1 [2|\partial u(\eta)|^2 + \gamma |u(\eta)|^4 + 2|v(\eta)|^2] \, d\eta, \qquad x = (u, v) \in \mathfrak{B},$$

is a Liapunov function on $\mathfrak{B}$, and determine (or estimate) $\dot{V} : \mathfrak{B} \to \overline{\mathfrak{R}}$. (*Hint*: As $\omega I - A$ is not accretive for any finite ω, use Theorem 2.1 in conjunction with the restrictions $\{S_\alpha(t)\}_{t \geq 0}$, $\alpha \in \mathfrak{R}$, defined in Example III.6.2.) What can you determine about positive invariant sets, boundedness of positive orbits, and existence of equilibria?

If $V : \mathfrak{X} \to \overline{\mathfrak{R}}$ is not Fréchet differentiable, or if $\{S(t)\}_{t \geq 0}$ is not linear and $\mathfrak{B}$ is not a Hilbert space, then neither Theorem 2.1 nor Corollary 2.1 is of help in computing $\dot{V} : \mathfrak{X} \to \overline{\mathfrak{R}}$, and the following result (Ref. 3) may be useful.

Theorem 2.2. Let $\{S(t)\}_{t \geq 0}$ be a dynamical system generated by a product formula on a metric space $\mathfrak{X}$ (see Definition III.5.1). Let there exist lower semicontinuous functions $V : \mathfrak{X} \to \overline{\mathfrak{R}}$, $U : \mathfrak{R}^+ \times \mathfrak{X} \to \overline{\mathfrak{R}}$, such that $V(x) > -\infty$ for all $x \in \mathfrak{X}$ and

$$V(x) \geq V(J_\lambda x) + \lambda U(\lambda, J_\lambda x) \tag{1}$$

for all $\lambda \in (0, \lambda_0)$, some $\lambda_0 > 0$, and all $x \in \mathfrak{X}$ with $V(x) < \infty$. Then

$$\dot{V}(x) \equiv \liminf_{t \searrow 0} \frac{1}{t} [V(S(t)x) - V(x)] \leq -U(0, x)$$

for every $x \in \mathfrak{X}$ with $V(x) < \infty$, and V is a Liapunov function on the set $\{x \in \mathfrak{X} \,|\, V(x) < \infty,\ U(0, x) \geq 0\}$.

Proof. For $x \in \mathcal{X}$ with $V(x) < \infty$, $t \in (0, \lambda_0)$, and $n = 1, 2, \ldots,$ we see that

$$\begin{aligned}\frac{1}{t}\left[V(J_{t/n}^{n}x) - V(x)\right] &= \frac{1}{t}\sum_{m=1}^{n}\left[V(J_{t/n}^{m}x) - V(J_{t/n}^{m-1}x)\right] \\ &\leqslant -\frac{1}{n}\sum_{m=1}^{n} U\left(\frac{t}{n}, J_{t/n}^{m}x\right) \\ &\leqslant -\inf_{m}\left\{U\left(\frac{t}{n}, J_{t/n}^{m}x\right)\middle| m = 1, 2, \ldots, n\right\} \\ &\leqslant -\inf_{m,\tau}\left\{U\left(\frac{\tau}{m}, J_{\tau/m}^{m}x\right)\middle| 0 < \tau \leqslant t; m = 1, 2, \ldots, n\right\},\end{aligned}$$

where $J_{\lambda}^{0}x = x$. Since $\{S(t)\}_{t \geqslant 0}$ is generated and V is lower semicontinuous,

$$\begin{aligned}\frac{1}{t}\left[V(S(t)x) - V(x)\right] &\leqslant \liminf_{n\to\infty}\frac{1}{t}\left[V(J_{t/n}^{n}x) - V(x)\right] \\ &\leqslant -\inf_{m,\tau}\left\{U\left(\frac{\tau}{m}, J_{\tau/m}^{m}x\right)\middle|\ 0 < \tau \leqslant t,\ m = 1, 2, \ldots\right\},\end{aligned}$$

and it follows that

$$\begin{aligned}\dot{V}(x) &\equiv \liminf_{t\searrow 0}\frac{1}{t}\left[V(S(t)x) - V(x)\right] \\ &\leqslant -\limsup_{t\searrow 0}\left[\inf_{m,\tau}\left\{U\left(\frac{\tau}{m}, J_{\tau/m}^{m}x\right)\middle|\ 0 < \tau \leqslant t,\ m = 1, 2, \ldots\right\}\right].\end{aligned}$$

Denoting the last term by $-f(x)$, $f: \mathcal{X} \to \overline{\mathcal{R}}$, it follows that for each $\varepsilon > 0$ there exists a sequence $\{\tau_k, m_k\}_{k=1,2,\ldots}$ depending on x and ε, such that m_k is a positive integer, $\tau_k > \tau_{k+1} > 0$, $\tau_k \to 0$ as $k \to \infty$, and

$$f(x) + \varepsilon \geqslant \liminf_{k\to\infty} U\left(\frac{\tau_k}{m_k}, J_{\tau_k/m_k}^{m_k}x\right).$$

If the sequence $\{m_k\}_{k=1,2,\ldots}$ is bounded, then the lower semicontinuity of U and the fact that $J_{\lambda}^{m}x \to x$ as $\lambda \searrow 0$, uniformly in $m = 1, 2, \ldots, n$ for finite n, together imply that $f(x) + \varepsilon \geqslant U(0, x)$. On the other hand, if the sequence $\{m_k\}_{k=1,2,\ldots}$ is not bounded, then there exists a subsequence $\{t_p, n_p\}_{p=1,2,\ldots}$ of $\{\tau_k, m_k\}_{k=1,2,\ldots}$ such that n_p is a positive integer, $n_p \to \infty$ as $p \to \infty$, $t_p > t_{p+1} > 0$, $t_p \to 0$ as $p \to \infty$, and

$$f(x) + \varepsilon \geqslant \liminf_{p\to\infty} U\left(\frac{t_p}{n_p}, J_{t_p/n_p}^{n_p}x\right).$$

Then, since $J^n_{t/n}x \to S(t)x$ as $n \to \infty$, uniformly on compact t-intervals, it again follows from the lower semicontinuity of U that $f(x) + \varepsilon \geqslant U(0, x)$. As $\varepsilon > 0$ was arbitrary, we obtain $\dot{V}(x) \leqslant -f(x) \leqslant -U(0, x)$ for all $x \in \mathfrak{X}$ such that $V(x) < \infty$. The proof is complete. □

Defining $\dot{V}(x) = 0$ for all $x \in \mathfrak{X}$ such that $V(x) = +\infty$, we see from Theorem 2.2 that V is a Liapunov function for $\{S(t)\}_{t \geqslant 0}$ on the set $\mathfrak{S} \equiv \{x \in \mathfrak{X} \mid U(0, x) \geqslant 0,\ V(x) < \infty\} \cup \{x \in \mathfrak{X} \mid V(x) = \infty\}$; moreover, $\dot{V}(x) \leqslant -W(x)$ for all $x \in \mathfrak{S}$, where $W : (\mathfrak{S} \subset \mathfrak{X}) \to \overline{\mathfrak{R}}^+$ is the lower semicontinuous function defined by

$$W(x) \equiv U(0, x) \qquad \text{for } x \in \mathfrak{S} \text{ with } V(x) < \infty,$$
$$W(x) \equiv 0 \qquad \text{for } x \in \mathfrak{S} \text{ with } V(x) = \infty.$$

If $\mathfrak{S} \supset \overline{\mathfrak{G}}_\alpha$, where $\overline{\mathfrak{G}}_\alpha$ is a disjoint component of the set $\{x \in \mathfrak{X} \mid V(x) \leqslant \alpha\}$ for some $\alpha < \infty$, then we see by Theorem 1.2 that $\overline{\mathfrak{G}}_\alpha$ is positive invariant and $\dot{V}(x) \leqslant -W(x) = -U(0, x)$ for all $x \in \overline{\mathfrak{G}}_\alpha$.

In Theorem 2.2, we also notice that if a multivalued operator $A : (\mathfrak{D}(A) \subset \mathfrak{B}) \to \mathfrak{B}$ generates $\{S(t)\}_{t \geqslant 0}$ in the sense of Theorem III.5.1, where the Banach space $\mathfrak{B} \supset \mathfrak{X} \equiv \mathrm{Cl}_{\mathfrak{B}}\mathfrak{D}(A)$ and $J_\lambda \equiv (I - \lambda A)^{-1}$, $\lambda \in (0, \lambda_0)$, then condition (1) is equivalent to

$$V(x - \lambda y) \geqslant V(x) + \lambda U(\lambda, x) \tag{2}$$

for all $\lambda \in (0, \lambda_0)$ and all $y \in Ax$, $x \in \mathfrak{D}(A)$, such that $x - \lambda y \in \mathfrak{X}$, $V(x - \lambda y) < \infty$. Condition (2) is usually more useful in applications.

Exercise 2.3. Prove Proposition 1.1 without making use of the estimate of Theorem III.5.1.

Example 2.2. Reconsider the nonlinear heat conduction problem of Example III.5.2, where $\mu I - f : \mathfrak{R} \to \mathfrak{R}$ is continuous and monotone for some $\mu \in \mathfrak{R}$, $f(0) = 0$, and

$$\mathfrak{X} \equiv \{x \in \mathcal{C}[0, 1] \mid x(0) = 0 = x(1)\}, \qquad \|x\| \equiv \max_\eta |x(\eta)|,$$

$$Ax(\eta) \equiv \partial^2 x(\eta) + f(x(\eta)), \quad 0 \leqslant \eta \leqslant 1, \qquad x \in \mathfrak{D}(A) \equiv \{x \in \mathfrak{X} \mid \partial^2 x \in \mathfrak{X}\}.$$

There we found that A generates a dynamical system $\{S(t)\}_{t \geqslant 0}$ on $\mathfrak{X}$, in the sense of Theorem III.5.1, and $\|S(t)x - S(t)y\| \leqslant e^{\mu t}\|x - y\|$ for all $x, y \in \mathfrak{X}$, $t \in \mathfrak{R}^+$.

Extending an idea of Ref. 14 as in Ref. 3, we define $V : \mathfrak{X} \to \overline{\mathfrak{R}}$ by

$$V(x) \equiv \int_0^1 \left[|\partial x(\eta)|^2 - 2\int_0^{x(\eta)} f(\xi)\, d\xi \right] d\eta \qquad \text{for } x \in \mathfrak{X} \cap \mathfrak{W}_2^1(0, 1),$$

$$V(x) \equiv \infty \qquad \text{for } x \notin \mathfrak{W}_2^1,\ x \in \mathfrak{X}.$$

It can be shown (Ref. 3) that V is lower semicontinuous and

$$V(x - \lambda Ax) = \int_0^1 \left[|\partial x(\eta) - \lambda \partial Ax(\eta)|^2 - 2\int_0^{x(\eta)-\lambda A(\eta)} f(\xi)\, d\xi \right] d\eta$$

$$= V(x) + \int_0^1 \left[2\lambda \partial^2 x(\eta) \cdot Ax(\eta) + \lambda^2 |\partial Ax(\eta)|^2 - 2\int_{x(\eta)}^{x(\eta)-\lambda Ax(\eta)} f(\xi)\, d\xi \right] d\eta$$

for all $\lambda > 0$, $x \in \mathfrak{D}(A)$. As $(\mu I - f) : \mathfrak{R} \to \mathfrak{R}$ is monotone,

$$-2\int_\alpha^{\alpha+\beta} f(\xi)\, d\xi = -2\mu \int_\alpha^{\alpha+\beta} \xi\, d\xi + 2\int_\alpha^{\alpha+\beta} (\mu I - f)(\xi)\, d\xi$$

$$\geqslant -\mu\left[(\alpha+\beta)^2 - \alpha^2 \right] + 2\beta(\mu I - f)(\alpha) = -\mu\beta^2 - 2\beta f(\alpha)$$

for $\alpha, \beta \in \mathfrak{R}$; hence, for all $\lambda > 0$ and $x \in \mathfrak{D}(A)$,

$$V(x - \lambda Ax) \geqslant V(x) + \left[2\lambda + \lambda^2(\pi^2 - \mu) \right] \int_0^1 |Ax(\eta)|^2\, d\eta,$$

and defining $W : \mathfrak{X} \to \overline{\mathfrak{R}}^+$ by

$$W(x) \equiv 2\int_0^1 |\partial^2 x(\eta) + f(x(\eta))|^2\, d\eta \qquad \text{for } x \in \mathfrak{X} \cap \mathfrak{W}_2^2(0, 1),$$

$$W(x) \equiv \infty \qquad \text{for } x \in \mathfrak{X},\ x \notin \mathfrak{W}_2^2(0, 1),$$

we see that for every $\varepsilon > 0$, $\varepsilon < 1$, there exists $\lambda_0(\varepsilon) > 0$ such that $V(x - \lambda Ax) \geqslant V(x) + \lambda(1 - \varepsilon)W(x)$ for all $\lambda \in (0, \lambda_0(\varepsilon))$, $x \in \mathfrak{D}(A)$. It can be shown that W is lower semicontinuous (Ref. 3); consequently, Theorem 2.2 shows that $\dot{V}(x) \leqslant -(1 - \varepsilon)W(x) \leqslant 0$ for all $x \in \mathfrak{X}$ such that $V(x) < \infty$. Letting $\varepsilon \searrow 0$, it follows that V is a Liapunov function on $\mathfrak{X} \cap \mathfrak{W}_2^1(0, 1)$, with $\dot{V}(x) \leqslant -W(x)$ for $x \in \mathfrak{X} \cap \mathfrak{W}_2^1(0, 1)$.

By Theorem 1.2, we see that the sets $\overline{\mathcal{G}}_\alpha \equiv \{x \in \mathfrak{X} \mid V(x) \leqslant \alpha\}$ and $\mathcal{G}_\alpha \equiv \{x \in \overline{\mathcal{G}}_\alpha \mid V(x) < \alpha\}$ are positive invariant for every $\alpha \in \mathfrak{R}$. As A is closed, x_e is an equilibrium if and only if $x_e \in \mathfrak{D}(A)$ and $Ax_e = 0$; this in turn is true if and only if $W(x_e) = 0$.

In the foregoing example, notice that we do not know if any of the motions are right-differentiable. Moreover, V is not Fréchet differentiable, or even continuous on $\mathfrak{X}$; in fact, V is not finite valued at some $x \in \mathfrak{X}$. We also notice that, for any $\alpha \in \mathfrak{R}$, the sets $\overline{\mathcal{G}}_\alpha$ and $\mathcal{G}_\alpha$ have empty interior; hence, our conclusions here are qualitatively different from those that might be obtained by the use of a continuous Liapunov function, where $\mathcal{G}_\alpha$ would be open.

Exercise 2.4. A is closed in Example 2.2. Show that there exists exactly one equilibrium (at $x = 0$) if $f : \mathfrak{R} \to \mathfrak{R}$ satisfies

$$\int_0^1 x(\eta) f(x(\eta))\, d\eta < \pi^2 \int_0^1 |x(\eta)|^2\, d\eta \qquad \text{for all nonzero } x \in \mathfrak{D}(A).$$

If V happens to be convex, the following result often provides a simple way of verifying condition (2) and, thereby, condition (1) of Theorem 2.2.

Proposition 2.1. For $\mathfrak{B}$ a Banach space, let a multivalued operator $A : (\mathfrak{D}(A) \subset \mathfrak{B}) \to \mathfrak{B}$ generate a dynamical system $\{S(t)\}_{t \geq 0}$ on $\mathfrak{X} \equiv \mathrm{Cl}_{\mathfrak{B}} \mathfrak{D}(A)$, in the sense of Theorem III.5.1. Let there exist lower semicontinuous functions $V : \mathfrak{B} \to \overline{\mathfrak{R}}$, $W : \mathfrak{X} \to \overline{\mathfrak{R}}$, such that V is convex, $V(x) > -\infty$ for all $x \in \mathfrak{B}$, and

$$\limsup_{\lambda \searrow 0} \frac{1}{\lambda} [V(x - \lambda y) - V(x)] \geq W(x) \tag{3}$$

for all $y \in Ax$, $x \in \mathfrak{D}(A)$. Then

$$\dot{V}(x) \equiv \liminf_{t \searrow 0} \frac{1}{t} [V(S(t)x) - V(x)] \leq -W(x)$$

for every $x \in \mathfrak{X}$, and V is a Liapunov function on the set $\{x \in \mathfrak{X} \mid V(x) < \infty, W(x) \geq 0\}$.

Proof. For all $x, z \in \mathfrak{B}$, $0 < h < 1$, we have

$$V(x + hz) - V(x) \leq h[V(x + z) - V(x)],$$

by the assumed convexity of $V : \mathfrak{B} \to \overline{\mathfrak{R}}$; hence, for every $x \in \mathfrak{D}(A)$, $y \in Ax$, $\lambda > 0$,

$$\begin{aligned} V(x - \lambda y) - V(x) &\geq \limsup_{h \searrow 0} \frac{\lambda}{\lambda h} [V(x - h\lambda y) - V(x)] \\ &\geq \lambda W(x), \end{aligned}$$

where we have used condition (3). Having satisfied condition (2), we see by Theorem 2.2 that the proof is complete. □

We notice that if V is Fréchet differentiable, as well as convex, then condition (3) is equivalent to

$$V_x'(y) \leq -W(x) \qquad \text{for all } x \in \mathfrak{D}(A), y \in Ax. \tag{4}$$

Exercise 2.5. Reconsider the dynamical system $\{S(t)\}_{t \geq 0}$ of Example 1.1, where $\mathcal{X} \equiv \{x \in \mathcal{C}[0, 1] | x(0) = 0 = x(1)\}$ and $\tilde{V} : \mathcal{X} \to \mathcal{R}$ is the continuous convex function defined there. Defining a lower semicontinuous function $\tilde{W} : \mathcal{X} \to \overline{\mathcal{R}}^+$ by

$$\tilde{W}(x) \equiv 2\int_0^1 |\partial x(\eta)|^2 \, d\eta \qquad \text{for } x \in \mathcal{X} \cap \mathcal{W}_2^1(0, 1),$$

$$\tilde{W}(x) \equiv \infty \qquad \text{for } x \in \mathcal{X}, x \notin \mathcal{W}_2^1(0, 1),$$

show that $\dot{\tilde{V}}(x) \leq -\tilde{W}(x)$ for every $x \in \mathcal{X}$.

3. Stability and Liapunov's Direct Method

For $\mathcal{X}$ a metric space with metric $d : \mathcal{X} \times \mathcal{X} \to \mathcal{R}^+$, we denote by $\mathcal{S}_r(x)$, $r > 0$, the open ball $\{y \in \mathcal{X} | d(y, x) < r\}$ for $x \in \mathcal{X}$. We also define a function $\hat{d} : \mathcal{X} \times 2^{\mathcal{X}} \to \mathcal{R}^+$ by

$$\hat{d}(y, \mathcal{P}) \equiv \inf_{x \in \mathcal{P}} d(y, x), \qquad y \in \mathcal{X}, \mathcal{P} \subset \mathcal{X};$$

we denote by $\hat{\mathcal{S}}_r(\mathcal{P})$, $r > 0$, the open set $\{y \in \mathcal{X} | \hat{d}(y, \mathcal{P}) < r\}$ for $\mathcal{P} \subset \mathcal{X}$.

Definition 3.1. Let $\{S(t)\}_{t \geq 0}$ be a dynamical system on a metric space $\mathcal{X}$. A particular motion $S(\cdot)x_0 : \mathcal{R}^+ \to \mathcal{X}$ is *stable* if, for each $\varepsilon > 0$, there exists $\delta(\varepsilon) > 0$ such that $y \in \mathcal{S}_\delta(x_0)$ implies $d(S(t)y, S(t)x_0) < \varepsilon$ for all $t \in \mathcal{R}^+$; a motion is *unstable* if it is not stable. The motion $S(\cdot)x_0$ is *asymptotically stable* if it is stable and there exists $\delta > 0$ such that $y \in \mathcal{S}_\delta(x_0)$ implies $d(S(t)y, S(t)x_0) \to 0$ as $t \to \infty$; this motion is *exponentially stable* if there exist $\delta > 0$, $\alpha(\delta) > 0$, $M(\delta) < \infty$, such that $y \in \mathcal{S}_\delta(x_0)$ implies $d(S(t)y, S(t)x_0) \leq Me^{-\alpha t}d(y, x_0)$ for all $t \in \mathcal{R}^+$. Either of the latter two properties is said to be *global* if δ can be chosen arbitrarily large.

An equilibrium x_e can be identified with the motion $S(\cdot)x_e$, since $S(t)x_e = x_e$ for all $t \in \mathcal{R}^+$; the stability properties of x_e are defined to be those of this motion. We see that an equilibrium x_e is stable if and only if, for each $\varepsilon > 0$, there exists $\delta(\varepsilon) > 0$ such that $y \in \mathcal{S}_\delta(x_e)$ implies $\gamma(y) \subset \mathcal{S}_\varepsilon(x_e)$. However, an equilibrium x_e can also be viewed as a subset $\{x_e\}$ of $\mathcal{X}$, and an equivalent definition of the stability properties of an equilibrium x_e can be arrived at by the following definition of the stability properties of an arbitrary set.

Definition 3.2. Let $\{S(t)\}_{t \geq 0}$ be a dynamical system on a metric space $\mathcal{X}$. A set $\mathcal{P} \subset \mathcal{X}$ is *stable* if, for each $\varepsilon > 0$, there exists $\delta(\varepsilon) > 0$ such that

$y \in \hat{\mathbb{S}}_\delta(\mathcal{P})$ implies $\gamma(y) \subset \hat{\mathbb{S}}_\epsilon(\mathcal{P})$; a set is unstable if it is not stable. The set $\mathcal{P}$ is *asymptotically stable* if it is stable and there exists $\delta > 0$ such that $y \in \hat{\mathbb{S}}_\delta(\mathcal{P})$ implies $\hat{d}(S(t)y, \mathcal{P}) \to 0$ as $t \to \infty$; this set is *exponentially stable* if there exist $\delta > 0$, $\alpha(\delta) > 0$, $M(\delta) < \infty$, such that $y \in \hat{\mathbb{S}}_\delta(\mathcal{P})$ implies $\hat{d}(S(t)y, \mathcal{P}) \leqslant Me^{-\alpha t}\hat{d}(y, \mathcal{P})$ for all $t \in \mathcal{R}^+$. Either of the latter two properties is said to be *global* if δ can be chosen arbitrarily large.

It is easily verified that a stable set must have a positive invariant and stable closure. A motion $S(\cdot)x_0$ is said to be *orbitally stable* if the positive orbit $\gamma(x_0)$ is a stable set; it is apparent that an equilibrium is stable if and only if it is orbitally stable. Any stable motion must be orbitally stable, but the converse is not true in general. In Example III.6.1, all motions are orbitally stable, but the equilibrium $x_e = 0$ is the only stable motion; obviously, $\{0\}$ is also stable as a set, and it is easily verified that $\mathbb{S}_r(0)$ is a stable set for every $r > 0$.

Exercise 3.1. Let a multivalued operator $A : (\mathcal{D}(A) \subset \mathcal{B}) \to \mathcal{B}$, $\mathcal{B}$ a Banach space, generate a dynamical system $\{S(t)\}_{t \geqslant 0}$ on $\mathcal{X} \equiv \mathrm{Cl}_{\mathcal{B}}\mathcal{D}(A)$ in the sense of Theorem III.5.1. Show that every motion is stable (or exponentially stable) if $\omega \leqslant 0$ (respectively, if $\omega < 0$).

For linear dynamical systems, some apparently different ideas coincide, as demonstrated in the following two exercises.

Exercise 3.2. Consider a linear dynamical system on a Banach space $\mathcal{X}$. Show that the following statements are equivalent.

(i) At least one motion is stable.
(ii) The equilibrium $x_e = 0$ is stable.
(iii) Every positive orbit is bounded.
(iv) Every motion is stable.
(v) Every motion is orbitally stable.

Hint: Show (iv) ⇒ (v) ⇒ (i) ⇒ (ii) ⇒ (iii); show (iii) ⇒ (iv) by use of the resonance theorem (Ref. 15), a corollary to the uniform boundedness theorem.

Exercise 3.3. Consider a linear dynamical system on a Banach space $\mathcal{X}$. Show that the following statements are equivalent.

(i) At least one motion is asymptotically stable (A.S.) [respectively, exponentially stable (E.S.)].
(ii) The equilibrium $x_e = 0$ is A.S. [respectively, E.S.].
(iii) Every motion is globally A.S. [respectively, globally E.S.].

For nonlinear dynamical systems, such coincidences are not generally true; referring to Example III.6.3, we see that we may have all motions stable, even asymptotically stable, but there need be no bounded positive orbits. It also is clear that, in general, asymptotic stability does not imply exponential stability. However, for a linear dynamical system on $\mathcal{R}^n$, we saw in Exercise I.9.2. that asymptotic stability of $x_e = 0$ implies global exponential stability of $x_e = 0$ and, by linearity, of every motion. It is interesting that this conclusion is not even necessarily true for a *linear* dynamical system on an infinite-dimensional Banach space, as is demonstrated by the following example (Ref. 5).

Example 3.1. Consider the real Hilbert space l_2 of infinite sequences in $\mathcal{R}$, as described in Example II.1.2, with

$$\langle x, \hat{x} \rangle \equiv \lim_{n\to\infty} \sum_{p=1}^{n} \alpha_p \hat{\alpha}_p \qquad \text{for } x = \{\alpha_p\}, \hat{x} = \{\hat{\alpha}_p\}.$$

Define a linear operator $A : l_2 \to l_2$ by

$$Ax \equiv \left\{ -\frac{1}{p}\alpha_p \right\}, \qquad x = \{\alpha_p\} \in l_2,$$

noting that $\mathcal{D}(A) = l_2$, $-A$ is monotone (accretive), and $\mathcal{R}(I - \lambda A) = l_2$ for every $\lambda > 0$. It follows that A is the infinitesimal generator of a linear dynamical system $\{S(t)\}_{t \geqslant 0}$ on l_2 with $\|S(t)x\| \leqslant \|x\|$ for all $t \in \mathcal{R}^+$, $x \in \mathcal{X}$.

Actually, we see that

$$S(t)x = \left\{ e^{-t/p}\alpha_p \right\}, \qquad x = \{\alpha_p\} \in l_2, \qquad t \in \mathcal{R}^+,$$

implying that $S(t)x \to 0$ as $t \to \infty$ for every $x \in l_2$; hence, the equilibrium $x_e = 0$ is asymptotically stable. However, it is apparent that $x_e = 0$ is *not* exponentially stable.

Upon complexification of the foregoing example, we notice that all eigenvalues of A are negative real. If the space were finite dimensional, this property would be sufficient to ensure exponential stability; that this is not true in this example can be attributed to the fact that the set $\{-1/p\}$ of eigenvalues has 0 as a limit point in $\hat{\mathcal{C}}$.

Unfortunately, there are discrepancies between eigenvalue analysis and stability analysis that are much more difficult to understand (Ref. 16). In general, for a linear dynamical system on an infinite-dimensional complex Banach space, it is impossible to state conditions sufficient for stability (or asymptotic stability, or exponential stability) solely in terms of the eigenvalues of the infinitesimal generator (Ref. 16). This fact casts considerable doubt on all claims of "stability" that are based on "separation of variables" in linear partial differential equations; such claims are quite common in the applied sciences but would be difficult to refute because the meaning of stability is seldom made precise in unsophisticated analyses.

Unlike sufficient conditions, a necessary condition for stability (or asymptotic stability, or exponential stability) can be stated solely in terms of eigenvalues, when dealing with a linear dynamical system.

Proposition 3.1. Let $\{S(t)\}_{t\geq 0}$ be a linear dynamical system on a complex Banach space $\mathfrak{X}$, with infinitesimal generator $A : (\mathfrak{D}(A) \subset \mathfrak{X}) \to \mathfrak{X}$. If the equilibrium $x_e = 0$ is stable, then no eigenvalue of A has positive real part. If $x_e = 0$ is asymptotically stable, then every eigenvalue has negative real part. If $x_e = 0$ is exponentially stable, with exponent $-\alpha t$, some $\alpha > 0$, then no eigenvalue of A has real part greater than $-\alpha$.

Exercise 3.4. Prove Proposition 3.1.

In Chapter I we saw that the Liapunov approach was essential to the investigation of stability of equilibria for a nonlinear dynamical system on $\mathfrak{R}^n$ (or on $\hat{\mathcal{C}}^n$); however, for a linear dynamical system on $\hat{\mathcal{C}}^n$, eigenvalue analysis furnishes an alternative approach. Our comments indicate that this alternative may not exist even for a linear dynamical system when the space has infinite dimension; hence, the Liapunov approach now assumes even greater importance than was apparent in Chapter I.

The next two theorems summarize Liapunov's Direct Method for investigating stability of an equilibrium of a dynamical system $\{S(t)\}_{t\geq 0}$ on a metric space $\mathfrak{X}$; here, $\mathfrak{F}_r$ denotes the set of all monotone functions $f : ([0, r) \subset \mathfrak{R}) \to \mathfrak{R}$ with $f(0) = 0$, $f(\eta) > 0$ for all $\eta \in (0, r)$. We assume the existence of at least one equilibrium x_e.

Theorem 3.1. If $V : \mathfrak{X} \to \mathfrak{R}$ is a continuous Liapunov function on the open ball $\mathcal{S}_r(x_e)$, some $r > 0$, such that

$$V(x) \geq V(x_e) + f(d(x, x_e)) \qquad \text{for all } x \in \mathcal{S}_r(x_e),$$

some $f \in \mathfrak{F}_r$, then the equilibrium x_e is stable. If, in addition, $\dot{V}(x) \leq -g(d(x, x_e))$ for all $x \in \mathcal{S}_r(x)$, some $g \in \mathfrak{F}_r$, then x_e is asymptotically stable. If, in addition, there exist real numbers $c_2 \geq c_1 > 0$, $\mu > 0$, $k > 0$, such that

$$\dot{V}(x) \leq -\mu[V(x) - V(x_e)],$$

$$c_1[d(x, x_e)]^k \leq V(x) - V(x_e) \leq c_2[d(x, x_e)]^k,$$

for all $x \in \mathcal{S}_r(x_e)$, then x_e is exponentially stable with exponent $-\mu t/k$.

Proof. Given any $\varepsilon > 0$, we may suppose without loss of generality that $\varepsilon < r$. Defining $\alpha(\varepsilon) \equiv \inf_{\|x - x_e\| = \varepsilon} V(x)$ and noting that $\alpha \geq V(x_e) + f(\varepsilon)$, we see that $\mathcal{S}_\varepsilon(x_e)$ contains a disjoint component $\mathcal{G}_\alpha$ of the set

$\{x \in \mathcal{X} \mid V(x) < \alpha\}$ such that $x_e \in \mathcal{G}_\alpha$. As $V : \mathcal{X} \to \mathcal{R}$ is continuous, $\mathcal{G}_\alpha$ is open and there must exist $\delta(\varepsilon) > 0$ such that $\mathcal{S}_\delta(x_e) \subset \mathcal{G}_\alpha$. By Corollary 1.1, $\mathcal{G}_\alpha$ is positive invariant; hence, given any $x_0 \in \mathcal{S}_\delta(x_e) \subset \mathcal{G}_\alpha$, we have $\gamma(x_0) \subset \mathcal{G}_\alpha \subset \mathcal{S}_\varepsilon(x_e)$, and x_e is seen to be stable.

With α and δ defined as before, and fixing ε, $0 < \varepsilon < r$, consider any fixed $x_0 \in \mathcal{S}_\delta(x_e) \subset \mathcal{G}_\alpha$. Theorem 1.1 implies that $V(S(\cdot)x_0)$ is nonincreasing on $\mathcal{R}^+$ and

$$V(S(t)x_0) \leqslant V(x) + \int_0^t \dot{V}(S(\tau)x_0)\, d\tau, \qquad t \in \mathcal{R}^+.$$

By section 6.8 of Ref. 4 it follows that $V(S(t)x_0) \to \inf_{t \in \mathcal{R}^+} V(S(t)x_0) \equiv \beta(x_0) \in \overline{\mathcal{R}}$ as $t \to \infty$; clearly, $\alpha \geqslant \beta \geqslant V(x_e)$ because $V(x) \geqslant V(x_e)$ for all $x \in \mathcal{G}_\alpha \subset \mathcal{S}_r(x_e)$. Either $\beta = V(x_e)$ or $\gamma(x_0) \cap \{x \in \mathcal{G}_\alpha \mid V(x) < \beta\}$ is empty with $\beta > V(x_e)$; in the latter case, continuity of V implies the existence of $\nu > 0$ such that $\gamma(x_0) \cap \mathcal{S}_\nu(x_e)$ is empty. Assuming that $\beta \neq V(x_e)$ and $\dot{V}(x) \leqslant -g(d(x, x_e))$ for all $x \in \mathcal{G}_\alpha \subset \mathcal{S}_r(x)$, we see that $g(\nu) > 0$ and

$$V(x_e) < V(S(t)x_0) \leqslant V(x_0) - \int_0^t g(\nu)\, dt$$

$$= V(x_0) - tg(\nu) \to -\infty \qquad \text{as } t \to \infty,$$

which is impossible; hence, $\beta = V(x_e)$ and $V(S(t)x_0) \to V(x_e)$ as $t \to \infty$. This implies that $f(d(S(t)x_0, x_e)) \to 0$ as $t \to \infty$; therefore, $S(t)x_0 \to x_e$ as $t \to \infty$, and x_e is seen to be asymptotically stable.

Finally, assuming the existence of suitable real numbers $c_2 \geqslant c_1 > 0$, $\mu > 0$, $k > 0$, and defining $\tilde{V}(x) \equiv V(x) - V(x_e)$, we see that $\dot{\tilde{V}} = \dot{V}$ and apply Theorem 1.1 to obtain

$$c_1[d(S(t)x_0, x_e)]^k \leqslant \tilde{V}(S(t)x_0) \leqslant e^{-\mu t}\tilde{V}(x_0) \leqslant e^{-\mu t}c_2[d(x_0, x_e)]^k$$

for every $x_0 \in \mathcal{G}_\alpha \subset \mathcal{S}_r(x_e)$; hence $d(S(t)x_0, x_e) \leqslant (c_2/c_1)^{-\mu t/k} d(x_0, x_e)$ for $x_0 \in \mathcal{G}_\alpha$, $t \in \mathcal{R}^+$, and the proof is complete. □

We now see that Theorem I.8.1 follows directly from Theorem 3.1, using the local compactness of $\mathcal{R}^n$ to show the existence of suitable $f, g \in \mathcal{F}_r$, some $r > 0$, under the assumptions of Theorem I.8.1.

Referring to the proof of Theorem 3.1, we see that we could state a completely similar theorem regarding the stability of a given set $\mathcal{P} \subset \mathcal{X}$; we need only replace $\mathcal{S}_r(x_e)$, $d(x, x_e)$, and $V(x_e)$ by $\hat{\mathcal{S}}_r(\mathcal{P})$, $\hat{d}(x, \mathcal{P})$, and $\sup_{x \in \mathcal{P}} V(x)$, respectively, in Theorem 3.1. However, notice that the conditions of the resulting theorem would require the closure of $\mathcal{P}$ to be a disjoint component of $\{x \in \mathcal{X} \mid V(x) \leqslant \alpha\}$ for some $\alpha \in \mathcal{R}$. We shall not go into such extensions of Liapunov's Direct Method.

Theorem 3.1 is the stability part of Liapunov's Direct Method; the following theorem provides an instability result.

Theorem 3.2. Let $V : \mathcal{X} \to \mathcal{R}$ be a continuous Liapunov function on $\mathcal{G} \equiv \{x \in \mathcal{S}_r(x_e) | V(x) < V(x_e)\}$, some $r > 0$, with $\dot{V}(x) \leq -g(V(x_e) - V(x))$ for all $x \in \mathcal{G}$, some $g \in \mathcal{F}_r$. If $\mathcal{G} \cap \mathcal{S}_\varepsilon(x_e)$ is nonempty for every $\varepsilon > 0$, then the equilibrium x_e is unstable.

Proof. As V is continuous, we may choose some $\delta \in (0, r)$ such that $V(x) > V(x_e) - 1$ for all $x \in \mathcal{S}_\delta(x_e)$. If x_e is stable, there must exist $\varepsilon > 0$ such that $\gamma(y) \subset \mathcal{S}_\delta(x_e)$ for all $y \in \mathcal{S}_\varepsilon(x_e)$; hence, applying Theorem 1.1, our assumptions imply that $\gamma(y) \subset \mathcal{G} \cap \mathcal{S}_\delta(x_e)$ for all $y \in \mathcal{G} \cap \mathcal{S}_\varepsilon(x_e)$, and

$$
\begin{aligned}
V(S(t)y) &\leq V(y) + \int_0^t \dot{V}(S(\tau)y)\, d\tau \\
&\leq V(y) - \int_0^t g(V(x_e) - V(S(\tau)y))\, d\tau \\
&\leq V(y) - \int_0^t g(V(x_e) - V(y))\, d\tau \\
&\to -\infty \qquad \text{as } t \to \infty, y \in \mathcal{G} \cap \mathcal{S}_\varepsilon(x_e).
\end{aligned}
$$

This is a contradiction, because $V(S(t)y) > V(x_e) - 1$ for all $t \in \mathcal{R}^+$ if $\gamma(y) \subset \mathcal{S}_\varepsilon(x_e)$. It follows that x_e is not stable, and the proof is complete. □

The assumption made on $\dot{V}$ in Theorem 3.2 is implied by the simpler but stronger condition $\dot{V}(x) \leq -f(d(x, x_e))$ for all $x \in \mathcal{S}_r(x_e)$, some $f \in \mathcal{F}_r$; hence Theorem I.8.2 follows directly from Theorem 3.2 with this stronger condition, using the local compactness of $\mathcal{R}^n$ to show the existence of a suitable $f \in \mathcal{F}_r$ under the assumptions of Theorem I.8.2.

Example 3.2. Again consider the dry friction problem of Example III.5.1. We recall that the multivalued operator $A : \mathcal{R}^2 \to \mathcal{R}^2$ defined there generates a dynamical system $\{S(t)\}_{t \geq 0}$ on $\mathcal{R}^2$, a Hilbert space, and the canonical restriction is

$$
\begin{aligned}
A^0x &\equiv (v, -u - \mu v/|v|) \qquad \text{if } v \neq 0, \\
A^0x &\equiv (0, 0) \qquad \text{if } v = 0, |u| \leq \mu, \\
A^0x &\equiv (0, -u + \mu u/|u|) \qquad \text{if } v = 0, |u| > \mu,
\end{aligned}
$$

for every $x = (u, v) \in \mathcal{R}^2$. As A is closed, Theorem III.5.2 shows that $\mathcal{Q}_e \equiv \{(u, v) \in \mathcal{R}^2 | v = 0, |u| \leq \mu\}$ is the set of all equilibria. As every neighborhood of any equilibrium contains another equilibrium, it is apparent that no equilibrium is asymptotically stable.

Consider some equilibrium $x_e = (u_0, 0)$, some $u_0 \in [-\mu, \mu]$, and define a Fréchet differentiable function $V : \mathcal{R}^2 \to \mathcal{R}$ as $V(x) \equiv (u - u_0)^2 + v^2$ for $x = (u, v) \in \mathcal{R}^2$. Applying Theorem 2.1, we find that $\dot{V}(x) = -2\mu|v| - 2u_0 v \leq 0$ for every $x = (u, v) \in \mathcal{R}^2$. Hence, V is a Liapunov function on all of $\mathcal{R}^2$, and Theorem 1.2 shows that the bounded set $\{x \in \mathcal{R}^2 | V(x) \leq \alpha\}$ is positive invariant for each

$\alpha \in \mathcal{R}$, implying that all positive orbits are bounded. Moreover, Theorem 3.1 shows $x_e = (u_0, 0)$ to be stable; hence, every equilibrium is stable, and from the compactness of $\mathcal{Q}_e$ it follows that $\mathcal{Q}_e$ is a stable set.

Although no $x_e \in \mathcal{Q}_e$ is asymptotically stable, the results of Section 4 will show $\mathcal{Q}_e$ to be an asymptotically stable set.

Example 3.3. Again consider the scalar differential–difference equation of Example III.3.3,

$$\dot{y}(t) = \alpha y(t) + \beta y(t-\tau), \qquad t \geq 0,$$

$$y(-\eta) \equiv u_0(\eta), \qquad 0 \leq \eta \leq \tau,$$

for given $\alpha, \beta, \tau \in \mathcal{R}$, $\tau > 0$, and $u_0(\cdot) : [0, \tau] \to \mathcal{R}$. In Example III.3.3 we found that this equation led to a linear dynamical system $\{S(t)\}_{t \geq 0}$ on $\mathcal{X} \equiv \mathcal{R} \times \mathcal{L}_2(0, \tau)$, with infinitesimal generator

$$Ax \equiv (\alpha y + \beta u(\tau), -\partial u),$$

$$x = (y, u) \in \mathcal{D}(A) \equiv \{(y, u) \in \mathcal{R} \times \mathcal{W}_2^1(0, \tau) | u(0) = y\}.$$

Moreover, we found that $\|S(t)x\| \leq e^{\omega t}\|x\|$ for all $t \in \mathcal{R}^+$, $x \in \mathcal{X}$, where $\omega \equiv \max\{0, \alpha + |\beta|\}$ and

$$\|x\|^2 \equiv y^2 + |\beta| \int_0^\tau |u(\eta)|^2 \, d\eta, \qquad x = (y, u) \in \mathcal{X},$$

assuming $\beta \neq 0$.

Let us now make the additional assumption that $\alpha \leq -|\beta|$. Our exponential estimate now shows that $x_e = 0$ is a stable equilibrium, regardless of the time delay $\tau > 0$. Alternatively, the same sufficient condition follows from Theorem 3.1 after defining $V(x) \equiv \|x\|^2$, $x \in \mathcal{X}$, and noting that by Theorem 2.1,

$$\dot{V}(x) = V_x' Ax = 2\langle x, Ax\rangle = -|\beta|[u(\tau) - y]^2 + 2(\alpha + |\beta|)y^2$$

for all $x = (y, v) \in \mathcal{D}(A)$; therefore $\dot{V}(x) \leq 0$ for all $x \in \mathcal{X}$ if $\alpha \leq -|\beta|$. It can be shown that the condition $\alpha \leq -|\beta|$ is necessary, as well as sufficient, if the equilibrium is to be stable for *every* $\tau > 0$; necessity follows from eigenvalue analysis of A.

We have as yet obtained no sufficient conditions for exponential stability, or even asymptotic stability, of $x_e = 0$. To this end, let us attempt to generate a suitable Liapunov function in a manner suggested by Theorem III.4.2; thus we define

$$\tilde{V}(x) = \langle y, B_{11}y + B_{12}u\rangle_{\mathcal{R}} + \langle u, B_{21}y + B_{22}u\rangle_{\mathcal{L}_2}, \qquad x = (y, u) \in \mathcal{X},$$

where the linear operators $B_{11} : \mathcal{R} \to \mathcal{R}$, $B_{12} : \mathcal{L}_2 \to \mathcal{R}$, $B_{21} : \mathcal{R} \to \mathcal{L}_2$, $B_{22} : \mathcal{L}_2 \to \mathcal{L}_2$ are assumed to be bounded in order to ensure continuity of $\tilde{V} : \mathcal{X} \to \mathcal{R}$, a property needed in Theorem 3.1. For simplicity, let us choose $B_{12} = 0, B_{21} \equiv 0$, and

$$B_{11}y \equiv y, \qquad y \in \mathcal{R},$$

$$B_{22}u(\eta) = g(\eta)u(\eta), \qquad 0 \leq \eta \leq \tau, \quad u \in \mathcal{L}_2(0, \tau),$$

where $g \in \mathcal{C}[0, \tau]$. Then $\tilde{V}$ is Fréchet differentiable and, by Theorem 2.1,

$$\dot{\tilde{V}}(x) = \tilde{V}_x' Ax = 2y(\alpha y + \beta u(\tau)) - 2\int_0^\tau g(\eta)\cdot u(\eta)\cdot \partial u(\eta)\, d\eta$$

$$= 2\alpha y^2 + 2\beta y \cdot u(\tau) - \int_0^\tau \frac{d}{d\eta}[g(\eta)|u(\eta)|^2]\, d\eta + \int_0^\tau g'(\eta)|u(\eta)|^2\, d\eta$$

$$= (2\alpha + g(0))y^2 + 2\beta y \cdot u(\tau) - g(\tau)|u(\tau)|^2 + \int_0^\tau g'(\eta)|u(\eta)|^2\, d\eta$$

for all $x = (y, v) \in \mathfrak{D}(A)$. As we wish to have $\dot{\tilde{V}}(x) \leqslant 0$ wherever possible, an inspection of $\tilde{V}(x)$ suggests that we choose $g(\eta) \equiv \gamma e^{-2\delta\eta}$, $0 \leqslant \eta \leqslant \tau$, for some $\gamma > 0$, $\delta \geqslant 0$; hence,

$$\tilde{V}(x) = y^2 + \gamma \int_0^1 |e^{-\delta\eta} u(\eta)|^2\, d\eta, \qquad x = (y, v) \in \mathfrak{X},$$

and

$$\dot{\tilde{V}}(x) = (2\alpha + \gamma)y^2 + 2\beta y\cdot u(\tau) - \gamma e^{-2\delta\tau}|u(\tau)|^2 - 2\delta\gamma\int_0^\tau |e^{-\delta\eta}u(\eta)|^2\, d\eta$$

$$= -2\delta V(x) - \gamma\left[e^{-\delta\tau}u(\tau) - (\beta e^{\delta\tau}/\gamma)y\right]^2 + (2\delta + 2\alpha + \gamma + \beta^2 e^{2\delta\tau}/\gamma)y^2$$

for all $x = (y, u) \in \mathfrak{D}(A)$.

By Theorem 2.1 we see that we shall have

$$\dot{\tilde{V}}(x) \leqslant -2\delta V(x) \leqslant 0,$$

$$(\min[1, \gamma e^{-2\delta\tau}])\|x\|^2 \leqslant V(x) \leqslant (\max[1, \gamma])\|x\|^2,$$

for all $x \in \mathfrak{X}$, provided that we can choose $\gamma > 0$, $\delta \geqslant 0$, such that

$$0 \geqslant 2(\alpha + \delta)\gamma + \gamma^2 + \beta^2 e^{2\delta\tau}.$$

Such a choice is possible if $-\alpha > |\beta|$, for then we may choose $\delta > 0$ such that $\delta = -\alpha - |\beta| e^{\delta\tau}$, and define $\gamma = -\alpha - \delta = |\beta| e^{\delta\tau}$, thereby obtaining

$$2(\alpha + \delta)\gamma + \gamma^2 + \beta^2 e^{2\delta\tau} = -2\gamma^2 + \gamma^2 + \gamma^2 = 0.$$

Hence, if $\alpha < -|\beta|$, then for every $\tau > 0$ the equilibrium at $x = 0$ is exponentially stable with exponent $-\delta t$, where $\delta = -\alpha - |\beta| e^{\delta\tau} > 0$.

A much more general stability analysis, valid for all $\alpha, \beta \in \mathfrak{R}$, $\beta \neq 0$, $\tau > 0$, has been performed in Ref. 17 by using the same arguments and more complicated choices for the linear operators B_{11}, B_{12}, B_{21}, B_{22}.

Exercise 3.5. Referring to Example 3.3, choose $g(\eta) \equiv a - b\eta$, and show that $a, b \in \mathfrak{R}$ can be chosen so as to obtain another Liapunov function which also proves exponential stability for $\alpha < -|\beta|$.

Often, as seen in Section I.10, we are interested in the extent to which "small perturbations" of a generating operator change the stability properties of a given equilibrium. The following result extends Theorem I.10.1.

Theorem 3.3. For $\mathfrak{B}$ a Banach space, let a multivalued operator $A : (\mathfrak{D}(A) \subset \mathfrak{B}) \to \mathfrak{B}$ and a continuous operator $F : \mathfrak{B} \to \mathfrak{B}$ be such that $A + F$ generates a dynamical system $\{S(t)\}_{t\geqslant 0}$ on $\mathfrak{X} \equiv \mathrm{Cl}_{\mathfrak{B}}\mathfrak{D}(A)$, in the sense of Theorem III.5.1. Let there exist an equilibrium $x_0 \in \mathfrak{D}(A)$ such that F is Fréchet differentiable at x_0 and $-\alpha I - A - F'_{x_0}$ is $\|\cdot\|_e$-accretive for some equivalent norm $\|\cdot\|_e$, some $\alpha > 0$. Then x_0 is exponentially stable.

Proof. As $-\alpha I - A - F'_{x_0}$ is $\|\cdot\|_e$-accretive,

$$\|(1 - \mu\alpha)(x - \hat{x}) - \mu(y + F'_{x_0}x) + \mu(\hat{y} + F'_{x_0}\hat{x})\|_e \geqslant \|x - \hat{x}\|_e$$

for all $x, \hat{x} \in \mathfrak{D}(A)$, $y \in Ax$, $\hat{y} \in A\hat{x}$, $\mu > 0$; hence,

$$\|x - \hat{x} - \lambda(y + F'_{x_0}x) + \lambda(\hat{y} + F'_{x_0}\hat{x})\|_e \geqslant (1 + \alpha\lambda)\|x - \hat{x}\|_e$$

for all $\lambda > 0$. Defining $V(x) \equiv \|x - x_0\|_e$, $x \in \mathfrak{X}$, and noting that $0 \in Ax_0 + Fx_0$, we set $y_0 = -Fx_0 \in Ax_0$ and obtain

$$\begin{aligned} V(x - \lambda(y + Fx)) &= \|x - x_0 - \lambda(y + F'_{x_0}x) \\ &\quad + \lambda(y_0 + F'_{x_0}x_0) - \lambda[Fx - F'_{x_0}(x - x_0) + y_0]\|_e \\ &\geqslant (1 + \alpha\lambda)\|x - x_0\|_e - \lambda\|Fx - Fx_0 + F'_{x_0}(x - x_0)\|_e \end{aligned}$$

for all $x \in \mathfrak{D}(A)$, $y \in Ax$, $\lambda > 0$. By Theorem 2.2 we find that

$$\dot{V}(x) \leqslant -\alpha\|x - x_0\|_e + \|F(x) - Fx_0 - F'_{x_0}(x - x_0)\|_e$$

for all $x \in \mathfrak{X}$. Hence, given any $\beta \in (0, \alpha)$, there exists $r(\beta) > 0$ such that $\dot{V}(x) \leqslant -\beta\|x - x_0\|_e = -\beta V(x)$ for all $x \in \mathbb{S}_r(x_0)$. Applying Theorem 3.1 completes the proof. □

It is not difficult to see that Theorem 3.3 extends Theorem I.10.1; if A is a linear operator and $A + F'_{x_0}$ is the infinitesimal generator of a linear dynamical system $\{\hat{S}(t)\}_{t\geqslant 0}$ on $\mathfrak{B}$, having an exponentially stable equilibrium at $x = 0$, then $\|\hat{S}(t)x\| \leqslant Me^{-\alpha t}\|x\|$ for all $x \in \mathfrak{X}$, $t \in \mathfrak{R}^+$, some $M \geqslant 1$, $\alpha > 0$. But then Theorem III.3.2 implies the existence of an equivalent norm $\|\cdot\|_e$ such that $-\alpha I - A - F'_{x_0}$ is $\|\cdot\|_e$-accretive; hence, the condition of Theorem 3.3 is met, and the conclusion of Theorem 3.3 implies that of Theorem I.10.1.

In the proof of Theorem 3.3, notice that $V(x) \equiv \|x - x_0\|_e$ was defined so as to be a Liapunov function (on all of $\mathfrak{X}$) for a "nearby" dynamical system generated by $A + F'_{x_0}$, and our proof proceeded by showing that V was also a Liapunov function (on $\mathcal{S}_r(x_0)$) for the given dynamical system $\{S(t)\}_{t\geq 0}$ generated by $A + F$. This device is often useful in constructing Liapunov functions for nonlinear systems; that is, V is often constructed on the basis of a "local approximation" of the true generator. Notice also, in the proof of Theorem 3.3, that V can be used with Theorem 1.2 to obtain an assured "domain of attraction" $\mathcal{G}_\alpha \subset \mathcal{S}_r(x_0)$ for the equilibrium x_0 of $\{S(t)\}_{t\geq 0}$.

For dynamical systems described by linear partial differential equations, we mentioned earlier the inadequacy of separation of variables as a means of proving stability, unless it can be shown that there exist "enough" separated solutions, i.e., unless it can be shown that the infinitesimal generator has enough eigenvectors. Nevertheless, some people seem to prefer this "method" of stability analysis, believing that it is easier to find eigenvalues and eigenvectors than to construct useful Liapunov functions. The following exercises demonstrate that this belief is wrong, and that this method is nothing more than a very laborious way of constructing Liapunov functions.

Exercise 3.6. Let $A : (\mathfrak{D}(A) \subset \mathfrak{X}) \to \mathfrak{X}$, $\mathfrak{X}$ a complex Hilbert space, generate a linear dynamical system $\{S(t)\}_{t\geq 0}$ on $\mathfrak{X}$. Let $\{g_n\}$ be any finite family of independent eigenvectors of A, i.e., $0 \neq g_n \in \mathfrak{D}(A)$ and $Ag_n = \alpha_n g_n$, some $\alpha_n \in \mathcal{C}$, for $n = 1, 2, \ldots, N$. Define $\mathcal{E} \equiv \{x \in \mathfrak{X} | x = \sum_{n=1}^N \varphi_n g_n$ for some $\{\varphi_n\} \in \mathcal{C}^N\}$ and let $\{g_n^*\}$ denote a biorthogonal family in $\mathcal{E}$, i.e., $g_n^* \in \mathcal{E}$ and $\langle g_n^*, g_m\rangle = \delta_{nm}$ for $n, m, = 1, 2, \ldots, N$ (this family is easy to construct in terms of $\{g_n\}$). Define $\beta_n \equiv 1$ if $\mathrm{Re}(\alpha_n) < 0$, $\beta_n \equiv -1$ if $\mathrm{Re}(\alpha_n) > 0$, and

$$V(x) \equiv \sum_{n=1}^N \beta_n |\langle x, g_n^*\rangle|^2, \qquad x \in \mathfrak{X}.$$

(a) Show that V is a continuous Liapunov function on $\mathcal{E}$, with

$$\dot{V}(x) = -2 \sum_{n=1}^N |\mathrm{Re}(\alpha_n)|\, |\langle x, g_n^*\rangle|^2 < 0, \qquad x \in \mathcal{E}.$$

(b) Show that $\mathcal{E}$ is a positive invariant linear manifold, with $S(t)x = \sum_{n=1}^N \exp\{\alpha_n t\}\varphi_n g_n$ for every $x = \sum_{n=1}^N \varphi_n g_n \in \mathcal{E}$, $t \in \mathfrak{R}^+$.

(c) If $\mathrm{Re}(\alpha_n) < 0$ for all $n = 1, 2, \ldots, N$, show that $0 < V(S(t)x) < e^{\delta t}V(x)$ for all $t \in \mathfrak{R}^+$, $x \in \mathcal{E}$, where $\delta \equiv 2 \max_{1<n<N} \mathrm{Re}(\alpha_n) < 0$.

(d) If $\text{Re}(\alpha_p) > 0$ for some p, show that $\tilde{V}(x) \equiv -|\langle x, g_p^* \rangle|^2$ proves instability by Liapunov's Direct Method. *Hint:* Let $\hat{S}(t)$ be the restriction of $S(t)$ to $\hat{\mathfrak{X}} \equiv (\mathcal{E}, \|\cdot\|)$, note that $\{\hat{S}(t)\}_{t\geq 0}$ is a dynamical system on $\hat{\mathfrak{X}}$, and show that $\hat{\mathfrak{X}}$-instability implies $\mathfrak{X}$-instability (but not conversely).

Exercise 3.7. With A and $\mathfrak{X}$ defined as in Exercise 3.6, suppose that $\text{Re}(\alpha_n) \leq 0$ for all $n = 1, 2, \ldots,$ and that $\{g_n\}$ is a countable (not necessarily finite) family of orthonormal eigenvectors spanning $\mathfrak{X}$, i.e., $\langle g_m, g_n \rangle = \delta_{mn}$, and each $x \in \mathfrak{X}$ can be expressed as $x = \Sigma_n \varphi_n g_n$ for some $\{\varphi_n\} \in \mathcal{C}^\infty$. Show that $V(x) \equiv \|x\|^2$ is a Liapunov function on all of $\mathfrak{X}$, and that it proves stability by Liapunov's Direct Method. If $\text{Re}(\alpha_n) \leq \varepsilon < 0$ for all $n = 1, 2, \ldots,$ show that V proves exponential stability by Liapunov's Direct Method. *Hint:* First show that $\|x\|^2 = \Sigma_n |\langle x, g_n \rangle|^2$ for all $x \in \mathfrak{X}$.

Exercise 3.8. Consider a system formally described by

$$\frac{\partial}{\partial t}\theta(\eta, t) = \frac{\partial}{\partial \eta}\left[p(\eta)\frac{\partial}{\partial \eta}\theta(\eta, t)\right], \qquad t \geq 0,$$

$$\theta(0, t) = 0 = \frac{\partial}{\partial \eta}\theta(1, t), \qquad t \geq 0,$$

$$\theta(\eta, 0) = u_0(\eta), \qquad 0 \leq \eta \leq 1,$$

for some given $p \in \mathcal{C}^1[0, 1]$ such that $p(\eta) \geq \varepsilon > 0$, $0 \leq \eta \leq 1$.

(a) What can you prove by using separation of variables? Can you find even one separated solution "explicitly" for nonconstant $p(\cdot)$?

(b) Prove that this equation leads to a dynamical system on $\mathcal{L}_2(0, 1)$, $V(x) \equiv \|x\|^2_{\mathcal{L}_2}$ is a Liapunov function everywhere, and the equilibrium $x_e = 0$ is exponentially stable, with exponent $-\varepsilon\pi^2 t/4$.

Finally, we mention that the asymptotic stability and instability results of Liapunov's Direct Method involve stronger sufficient conditions on $\dot{V}$ than are absolutely necessary. In the following section we shall discuss LaSalle's Invariance Principle, which often provides a means of weakening these strong conditions on $\dot{V}$.

4. Positive Limit Sets and the Invariance Principle

We now inquire as to the asymptotic behavior of motions as $t \to \infty$. For example, given some $x \in \mathfrak{X}$, does there exist some $y(x) \in \mathfrak{X}$ such that $S(t)x \to y$ as $t \to \infty$, in the sense that $d(S(t)x, y) \to 0$ as $t \to \infty$? If so, we would like to locate y. Often there exists no such $y(x)$, and we may instead wish to ask whether, given $x \in \mathfrak{X}$, there exists a nonempty bounded set $\mathcal{S}(x) \subset \mathfrak{X}$ such that $S(t)x \to \mathcal{S}$ as $t \to \infty$, in the sense that $\hat{d}(S(t)x, \mathcal{S}) \to 0$ as $t \to \infty$. In this case, there may exist a *smallest closed* nonempty subset

$\hat{\mathcal{S}}(x) \subset \mathcal{S}$ that also has this property, and we would like to locate $\hat{\mathcal{S}}$ if it exists. We notice that a suitable nonempty bounded set $\mathcal{S}(x)$ exists if and only if $\gamma(x)$ is bounded; then any set $\mathcal{S} \supset \gamma(x)$ has the desired property. However, we would much prefer to locate the smallest possible closed nonempty set $\hat{\mathcal{S}}(x)$ having this property, if such a set does exist.

In discussing questions of this type, the following idea is quite useful.

Definition 4.1. Given $x \in \mathcal{X}$, the *positive limit set* $\Omega(x)$ is defined by the following property: $y \in \Omega(x)$ if there exists a sequence $\{t_n\} \subset \mathcal{R}^+$ such that $t_n \to \infty$ and $S(t_n)x \to y$ as $n \to \infty$ (Ref. 18).

It is easily seen that $\Omega(x)$ is the (possibly empty) set

$$\bigcap_{\tau \in \mathcal{R}^+} \left(\mathrm{Cl} \bigcup_{t > \tau} S(t)x \right) = \bigcap_{\tau \in \mathcal{R}^+} (\mathrm{Cl}\, \gamma(S(\tau)x)),$$

and it is obvious that $\Omega(x) \subset \mathrm{Cl}\, \gamma(x)$. Hence, $\Omega(x)$ is bounded (or precompact) if $\gamma(x)$ is bounded (respectively, precompact). $\Omega(x)$ may be empty if $\gamma(x)$ is not precompact, but $\Omega(x)$ may be nonempty even if $\gamma(x)$ is unbounded. It is apparent that $\Omega(x_e) = \{x_e\}$ if x_e is an equilibrium, and $\Omega(x) = \gamma(x) \equiv \cup_{0 \leqslant t \leqslant T} S(t)x$ if $S(\cdot)x$: $\mathcal{R}^+ \to \mathcal{X}$ is periodic with period T. The sketches of Figure 6 may be helpful in visualizing the positive limit set $\Omega(x)$ for motions in $\mathcal{X} \equiv \mathcal{R}^2$.

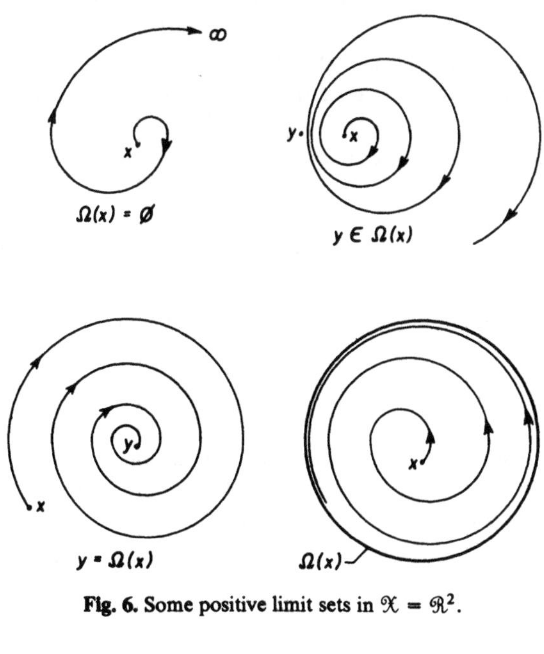

Fig. 6. Some positive limit sets in $\mathcal{X} = \mathcal{R}^2$.

Exercise 4.1. Let x_e be a globally asymptotically stable equilibrium of a dynamical system $\{S(t)\}_{t\geq 0}$ on a metric space $\mathfrak{X}$. Show that, for every $x \in \mathfrak{X}$, $\Omega(x) = \{x_e\}$, $\hat{d}(S(t)x, \Omega(x)) \to 0$ as $t \to \infty$, and $\gamma(x)$ is precompact. *Hint:* Given any sequence $\{t_n\} \subset \mathfrak{R}^+$, there exists a subsequence $\{t_m\}$ such that $t_{m+1} \geq t_m$ and either $t_m \to T < \infty$ or $S(t_m)x \to x_e$ as $m \to \infty$, using the assumption on x_e.

Proposition 4.1. $\Omega(x)$ is closed and positive invariant.

Proof. We first show that $\Omega(x)$ is positive invariant; to this end, consider $y \in \Omega(x)$ and a sequence $\{t_n\} \subset \mathfrak{R}^+$ such that $t_n \to \infty$ and $S(t_n)x \to y$ as $n \to \infty$. Given $\tau > 0$, we see by continuity of $S(\tau)$: $\mathfrak{X} \to \mathfrak{X}$ that $S(t_n + \tau)x = S(\tau)S(t_n)x \to S(\tau)y$ as $n \to \infty$; hence, $S(\tau)y \in \Omega(x)$ for every $\tau \in \mathfrak{R}^+$.

To show that $\Omega(x)$ is closed, consider a sequence $\{y_m\} \subset \Omega(x)$ such that $d(y_m, y_0) < 1/m$ for $m = 1, 2, \ldots,$ some $y_0 \in \mathfrak{X}$. To each y_m there corresponds some sequence $\{t_{nm}\}_{n=1,2,\ldots}$ such that $t_{nm} \to \infty$ as $n \to \infty$ and $d(S(t_{nm})x, y_m) < 1/n$. Hence,

$$d(S(t_{nn})x, y_0) \leq d(S(t_{nn})x, y_n) + d(y_n, y_0) \leq 2/n \to 0 \qquad \text{as } n \to \infty.$$

This implies that $y_0 \in \Omega(x)$, and the proof is complete. □

We notice that if $x \in \mathfrak{X}$ and nonempty $\mathcal{S} \subset \mathfrak{X}$ are such that $S(t)x \to \mathcal{S}$ as $t \to \infty$, then $\Omega(x) \subset \mathrm{Cl}\, \mathcal{S}$; if $\Omega(x)$ is nonempty and $S(t)x \to \Omega(x)$ as $t \to \infty$, then $\Omega(x)$ is the smallest closed set having this property, and we would like to locate $\Omega(x)$. Unfortunately, even if $\Omega(x)$ can be located, it is not necessarily true that $S(t)x \to \Omega(x)$ as $t \to \infty$; in fact, $\Omega(x)$ might even be empty. In a later result (Theorem 4.1) we shall show that $\Omega(x) \neq \varnothing$ and $S(t)x \to \Omega(x)$ if $\gamma(x)$ is precompact and $\mathfrak{X}$ is complete; moreover, in this case, $\gamma(x)$ is not only positive invariant but also invariant.

Definition 4.2. A set $\mathcal{S} \subset \mathfrak{X}$ is *invariant* under $\{S(t)\}_{t\geq 0}$ if there exists a mapping U: $\mathfrak{R} \times \mathcal{S} \to \mathcal{S}$ such that

$$U(0, x) = x, \qquad U(t+s, x) = S(t)\,U(s, x),$$

for all $x \in \mathcal{S}$, $s \in \mathfrak{R}$, $t \in \mathfrak{R}^+$.

It is easy to see that an invariant set $\mathcal{S}$ must be positive invariant; that is, we must have $S(t)\mathcal{S} \subset \mathcal{S}$ for all $t \in \mathfrak{R}^+$, where $S(t)\mathcal{S} \equiv \{y \in \mathfrak{X} \,|\, y = S(t)x, \text{ some } x \in \mathcal{S}\}$ for each $t \in \mathfrak{R}^+$. In fact, it is not difficult to show that $\mathcal{S}$ is invariant if and only if $S(t)\mathcal{S} = \mathcal{S}$ (in $2^{\mathfrak{X}}$) for all $t \in \mathfrak{R}^+$.

Exercise 4.2. Show that
(i) if x_e is an equilibrium, $\gamma(x_e) = \{x_e\}$ is invariant;
(ii) if a motion $S(\cdot)x_0$ is periodic, $\gamma(x_0)$ is invariant;
(iii) any countable union of invariant sets is invariant;
(iv) $\mathcal{X}$ is invariant if and only if $\mathcal{R}(S(t)) = \mathcal{X}$ for all $t > 0$.

We recall that if $\Omega(x)$ is nonempty and if $S(t)x \to \Omega(x)$ as $t \to \infty$, then $\Omega(x)$ is the smallest closed set having this property; hence, the following result is very important.

Theorem 4.1. If $\mathcal{X}$ is complete and $\gamma(x)$ is precompact, then $\Omega(x)$ is nonempty, compact, connected, and invariant; moreover, $S(t)x \to \Omega(x)$ as $t \to \infty$. In fact, $\Omega(x)$ is the smallest closed set that $S(t)x$ approaches; if $S(t)x \to \mathcal{S} \subset \mathcal{X}$ as $t \to \infty$, then $\Omega(x) \subset \text{Cl}\, \mathcal{S}$.

Proof. As $\mathcal{X}$ is complete and $\gamma(x)$ is precompact, Cl $\gamma(x)$ is compact. As $\Omega(x)$ is closed and contained in Cl $\gamma(x)$, it follows that $\Omega(x)$ is compact.

Given any sequence $\{t_n\} \subset \mathcal{R}^+$, $t_n \to \infty$ as $n \to \infty$, the precompactness of $\{S(t_n)x\} \subset \gamma(x)$ implies the existence of a subsequence $\{t_m\} \subset \{t_n\}$ such that $\{S(t_m)x\}$ is Cauchy; hence, by completeness of $\mathcal{X}$, there exists $y \in \mathcal{X}$ such that $S(t_m)x \to y$ as $m \to \infty$. It follows that $y \in \Omega(x)$ and $\Omega(x)$ is not empty.

Suppose that $\hat{d}(S(t)x, \Omega(x)) \not\to 0$ as $t \to \infty$; that is, suppose there exists $\varepsilon > 0$ and a sequence $\{t_n\} \subset \mathcal{R}^n$ with $t_n \geqslant n$ and $\hat{d}(S(t_n)x, \Omega(x)) > \varepsilon$ for all $n = 1, 2, \ldots$. Then, by precompactness of $\{S(t_n)x\} \subset \gamma(x)$, there exists a subsequence $\{t_m\}$ such that $\{S(t_m)x\}$ is Cauchy with $t_m \to \infty$ and $\hat{d}(S(t_m)x, \Omega(x)) \to 0$ as $m \to \infty$. As this contradicts our supposition, we see that $\hat{d}(S(t)x, \Omega(x)) \to 0$ as $t \to \infty$.

Suppose that $\Omega(x)$ is not connected; that is, suppose that $\Omega(x) = \Omega_1 \cup \Omega_2$, where $\Omega_1 \cap \Omega_2 = \emptyset$. As $\Omega(x)$ is compact, so are Ω_1 and Ω_2; hence

$$\inf_{\substack{y \in \Omega_1(x) \\ z \in \Omega_2(x)}} d(y, z) = \delta > 0.$$

Given $y \in \Omega_1$, $z \in \Omega_2$, there must exist sequences $\{t_n\} \subset \mathcal{R}^+$, $\{\tau_n\} \subset \mathcal{R}^+$, such that $\tau_n > t_n$, and $n \to \infty$ implies $t_n \to \infty$, $S(t_n x) \to y$, $S(\tau_m x) \to z$. Therefore, there exists N such that $\hat{d}(S(t_n)x, \Omega_1) < \delta/2$ and $\hat{d}(S(\tau_n), \Omega_2) < \delta/2$ for all $n > N$; however, by the continuity $S(\cdot)x$: $\mathcal{R}^+ \to \mathcal{X}$, this implies the existence of a sequence $\{s_n\}$, $\tau_n > s_n > t_n$, such that $\hat{d}(S(s_n)x, \Omega_1) \geqslant \delta/2$ and $\hat{d}(S(s_n)x, \Omega_2) \geqslant \delta/2$ for all $n > N$. This means that $\hat{d}(S(s_n)x, \Omega(x)) \geqslant \delta/2$ for all $n > N$, contradicting our result that $\hat{d}(S(t)x, \Omega(x)) \to 0$ as $t \to \infty$. Hence, $\Omega(x)$ is connected.

Finally, in order to show invariance of $\Omega(x)$, consider $y \in \Omega(x)$ and a sequence $\{t_n\} \subset \mathcal{R}^+$, $t_{n+1} > t_n$, such that $t_n \to \infty$ and $S(t_n)x \to y$ as $n \to \infty$. For each $n = 1, 2, \ldots,$ define ψ_n: $\mathcal{R} \to \mathcal{X}$ by

$$\psi_n(s) \equiv x \qquad \text{for all} \quad s < -t_n,$$

$$\psi_n(s) \equiv S(s + t_n)x \qquad \text{for all} \quad s \geqslant -t_n.$$

For each fixed $s \in \mathcal{R}$, we notice that $\{\psi_n(s)\} \subset \gamma(x)$ is precompact; hence, for each fixed $s \in \mathcal{R}$, there exist subsequences $\{t_m\}$ and $\{\psi_m(s)\}$ such that $t_m \to \infty$ as $m \to \infty$ and the limit $U(s, y) \equiv \lim_{m\to\infty}\psi_m(s)$ exists. Clearly, $U(0, y) = y$ and $U(s, y) = \lim_{m\to\infty} S(s + t_m)x \in \Omega(x)$ for every $s \in \mathcal{R}$. Moreover, for $t \in \mathcal{R}^+$, $s \in \mathcal{R}$, we see that

$$\begin{aligned} U(t + s, y) &= \lim_{m\to\infty} S(t + s + t_m)x = S(t) \lim_{m\to\infty} S(s + t_m)x \\ &= S(t)U(s, y). \end{aligned}$$

Repeating this argument for every $y \in \Omega(x)$, we obtain a function U: $\mathcal{R} \times \Omega(x) \to \Omega(x)$ satisfying the conditions of Definition 4.2. Hence, $\Omega(x)$ is invariant, and the proof is complete. □

Theorem 4.1 provides a means of describing the asymptotic behavior of a motion $S(\cdot)x$: $\mathcal{X} \to \mathcal{R}$, for given $x \in \mathcal{X}$, provided that $\gamma(x)$ is precompact, $\mathcal{X}$ is complete, and we can locate $\Omega(x)$. In most applications we simply choose the state space $\mathcal{X}$ to be complete, and often $\gamma(x)$ can be shown to be precompact by proving that it belongs to some precompact subset of $\mathcal{X}$. A means of doing this, when a suitable Liapunov function is available, may be provided by Theorem 1.2 if $x \in \overline{\mathcal{G}}_\alpha$ and $V(x) < \alpha$, with V and $\overline{\mathcal{G}}_\alpha$ as defined there; then $\gamma(x) \subset \overline{\mathcal{G}}_\alpha$, and this implies that $\gamma(x)$ is precompact if $\overline{\mathcal{G}}_\alpha$ is precompact. In Section 5 we shall discuss other means of establishing precompactness of positive orbits.

If $\mathcal{X}$ is complete and, for some given $x \in \mathcal{X}$, we can show $\gamma(x)$ to be precompact, Theorem 4.1 shows the *existence* of a smallest closed nonempty set $\hat{\mathcal{S}}$ such that $S(t)x \to \hat{\mathcal{S}}$ as $t \to \infty$; $\hat{\mathcal{S}}$ is the positive limit set $\Omega(x)$. Hence, we now desire some method of *locating* $\Omega(x)$ in the absence of explicit knowledge of the motion $S(\cdot)x$; such a method is provided by LaSalle's Invariance Principle (Refs. 19, 20) and its various extensions (Refs. 1, 3, 19, 21).

Theorem 4.2 (Invariance Principle). Let $\{S(t)\}_{t\geqslant 0}$ be a dynamical system on a metric space $\mathcal{X}$, and let V: $\mathcal{X} \to \mathcal{R}$ be a Liapunov function on a set $\mathcal{G} \subset \mathcal{X}$ such that $\dot{V}(x) \leqslant -W(x)$ for all $x \in \mathcal{G}$, where W: $(\overline{\mathcal{G}} \subset \mathcal{X}) \to \overline{\mathcal{R}}^+$ is lower semicontinuous and $V(y) > -\infty$ for all $y \in \overline{\mathcal{G}} \equiv \text{Cl}\,\mathcal{G}$.

If $\gamma(x) \subset \mathcal{G}$, then $\Omega(x) \subset \mathfrak{M}^+$, where $\mathfrak{M}^+$ is the largest positive invariant subset of

(a) $\mathfrak{M}_1 = \{z \in \bar{\mathcal{G}} \mid \dot{V}(z) = 0\}$ if V is continuous (in fact, $\Omega(x) \subset \mathfrak{M}^+ \cap V^{-1}(\beta)$ for some $\beta \in \mathfrak{R}$, where $V^{-1}(\beta) \equiv \{z \in \mathfrak{X} \mid V(z) = \beta\}$).

(b) $\mathfrak{M}_2 = \{z \in \bar{\mathcal{G}} \mid W(z) = 0\}$ if V is only lower semicontinuous.

If, in addition, $\mathfrak{X}$ is complete and $\gamma(x)$ is precompact, then $\hat{d}(S(t)x, \mathfrak{M}) \to 0$ as $t \to \infty$, where $\mathfrak{M}$ is the largest invariant subset of $\mathfrak{M}_1 \subset \mathfrak{M}_2$ if V is continuous, or of $\mathfrak{M}_2$ if V is only lower semicontinuous.

Proof. We recall that $\Omega(x)$ is closed and positive invariant. If $\mathfrak{X}$ is complete and $\gamma(x)$ is precompact, then $\Omega(x)$ is nonempty, compact, connected, and invariant; moreover, $\hat{d}(S(t)x, \Omega(x)) \to 0$ as $t \to \infty$. Assuming that $\gamma(x) \subset \bar{\mathcal{G}}$, we have $\Omega(x) \subset \mathrm{Cl}\,\gamma(x) \subset \bar{\mathcal{G}}$. If $\gamma(x)$ is not precompact, $\Omega(x)$ may be empty, in which case the theorem is obviously true but vacuous; hence, we shall assume that $\Omega(x)$ is nonempty. There now are several cases to be considered.

If $V(S(t)x) = \infty$ for all $t \in \mathfrak{R}^+$, Definition 1.1 implies that $W(S(t)x) \equiv 0$ on $\mathfrak{R}^+$; hence, by the lower semicontinuity and nonnegativity of W, $W(z) = 0$ for every $z \in \Omega(x)$ and result (b) applies.

If $V(x) = \infty$ but $V(S(t_0)x) < \infty$ for some $t_0 > 0$, we may replace x by $x_0 \equiv S(t_0)x$ and note that $\Omega(x_0) = \Omega(x)$, $V(x_0) < \infty$; hence, the proof of (b) for this case can be embedded in the proof for the following case.

If $V(x) < \infty$, Theorem 1.1 shows that $V(S(\cdot)x)$ is nonincreasing as well as finite valued on $\mathfrak{R}^+$. This implies that $V(S(t)x) \to \beta < \infty$ as $t \to \infty$, where $\beta = \inf_{t \in \mathfrak{R}^+} V(S(t)x)$ (Ref. 4); as $\mathrm{Cl}\,\gamma(x) \subset \bar{\mathcal{G}}$, $\Omega(x)$ has been assumed nonempty, and $V(y) > -\infty$ for every $y \in \bar{\mathcal{G}}$, lower semicontinuity of V implies $\beta > -\infty$. If $V\colon \mathfrak{X} \to \bar{\mathfrak{R}}$ is continuous, it follows from the definition of $\Omega(x)$ (Definition 4.1) that $V(z) = \beta$ for every $z \in \Omega(x)$; furthermore, as $\Omega(x)$ is positive invariant, $\dot{V}(z) = 0$ for every $z \in \Omega(x)$ and result (a) follows. On the other hand, if V is only lower semicontinuous, we note by Theorem 1.1 that $V(S(\cdot)x)\ \mathfrak{R}^+ \to \bar{\mathfrak{R}}$ is differentiable a.e. and

$$V(u(t, x)) \leqslant V(x) + \int_0^t \dot{V}(S(\tau)x)\, d\tau, \qquad t \in \mathfrak{R}^+.$$

Therefore, considering any sequence $\{t_n\} \subset \mathfrak{R}^+$ such that $t_n \to \infty$ and $S(t_n)x \to z \in \Omega(x)$ as $n \to \infty$, the uniqueness of the limit β implies that for any $T > 0$,

$$0 \geqslant \int_0^T \dot{V}(S(s + t_n)x)\, ds \geqslant \left[V(S(t_n + T)x) - V(S(t_n)x) \right] \to 0 \qquad \text{as} \quad n \to \infty.$$

As W is nonnegative and lower semicontinuous, we apply Fatou's lemma (Ref. 4) to obtain

$$\int_0^T W(S(s)z)\,ds \leqslant \int_0^T \Big[\liminf_{n\to\infty} W(S(s)S(t_n)x)\Big]\,ds$$
$$\leqslant \liminf_{n\to\infty} \int_0^T W(S(s+t_n)x)\,ds$$
$$\leqslant -\lim_{n\to\infty} \int_0^T \dot{V}(S(s+t_n)x)\,ds = 0.$$

Therefore, $W(S(s)z) = 0$ a.e. $s \in [0, T]$, and the lower semicontinuity of W now implies that $0 = W(S(0)z) = W(z)$; hence, $\Omega(x) \subset \{z \in \bar{\mathcal{G}} \mid W(z) = 0\}$ and the proof of (b) is complete. □

We notice that if $\mathfrak{X} = \mathfrak{R}^n$, then $\mathfrak{X}$ is locally compact and $\gamma(x)$ is precompact if and only if it is bounded. In Chapter I we considered only continuous Liapunov functions (Definition I.8.4), and therefore Theorem I.11.1 now follows directly from (a) of Theorem 4.2.

Theorem 4.2 provides a simple means of locating positive limit sets; moreover, it provides a means of studying asymptotic behavior of a motion with precompact positive orbit $\gamma(x)$, for then we know that $S(t)x \to \Omega(x)$ as $t \to \infty$. Obviously, $\gamma(x)$ must be bounded if it is precompact, but the converse is not generally true unless $\mathfrak{X}$ is a locally compact metric space. In the following examples the precompactness question is answered by using a suitable Liapunov function with Theorem 1.2.

Example 4.1. Again consider the dry friction problem of Example III.5.1, where a dynamical system $\{S(t)\}_{t\geqslant 0}$ is generated on $\mathfrak{R}^2$ by the multivalued operator $A: \mathfrak{R}^2 \to \mathfrak{R}^2$ defined by

$$Ax \equiv \left\{ (v, -u-w) \in \mathfrak{R}^2 \,\middle|\, \begin{array}{ll} w = \mu v/|v| & \text{if } v \neq 0 \\ w \in [-\mu, \mu] & \text{if } v = 0 \end{array} \right\}, \qquad x = (u, v) \in \mathfrak{R}^2.$$

There we found that $\{S(t)\}_{t\geqslant 0}$ is nonexpansive.

In Example 3.2 we noticed that $\mathfrak{Q}_e \equiv \{(u, v) \in \mathfrak{R}^2 \mid v = 0, |u| \leqslant \mu\}$ is the set of all equilibria; moreover, every equilibrium is stable, none is asymptotically stable, and all positive orbits are bounded. In addition, the continuous function $V: \mathfrak{X} \to \mathfrak{R}$ defined by

$$V(x) = \|x\|^2 = u^2 + v^2, \qquad x = (u, v) \in \mathfrak{R}^2,$$

is a Liapunov function on all of $\mathfrak{R}^2$, with $\dot{V}(x) = -2\mu|v|$ for all $x = (u, v) \in \mathfrak{R}^2$.

As

$$\frac{d^+}{dt} S(t)x = A^0 S(t)x$$

for a.e. $t \in \mathcal{R}^+$, $x \in \mathcal{R}^2$, where A^0 is defined in Example 3.2, we see that the largest positive invariant set in

$$\mathfrak{M}_1 \equiv \{x \in \mathcal{R}^2 | \dot{V}(x) = 0\} = \{(u, v) \in \mathcal{R}^2 | v = 0\}$$

is $\mathfrak{M}^+ \equiv \{(u, v) \in \mathcal{R}^2 | v = 0, |u| \leqslant \mu\} = \mathcal{Q}_e$. Noting that

$$\mathfrak{M}^+ \cap V^{-1}(\beta) = \begin{cases} \varnothing & \text{if } \beta < 0 \text{ or } \beta > \mu^2, \\ \{(-\sqrt{\beta}, 0), (\sqrt{\beta}, 0)\} & \text{if } 0 \leqslant \beta \leqslant \mu^2, \end{cases}$$

we see by (a) of Theorem 4.2 that $\Omega(x) \subset \mathcal{Q}_e$, and $\Omega(x)$ contains at most two points, for every $x \in \mathcal{R}^+$. Moreover, since boundedness of $\gamma(x)$ in $\mathcal{R}^2$ implies precompactness in $\mathcal{R}^2$, $\Omega(x)$ is nonempty and connected; hence, Theorem 4.2 implies that, for each $x \in \mathcal{R}^2$, there exists $x_e \in \mathcal{Q}_e$ such that $S(t)x \to x_e$ as $t \to \infty$.

Exercise 4.3. Consider a dry friction problem with a nonlinear spring, described by

$$0 \in \ddot{u}(t) + f(\dot{u}(t)) + (u(t))^3, \qquad t \geqslant 0,$$

$$(u(0), \dot{u}(0)) = (u_0, v_0) \in \mathcal{R}^2,$$

where f: $\mathcal{R} \to 2^{\mathcal{R}}$ is defined by

$$f(v) \equiv \mu v/|v| \qquad \text{if } v \neq 0,$$

$$f(v) \equiv [-\mu, \mu] \qquad \text{if } v = 0,$$

for some $\mu > 0$. Show that this leads to a dynamical system on $\mathcal{R}^2$, and duplicate the analysis of Example 4.1. *Hint:* First apply Theorem III.6.1 with $V(x) \equiv 2v^2 + u^4$ for all $x = (u, v) \in \mathcal{R}^2$.

Example 4.2. Recalling Example 3.1, consider the dynamical system $\{S(t)\}_{t \geqslant 0}$ on l_2 having the infinitesimal generator

$$Ax \equiv \{-p^{-1}\alpha_p\}, \qquad x = \{\alpha_p\} \in l_2.$$

We found that $\|S(t)x\| \leqslant \|x\|$ for all $t \in \mathcal{R}^+$, $x \in \mathcal{X}$, where

$$\|x\|^2 \equiv \lim_{n \to \infty} \sum_{p=1}^{n} \alpha_p^2, \qquad x = \{\alpha_p\} \in l_2.$$

The exponential estimate implies that the equilibrium $x_e = 0$ is stable and all positive orbits are bounded; alternatively, these conclusions follow upon defining $V(x) = \|x\|^2$, obtaining

$$\dot{V}(x) = -2 \sum_{p=1}^{\infty} \alpha_p^2/p, \qquad x = \{\alpha_p\} \in l_2,$$

from Theorem 2.1, and applying Theorems 1.2 and 3.1.

Noting that $V: l_2 \to \mathcal{R}$ is continuous and

$$\frac{d}{dt} S(t)x = AS(t)x, \qquad t \in \mathcal{R}^+,$$

for all $t \in \mathcal{R}$, every $x \in \mathcal{D}(A) = l_2$, we see that largest positive invariant set contained in $\{x \in \mathcal{X} \mid \dot{V}(x) = 0\} = \{0\}$ is $\mathfrak{M}^+ = \{0\}$; hence, the Invariance Principle (Theorem 4.2) implies that $\Omega(x) \subset \{0\}$ for every $x \in \mathcal{X}$. Unfortunately, we cannot (yet) conclude that $\Omega(x) = \{0\}$ and that $S(t) \to 0$ as $t \to \infty$, because we only know that $\gamma(x)$ is bounded; we do not (yet) know that $\gamma(x)$ is precompact. Bounded sets in l_2 are not necessarily precompact, i.e., l_2 is not a locally compact space.

To study the precompactness question, define $\tilde{V}: \mathcal{X} \to \overline{\mathcal{R}}$ as

$$\tilde{V}(x) = \liminf_{n \to \infty} \sum_{p=1}^{n} p\alpha_p^2, \qquad x = \{\alpha_p\} \in l_2.$$

$\tilde{V}$ is lower semicontinuous and $\tilde{V}(x) \geq 0$ for all $x \in l_2$; moreover, for $\lambda > 0$, $x = \{\alpha_p\}$, $\tilde{V}(x) < \infty$, we see that

$$\tilde{V}(x - \lambda Ax) = \liminf_{n \to \infty} \sum_{p=1} (p + 2\lambda + \lambda^2/p)\alpha_p^2$$

$$\geq \tilde{V}(x) + 2\lambda V(x),$$

and it follows from Theorem 2.2 that $\tilde{V}$ is a Liapunov function on all of $\mathcal{X}$, with

$$\dot{\tilde{V}}(x) \leq -2V(x) \qquad \text{for all } x \text{ with } \tilde{V}(x) < \infty,$$

$$\dot{\tilde{V}}(x) \equiv 0 \qquad \text{for all } x \text{ with } \tilde{V}(x) = \infty.$$

It can be shown that $\{x \in l_2 \mid \tilde{V}(x) \leq \alpha\}$ is compact for any $\alpha < \infty$, and Theorem 1.2 shows this set to be positive invariant as well. Hence, for every x such that $\tilde{V}(x) < \infty$, $\gamma(x)$ is precompact and $S(t)x \to \Omega(x) = \{0\}$ as $t \to \infty$. As $\{x \in l_2 \mid \tilde{V}(x) < \infty\}$ is dense, and as all motions are stable (by linearity), it now follows that $S(t)x \to 0$ as $t \to \infty$, for every $x \in l_2$. As $x_e = 0$ is stable, we see that $x_e = 0$ is a (globally) asymptotically stable equilibrium.

Notice that, unlike V, the Liapunov function $\tilde{V}$ cannot be used in Liapunov's Direct Method to prove stability of $x_e = 0$, because $\tilde{V}$ is not continuous. Notice also that neither $\tilde{V}$ nor V satisfies the sufficient conditions of Theorem 3.1 for asymptotic stability.

Example 4.3. Again consider the nonlinear heat conduction problem of Example III.5.2, where $f(0) = 0$, $\mu I - f: \mathcal{R} \to \mathcal{R}$ is continuous and monotone for some $\mu \in \mathcal{R}$, and

$$\mathcal{X} \equiv \{x \in \mathcal{C}[0, 1] \mid x(0) = 0 = x(1)\}, \qquad \|x\| \equiv \max_{\eta} |x(\eta)|,$$

$$Ax(\eta) \equiv \partial^2 x(\eta) + f(x(\eta)), \qquad 0 \leq \eta \leq 1, \quad x \in \mathcal{D}(A) \equiv \{x \in \mathcal{X} \mid \partial^2 x \in \mathcal{X}\}.$$

We found that A generates a dynamical system $\{S(t)\}_{t\geq 0}$ on $\mathfrak{X}$, in the sense of Theorem III.5.1, with $\|S(t)x - S(t)y\| \leq e^{\mu t}\|x - y\|$ for all $x, y \in \mathfrak{X}$, $t \in \mathfrak{R}^+$.

In Example 2.2 we found that the lower semicontinuous function $V: \mathfrak{X} \to \overline{\mathfrak{R}}$, defined by

$$V(x) \equiv \int_0^1 \left[|\partial x(\eta)|^2 - 2\int_0^{x(\eta)} f(\xi)\, d\xi \right] d\eta, \qquad x \in \mathfrak{X} \cap \mathfrak{W}_2^1(0, 1),$$

$$V(x) \equiv \infty, \qquad x \notin \mathfrak{W}_2^1, x \in \mathfrak{X},$$

is a Liapunov function on $\mathfrak{X}$, with $\dot{V}(x) \leq -W(x)$ for all $x \in \mathfrak{X}$, where the lower semicontinuous function $W: \mathfrak{X} \to \overline{\mathfrak{R}}^+$ is defined by

$$W(x) \equiv 2\int_0^1 |\partial^2 x(\eta) + f(x(\eta))|^2\, d\eta, \qquad x \in \mathfrak{X} \cap \mathfrak{W}_2^2(0, 1),$$

$$W(x) \equiv \infty \qquad \text{for } x \notin \mathfrak{W}_2^2(0, 1), x \in \mathfrak{X}.$$

We also saw that $\overline{\mathcal{G}}_\alpha \equiv \{x \in \mathfrak{X} \mid V(x) \leq \alpha\}$ is positive invariant for each $\alpha \in \mathfrak{R}$.

Let us now assume that $\mu < \pi^2$. If $W(x_0) = 0$, we must have $x_0 \in \mathfrak{X} \cap \mathfrak{W}_2^2(0, 1)$, $Ax_0 = 0$; hence,

$$\begin{aligned} \mu\int_0^1 |x_0(\eta) - 0|^2\, d\eta &\geq \int_0^1 [x_0(\eta) - 0][f(x_0(\eta)) - 0]\, d\eta \\ &= -\int_0^1 x_0(\eta) \cdot \partial^2 x_0(\eta)\, d\eta \\ &= \int_0^1 |\partial x_0(\eta)|^2\, d\eta \geq \pi^2 \int_0^1 |x_0(\eta)|^2\, d\eta, \end{aligned}$$

and $\mu < \pi^2$ implies that $x_0 = 0$. Therefore, for all sufficiently large $\alpha \in \mathfrak{R}$, we see that the largest positive invariant set in $\{x \in \overline{\mathcal{G}}_\alpha \mid W(x) = 0\} = \{0\}$ is $\mathfrak{M}^+ = \{0\}$. Hence, by Theorem 4.2, $\Omega(x) \subset \{0\}$ for every $x \in \mathfrak{X}$ such that $V(x) < \infty$, i.e., for every $x \in \mathfrak{X} \cap \mathfrak{W}_2^1(0, 1)$. To say more, we need to know if $\gamma(x)$ is precompact.

If $x \in \mathfrak{X}$ is such that $V(x) \leq \alpha \in \mathfrak{R}$, then $x \in \mathfrak{X} \cap \mathfrak{W}_2^1(0, 1)$ and

$$\begin{aligned} \int_0^1 |\partial x(\eta)|^2\, d\eta &= V(x) + 2\int_0^1 \left(\int_0^{x(\eta)} f(\xi)\, d\xi \right) d\eta \\ &\leq \alpha + \int_0^1 \mu |x(\eta)|^2\, d\eta \leq \alpha + \frac{\mu}{\pi^2}\int_0^1 |\partial x(\eta)|^2\, d\eta; \end{aligned}$$

therefore,

$$\begin{aligned} \|x\|^2_{\mathfrak{W}_2^1} &\equiv \int_0^1 (|x(\eta)|^2 + |\partial x(\eta)|^2)\, d\eta \\ &\leq (\pi^{-2} + 1)\int_0^1 |\partial x(\eta)|^2\, d\eta \leq \frac{(1 + \pi^2)\alpha}{\pi^2 - \mu} \end{aligned}$$

for all $x \in \bar{\mathcal{G}}_\alpha$. Hence, $\bar{\mathcal{G}}_\alpha$ is a $\|\cdot\|_{\mathfrak{W}_2^1}$-bounded set, and this can be shown to imply $\|\cdot\|_{\mathcal{C}}$-precompactness of $\bar{\mathcal{G}}_\alpha$ (Ref. 21). Hence, $\gamma(x)$ is precompact for every $x \in \mathfrak{X} \cap \mathfrak{W}_2^1(0, 1)$, and Theorem 4.1 now implies that $S(t)x \to \Omega(x) = \{0\}$ as $t \to \infty$ for every $x \in \mathfrak{X} \cap \mathfrak{W}_2^1(0, 1)$.

Notice that, using only this Liapunov function, we cannot conclude asymptotic stability (or even stability) of the (unique) equilibrium $x_e = 0$; V is not continuous, and Theorem 3.1 cannot be used. There is no other way to use this V to prove stability of $x_e = 0$; every $\bar{\mathcal{G}}_\alpha$ has empty interior and, hence, no $\bar{\mathcal{G}}_\alpha$ contains an open neighborhood of $x_e = 0$.

Exercise 4.4. Referring to Example 4.3 and assuming, as there, that $\mu < \pi^2$, use the function $\tilde{V}$: $\mathfrak{X} \to \mathfrak{R}$, defined by

$$\tilde{V}(x) \equiv \int_0^1 |x(\eta)|^2 \, d\eta, \qquad x \in \mathfrak{X},$$

to show that $\Omega(x) \subset \{0\}$ for every $x \in \mathfrak{X}$. Note that, unlike V, $\tilde{V}$ cannot be used in Theorem 1.2 to demonstrate precompactness of positive orbits. Note that, like V, $\tilde{V}$ cannot be used in Theorem 3.1 to prove stability of $x_e = 0$. Why not?

Exercise 4.5. Referring to Example 4.3 and without assuming $\mu < \pi^2$, show that $x_e = 0$ is stable if $\xi f(\xi) \leqslant 0$ for all $\xi \in [-\alpha, \alpha]$, some $\alpha > 0$. Show that all positive orbits are bounded if $\xi f(\xi) \leqslant 0$ for all $\xi \notin [-\beta, \beta]$, some $\beta > 0$. *Hint:* Define $\hat{V}(x) \equiv \|x\|$, $x \in \mathfrak{X}$, and show that $\dot{\hat{V}}(x) \leqslant 0$ for $\|x\| \leqslant \alpha$ under the first assumption, whereas $\dot{\hat{V}}(ix) \leqslant 0$ for $\|x\| \geqslant \beta$ under the second assumption.

In the following section we shall delve further into the orbital precompactness question, a question that must be answered if one wishes to use the Invariance Principle to investigate the asymptotic behavior of motions. However, let us notice that it may sometimes be easier to find a "better" Liapunov function than to employ the Invariance Principle.

Example 4.4. Consider the damped wave equation

$$\frac{\partial^2}{\partial t^2} u(\eta, t) + 2\alpha \frac{\partial}{\partial t} u(\eta, t) - \frac{\partial^2}{\partial \eta^2} u(\eta, t) = 0, \qquad 0 \leqslant \eta \leqslant 1, \quad t \geqslant 0,$$

$$u(0, t) = 0 = u(1, t), \qquad t \geqslant 0,$$

where $\alpha > 0$. Defining

$$\hat{\mathfrak{W}}_2^n(0, 1) \equiv \{u \in \mathfrak{W}_2^{2n}(0, 1) | \partial^m u(0) = 0 = \partial^m u(1), m = 0, 2, 4, \ldots, \leqslant n - 1\},$$

and

$$\mathfrak{X} \equiv \hat{\mathfrak{W}}_2^1(0, 1) \times \mathfrak{L}_2(0, 1),$$

$$\|x\|^2 \equiv \int_0^1 [|\partial u(\eta)|^2 + |v(\eta)|^2] \, d\eta, \qquad x = (u, v) \in \mathfrak{X},$$

$$Ax \equiv (v, -2\alpha v + \partial^2 u), \qquad x = (u, v) \in \mathfrak{D}(A) \equiv \hat{\mathfrak{W}}_2^2(0, 1) \times \hat{\mathfrak{W}}_2^1(0, 1),$$

we find by Proposition III.5.1 and Example III.4.3 that $A: (\mathcal{D}(A) \subset \mathcal{X}) \to \mathcal{X}$ is the infinitesimal generator of a linear dynamical system $\{S(t)\}_{t \geq 0}$ on $\mathcal{X}$. Moreover, $-A$ is accretive and $\|S(t)x\| \leq \|x\|$ for all $t \in \mathcal{R}^+$, $x \in \mathcal{X}$; hence, $x_e = 0$ is a stable equilibrium and all positive orbits are bounded.

We might now wish to use the Invariance Principle, noting that $V(x) \equiv \|x\|^2$ is a continuous Liapunov function on $\mathcal{X}$ and

$$\dot{V}(x) = V_x' Ax = -4\alpha \int_0^1 |v(\eta)|^2 \, d\eta, \qquad x \in \mathcal{D}(A),$$

$$\dot{V}(x) \leq -4\alpha \int_0^1 |v(\eta)|^2 \, d\eta \equiv -W(x), \qquad x \in \mathcal{X},$$

where we have used Theorem 2.1. If $\mathcal{M}^+$ is the largest positive invariant set in $\mathcal{M}^2 \equiv \{x \in \mathcal{X} \mid W(x) = 0\} = \{(u, v) \in \mathcal{X} \mid v = 0\}$, then the fact that $S(t)x_0$ is a strong solution of $\dot{x}(t) = Ax(t)$ for $x_0 \in \mathcal{D}(A)$ implies that $\mathcal{M}^+ \cap \mathcal{D}(A) = \{0\}$; in fact, one can write a distributional evolution equation (Ref. 15) that is satisfied by every motion, thereby obtaining $\mathcal{M}^+ = \{0\}$. Then a result of the following section (Theorem 5.2) can be used to show that every bounded positive orbit must be precompact, and the Invariance Principle can be used to show that $S(t)x \to 0$ as $t \to \infty$, for every $x \in \mathcal{X}$; thus, $x_e = 0$ is globally asymptotically stable.

However, rather than making this long and rather complicated argument, it is much easier to simply define

$$\tilde{V}(x) \equiv \|x\|^2 + 2\alpha \int_0^1 [v(\eta) \cdot u(\eta) + \alpha |u(\eta)|^2] \, d\eta, \qquad x = (u, v) \in \mathcal{X},$$

and obtain

$$\dot{\tilde{V}}(x) = \tilde{V}_x' Ax = -2\alpha \int_0^1 [|v(\eta)|^2 + |\partial u(\eta)|^2] \, d\eta, \qquad x = (u, v,) \in \mathcal{D}(A),$$

$$\dot{\tilde{V}}(x) \leq -2\alpha \|x\|^2, \qquad x = (u, v) \in \mathcal{X},$$

by using Theorem 2.1. Noting that there exist $c_2 \geq c_1 > 0$, $\delta > 0$, depending on $\alpha > 0$, such that $c_1\|x\|^2 \leq \tilde{V}(x) \leq c_2\|x\|^2$ and $\dot{\tilde{V}}(x) \leq -2\delta \tilde{V}(x)$ for all $x \in \mathcal{X}$, we find by Theorem 3.1 that $x_e = 0$ is globally exponentially stable with exponent $-\delta t$. Note that this is an even better result than we obtained by using the Invariance Principle with the original Liapunov function V. Also note that $\tilde{V}$ is easy to construct by use of the approach suggested by Theorem III.4.2.

5. Orbital Precompactness and Use of the Invariance Principle

The Invariance Principle provides a very effective way of investigating asymptotic behavior of motions, provided that these motions have precompact positive orbits and a suitable Liapunov function is known. In Examples 4.1–4.3, we saw how a Liapunov function V can be used to establish precompactness of positive orbits by using Theorem 1.2, provided that $\bar{\mathcal{G}}_\alpha \equiv \{x \in \mathcal{X} \mid V(x) \leq \alpha\}$ is precompact for some $\alpha \in \mathcal{R}$. Unfortunately, this approach to precompactness may not be as simple as it seems. If V is a Liapunov function useful in Liapunov's Direct Method (Theorems 3.1, 3.2), then V must be continuous on $\mathfrak{X}$. However, a continuous Liapunov function generally proves only boundedness of positive orbits, at best, and this does not imply precompactness of these positive orbits unless $\mathfrak{X}$ is a locally compact metric space; we recall that a Banach space is locally compact if and only if it has finite dimension. We encountered this difficulty in Examples 4.2 and 4.3, where we used discontinuous Liapunov functions with Theorem 1.2 to assure precompactness of positive orbits.

As useful Liapunov functions are often not particularly easy to find, we would prefer not to need more than one for each application, and we are interested in other approaches to the precompactness question when $\mathfrak{X}$ is not locally compact.

Often it is not too difficult to show that a positive orbit $\gamma(x)$ is bounded, for some given $x \in \mathfrak{X}$; our basic objective here is to find simple conditions sufficient to ensure precompactness of bounded positive orbits. That a bounded positive orbit need not be precompact is shown by the following example.

Example 5.1. Let $\{S(t)\}_{t\geq 0}$ be a dynamical system on

$$l_2 \equiv \left\{ \{\alpha_n\} \subset \mathfrak{R}^\infty \,\middle|\, \sum_{n=1}^{\infty} \alpha_n^2 < \infty \right\},$$

and suppose that some particular motion $S(\cdot)x_0 \colon \mathfrak{R}^+ \to l_2$ is such that

$$x_0 = \{1, 0, 0, \ldots\},$$

$$S(1)x_0 = \{0, 1, 0, 0, \ldots\},$$

$$S(2)x_0 = \{0, 0, 1, 0, 0, \ldots\},$$

$$\vdots$$

$$S(m-1)x_0 = \{\delta_{mn}\}_{n=1,2,\ldots}.$$

It is clear that $\gamma(x_0)$ might very well be bounded, since $\|S(t)x_0\| = 1$ for $t = 0, 1, 2, \ldots$; however, it is apparent that $\gamma(x_0)$ is not precompact, because the sequence $\{t_m\}$, $t_m \equiv m = 1, 2, \ldots,$ contains no subsequence $\{\tau_m\}$ for which $\{S(\tau_m)x_0\}$ is Cauchy.

The following lemma will be helpful in determining which bounded positive orbits happen to be precompact.

Lemma 5.1. For $\{S(t)\}_{t\geq 0}$ a dynamical system on a metric space $\mathcal{X}$, the sets

$$\mathcal{Q}_b \equiv \{x \in \mathcal{X} | \gamma(x) \text{ is bounded}\},$$

$$\mathcal{Q}_c \equiv \{x \in \mathcal{X} | \gamma(x) \text{ is precompact}\},$$

are positive invariant, with $\mathcal{Q}_c \subset \mathcal{Q}_b$. If all motions are stable, then $\mathcal{Q}_b$ and $\mathcal{Q}_c$ are closed. If $\{S(t)\}_{t\geq 0}$ is a linear dynamical system, $\mathcal{X}$ a Banach space, then $\mathcal{Q}_b$ and $\mathcal{Q}_c$ are closed linear manifolds, whether or not all motions are stable.

Proof. It is obvious that $\mathcal{Q}_c \subset \mathcal{Q}_b$, and positive invariance of $\mathcal{Q}_b$ and $\mathcal{Q}_c$ follows from $\gamma(S(t)x) \subset \gamma(x)$ for all $t \in \mathcal{R}^+$, $x \in \mathcal{X}$.

Assuming that all motions are stable, and considering any $y \in \text{Cl}\,\mathcal{Q}_b$, note that for each $\varepsilon > 0$ there exists $x \in \mathcal{Q}_b$ such that $d(S(t)y, S(t)x) < \varepsilon$ for all $t \in \mathcal{R}^+$; hence, $\hat{d}(S(t)y, \gamma(x)) < \varepsilon$ for all $t \in \mathcal{R}^+$, implying that $\gamma(y)$ is bounded and $y \in \mathcal{Q}_b$. Consequently, $\mathcal{Q}_b$ is closed.

Continuing to assume that all motions are stable, and considering any $z \in \text{Cl}\,\mathcal{Q}_c$, there exists a sequence $\{x_n\} \subset \mathcal{Q}_c$ such that $d(S(t)z, S(t)x_n) < 1/n$ for all $t \in \mathcal{R}^+$, $n = 1, 2, \ldots$. As $\gamma(x_n)$ is precompact, $n = 1, 2, \ldots$, Tychonov's theorem (Ref. 15) implies that any given sequence $\{\tau_m\} \subset \mathcal{R}^+$ admits a subsequence $\{t_m\}$ such that $\{S(t_m)x_n\}_{m=1,2,\ldots}$ is Cauchy, uniformly in $n = 1, 2, \ldots$. Noting that

$$\begin{aligned} d(S(t_p)y, S(t_m)y) &\leq d(S(t_p)y, S(t_p)x_n) \\ &\quad + d(S(t_p)x_n, S(t_m)x_n) + d(S(t_m)x_n, S(t_m)y) \\ &\leq 2/n + d(S(t_p)x_n, S(t_m)x_n) \end{aligned}$$

and letting $n \to \infty$, it follows that $\{S(t_m)y\}$ is Cauchy; hence, $\gamma(y)$ is precompact and $y \in \mathcal{Q}_c$. Thus $\mathcal{Q}_c$ is closed.

Assuming instead that $\{S(t)\}_{t\geq 0}$ is linear, $\mathcal{X}$ a Banach space, we define

$$\mathcal{G}_1 + \mathcal{G}_2 \equiv \{x + y | x \in \mathcal{G}_1, y \in \mathcal{G}_2\},$$

$$\alpha\mathcal{G}_1 \equiv \{\alpha x | x \in \mathcal{G}_1\},$$

for any sets $\mathcal{G}_1, \mathcal{G}_2 \in \mathcal{X}$, and α in the field of $\mathcal{X}$. Since $S(t)(\alpha x + \beta y) = \alpha S(t)x + \beta S(t)y$, we see that $\gamma(\alpha x + \beta y) \subset \alpha\gamma(x) + \beta\gamma(y)$; consequently, $\mathcal{Q}_b$ and $\mathcal{Q}_c$ are linear manifolds.

Continuing to assume that $\{S(t)\}_{t \geq 0}$ is linear, $\mathcal{X}$ a Banach space, we define $\hat{S}(t)$ to be the restriction of $S(t)$ to $\hat{\mathcal{X}} \equiv \{\mathcal{Q}_b, \|\cdot\|_{\mathcal{X}}\}$, $t \in \mathcal{R}^+$. As $\mathcal{Q}_b$ is positive invariant and $\hat{\mathcal{X}}$ is a normed linear space, the uniform boundedness theorem (Ref. 15) implies the existence of $c \in \mathcal{R}^+$ such that $\|S(t)x\| \leq c\|x\|$ for all $t \in \mathcal{R}^+$, $x \in \hat{\mathcal{X}}$. Therefore, $\|S(t)x\| \leq c\|x\|$ for all $x \in \mathcal{Q}_b$, and the continuity of $S(t)$: $\mathcal{X} \to \mathcal{X}$ now implies that $\|S(t)x\| \leq c\|x\|$ for all $x \in \mathrm{Cl}\,\mathcal{Q}_b$; consequently, $\mathcal{Q}_b = \mathrm{Cl}\,\mathcal{Q}_b$ and $\mathcal{Q}_b$ is closed. Moreover, $\hat{\mathcal{X}}$ is a Banach space and $\{\hat{S}(t)\}_{t \geq 0}$ is a linear dynamical system on $\hat{\mathcal{X}}$ with $\hat{\gamma}(x) = \gamma(x)$ for all $x \in \hat{\mathcal{X}}$. As all positive orbits of $\{\hat{S}(t)\}_{t \geq 0}$ are bounded, an exercise in Section IV.3 shows that the motion $\hat{S}(\cdot)x = S(\cdot)x$ is $\hat{\mathcal{X}}$-stable (not the same as $\mathcal{X}$-stable) for every $x \in \hat{\mathcal{X}}$. Hence, an earlier conclusion of this lemma implies that $\mathcal{Q}_c$ is $\hat{\mathcal{X}}$-closed, since $\mathcal{Q}_c \subset \mathcal{Q}_b = \hat{\mathcal{X}}$. As $\mathcal{Q}_b$ is $\mathcal{X}$-closed, $\mathrm{Cl}_{\mathcal{X}}\mathcal{Q}_c = \mathrm{Cl}_{\hat{\mathcal{X}}}\mathcal{Q}_c$, and it follows that $\mathcal{Q}_c$ is $\mathcal{X}$-closed. The proof is complete. □

Exercise 5.1. For given x_0, suppose that there exists $T \in \mathcal{R}^+$ such that $S(T)x_0 \in \mathcal{Q}_b$. Show that $x_0 \in \mathcal{Q}_b$. Assuming that $S(T)x_0 \in \mathcal{Q}_c$, show that $x_0 \in \mathcal{Q}_c$.

Exercise 5.2. Consider any positive invariant set $\mathcal{G} \subset \mathcal{X}$. Show that $\mathrm{Cl}\,\mathcal{G} \subset \mathcal{Q}_b$ if $\mathcal{G}$ is bounded, and $\mathrm{Cl}\,\mathcal{G} \subset \mathcal{Q}_c$ if $\mathcal{G}$ is precompact.

Exercise 5.3. In some applications, it can be shown that $S(T)$: $\mathcal{X} \to \mathcal{X}$ is a compact operator for some $T \in \mathcal{R}^+$. Show that, in such cases, $\mathcal{Q}_c = \mathcal{Q}_b$. *Hint:* Show that $\gamma(x) = (\cup_{0 \leq t < T} S(t)x) \cup (S(T)\gamma(x))$ for all $x \in \mathcal{X}$, where $S(T)\mathcal{G} \equiv \{y \in \mathcal{X} \mid y = S(T)x \text{ for some } x \in \mathcal{G}\}$ for $\mathcal{G} \subset \mathcal{X}$.

The following theorem provides three possible approaches to the precompactness question. One or more of these ideas have been used to obtain virtually all concrete results on orbital precompactness (Refs. 13, 21–25).

Theorem 5.1. Let $\{S(t)\}_{t \geq 0}$ be a dynamical system on a metric space $\mathcal{X}$.

(a) If $x_0 \in \mathcal{Q}_b$ and there exist operators P: $\mathcal{X} \to \mathcal{X}$, Q: $\mathcal{X} \to \mathcal{X}$, such that P is compact and $\gamma(Qx_0) \subset P\gamma(x_0)$, then $Qx_0 \in \mathcal{Q}_c$.

(b) If $x_0 \in \mathcal{Q}_b$ and there exists a family $\{P_n\}$ of compact operators P_n: $\mathcal{X} \to \mathcal{X}$, $n = 1, 2, \ldots,$ such that $P_n S(t)x_0 \to S(t)x_0$ as $n \to \infty$, *uniformly* in $t \in \mathcal{R}^+$, then $x_0 \in \mathcal{Q}_c$.

(c) Assume that $\mathcal{X} \subset \mathcal{B}$, $\mathcal{B}$ a Banach space, where $\|\cdot\|_{\mathcal{B}}$ induces the metric. If there exists a family $\{\mathcal{P}_n\}$ of precompact sets in $\mathcal{B}$, and a sequence $\{\varepsilon_n\} \subset \mathcal{R}^+$, $\varepsilon_n \searrow 0$ as $n \to \infty$, such that $\gamma(x_0) \subset \hat{\mathcal{S}}_{\varepsilon_n}(\mathcal{P}_n)$, $n = 1, 2, \ldots,$ then $x_0 \in \mathcal{Q}_c$.

Proof. The proof of (a) is obvious, since $\gamma(x_0)$ is bounded, $\gamma(Qx_0) \subset P\gamma(x_0)$, and P is compact.

Considering (b), we see that $P_n\gamma(x_0)$ is precompact for every n, and without loss of generality we can assume that $d(P_nS(t)x_0, S(t)x_0) < 1/n$ for all $t \in \mathcal{R}^+$, every $n = 1, 2, \ldots$. Given any sequence $\{\tau_m\} \subset \mathcal{R}^+$, Tychonov's theorem (Ref. 15) implies the existence of a subsequence $\{t_m\}$ such that, for each n, $d(P_nS(t_{m+1})x_0, P_nS(t_m)x_0) < 1/m$ for all $m = 1, 2, \ldots$. Noting that

$$\begin{aligned} d(S(t_{m+1})x_0, S(t_m)x_0) &\leqslant d(S(t_{m+1})x_0, P_nS(t_{m+1})x_0) \\ &\quad + d(P_nS(t_{m+1})x_0, P_nS(t_m)x_0) + d(P_nS(t_m)x_0, S(t_m)x_0) \\ &\leqslant \frac{1}{n} + \frac{1}{m} + \frac{1}{n}, \end{aligned}$$

and letting $n \to \infty$, we see that $\{S(t_m)x_0\}$ is a Cauchy sequence, hence, $x_0 \in \mathcal{Q}_c$.

In (c) we assume that $\mathcal{B}$ is a Banach space, and $\mathcal{X} \subset \mathcal{B}$ with the metric induced by the norm. To prove (c), we employ the α-measure of noncompactness defined by Kuratowski (Ref. 26): If $\mathcal{G}$ is a bounded set in $\mathcal{B}$, then $\alpha[\mathcal{G}]$ is the infimum of $\varepsilon > 0$ such that $\mathcal{G}$ can be covered by a finite number of sets of diameter no larger than ε. If $\mathcal{G}_1$ and $\mathcal{G}_2$ are bounded sets in $\mathcal{B}$, it follows from the definition that

(i) $\alpha[\mathcal{G}_1] = 0$ if and only if $\mathcal{G}_1$ is precompact;
(ii) $\mathcal{G}_1 \subset \mathcal{G}_2$ implies $\alpha[\mathcal{G}_1] \leqslant \alpha[\mathcal{G}_2]$;
(iii) $\alpha[\mathcal{G}_1 \cup \mathcal{G}_2] = \max\{\alpha[\mathcal{G}_1], \alpha[\mathcal{G}_2]\}$.

Moreover, defining $\mathcal{G}_1 + \mathcal{G}_2 \equiv \{x + y | x \in \mathcal{G}_1, y \in \mathcal{G}_2\}$, it has been shown (Ref. 27) that

(iv) $\alpha[\mathcal{G}_1 + \mathcal{G}_2] \leqslant \alpha[\mathcal{G}_1] + \alpha[\mathcal{G}_2]$.

Under the assumptions of (c), we see that $\hat{\mathcal{S}}_{\varepsilon_n}(\mathcal{P}_n) = \mathcal{P}_n + \mathcal{S}_{\varepsilon_n}(0)$, and $\alpha[\mathcal{P}_n] = 0$, $\alpha[\mathcal{S}_{\varepsilon_n}(0)] = 2\varepsilon_n$. Hence, employing (ii) and (iv), $\alpha[\gamma(x_0)] \leqslant \alpha[\hat{\mathcal{S}}_{\varepsilon_n}(\mathcal{P}_n)] = 2\varepsilon_n$, $n = 1, 2, \ldots$. As $\varepsilon_n \searrow 0$ for $n \to \infty$, $\alpha[\gamma(x_0)] = 0$ and (i) implies that $\gamma(x_0)$ is precompact. The proof is complete. □

Making use of Lemma 5.1 and Theorem 5.1, the following theorem provides some concrete results on our precompactness question for linear dynamical systems; this theorem extends some results of Ref. 23.

Theorem 5.2. Let a linear operator $A: (\mathcal{D}(A) \subset \mathcal{X}) \to \mathcal{X}$ be the infinitesimal generator of a linear dynamical system $\{S(t)\}_{t \geqslant 0}$ on a Banach space $\mathcal{X}$.

(a) If there exists a compact linear operator $P: \mathcal{X} \to \mathcal{X}$ such that

$$Px \in \mathcal{D}(A) \quad \text{and} \quad PAx = APx \qquad \text{for all } x \in \mathcal{D}(A),$$

then $x \in \mathcal{Q}_b$ implies $Px \in \mathcal{Q}_c$; further, if $\mathcal{Q}_b = \mathcal{X}$, then $\mathcal{Q}_c \supset \text{Cl}\,\mathcal{R}(P)$.

(b) If $J_\mu \equiv (I - \mu A)^{-1}$ is compact for some $\mu \in (0, \lambda_0)$, then $\mathcal{Q}_c = \mathcal{Q}_b$.

Proof. To show (a), recall that $S(t)x = \lim_{n\to\infty} J^n_{t/n}x$ for all $t \in \mathcal{R}^+$, $x \in \mathcal{X}$, and our assumption implies that $PJ_\lambda = J_\lambda P$ for all $\lambda \in (0, \lambda_0)$ because P is linear. As a compact operator must be bounded, and a bounded linear operator is continuous, we find that

$$S(t)Px = \lim_{n\to\infty} J^n_{t/n}Px = \lim_{n\to\infty} PJ^n_{t/n}x = PS(t)x$$

for all $t \in \mathcal{R}^+$, $x \in \mathcal{X}$, and (a) of Theorem 5.1 now implies that $Px \in \mathcal{Q}_c$ if $x \in \mathcal{Q}_b$. If $\mathcal{Q}_b = \mathcal{X}$, it also follows that $Px \in \mathcal{Q}_c$ for all $x \in \mathcal{X}$; hence, $\mathcal{R}(P) \subset \mathcal{Q}_c$, and we find that $\mathrm{Cl}\, \mathcal{R}(P) \subset \mathrm{Cl}\, \mathcal{Q}_c = \mathcal{Q}_c$ by Lemma 5.1.

To show (b), let $\{\hat{S}(t)\}_{t\geq 0}$ be the linear dynamical system on $\hat{\mathcal{X}} \equiv (\mathcal{Q}_b, \|\cdot\|_{\mathcal{X}})$ which was defined in the proof of Lemma 5.2, and note that the infinitesimal generator $\hat{A}$: $(\mathcal{D}(\hat{A}) \subset \mathcal{X}) \to \mathcal{X}$ of $\{\hat{S}(t)\}_{t\geq 0}$ is the restriction of A to $\mathcal{D}(A) \cap \mathcal{Q}_b$. As $\hat{\mathcal{X}}$ is a Banach space and $\{\hat{S}(t)\}_{t\geq 0}$ is linear, $I - \lambda\hat{A}$ must have range $\hat{\mathcal{X}}$ for all $\lambda \in (0, \lambda_0)$, and it follows that $\hat{J}_\lambda$ is the restriction of J_λ to $\mathcal{Q}_b$; hence, $\hat{J}_\mu$ is compact with $\mathcal{R}(\hat{J}_\mu) = \mathcal{D}(\hat{A})$ dense in $\hat{\mathcal{X}}$ for $\lambda \in (0, \lambda_0)$. Noting that $\hat{\gamma}(x) = \gamma(x)$ is bounded for $x \in \hat{\mathcal{X}}$, defining $\hat{P} = \hat{J}_\mu$, and applying (a) of this theorem to $\{\hat{S}(t)\}_{t\geq 0}$ on $\hat{\mathcal{X}}$, we find that $\hat{\gamma}(x) = \gamma(x)$ is precompact for all $x \in \mathrm{Cl}\, \mathcal{R}(\hat{J}_\mu) = \hat{\mathcal{X}} = \mathcal{Q}_b$; hence, $\mathcal{Q}_c \supset \mathcal{Q}_b$ and the proof is complete. □

Exercise 5.4. Show that all conclusions of Example 4.2 can be obtained without invoking the second Liapunov function $\tilde{V}$ to establish precompactness of positive orbits. *Hint:* A: $l_2 \to l_2$ is a compact linear operator with dense range.

Before applying Theorem 5.2 in some linear examples, let us consider a useful means of showing that a given operator P: $\mathcal{X} \to \mathcal{X}$ is a compact operator.

Definition 5.1. Let $\mathcal{X}$ and $\mathcal{Y}$ be metric spaces with $\mathcal{Y} \subset \mathcal{X}$. If every $d_{\mathcal{Y}}$-bounded set in $\mathcal{Y}$ is $d_{\mathcal{X}}$-precompact, we say that the *injection* $\hat{I}$: $\mathcal{Y} \to \mathcal{X}$ is *compact*, where $\hat{I}y \equiv y$ for $y \in \mathcal{Y}$. Equivalently, $\hat{I}$: $\mathcal{Y} \to \mathcal{X}$ is compact if every $d_{\mathcal{Y}}$-bounded sequence $\{y_n\} \subset \mathcal{Y}$ contains a $d_{\mathcal{X}}$-Cauchy subsequence.

It follows easily from Ascoli's theorem (Ref. 15) that the injection $\hat{I}$: $\mathcal{C}^n[0, 1] \to \mathcal{C}^{n-1}[0, 1]$ is compact, for every $n = 1, 2, \ldots$. From Sobolev's embedding theorem (Ref. 28), it also follows that the injection $\hat{I}$: $\mathcal{W}_2^n(0, 1) \to \mathcal{W}_2^{n-1}(0, 1)$ is compact for every $n = 1, 2, \ldots$. Such information is often useful in establishing that a given operator is compact.

Proposition 5.1. Given a function P: $\mathcal{X} \to \mathcal{X}$, $\mathcal{X}$ a metric space, let there exist a metric space $\mathcal{Y} \equiv (\mathcal{R}(P), d_{\mathcal{Y}})$ such that $d_{\mathcal{Y}}(Px, Py) \leq$

$\alpha d_{\mathcal{X}}(x, y)$ for all $x, y \in \mathcal{X}$, some $\alpha > 0$. If the injection $\hat{I}$: $\mathcal{Y} \to \mathcal{X}$ is compact, then P: $\mathcal{X} \to \mathcal{X}$ is a compact operator.

Proof. P takes $\mathcal{X}$-bounded sets into $\mathcal{Y}$-bounded sets, and $\mathcal{Y}$-bounded sets are $\mathcal{X}$-precompact by the compactness of the injection. The proof is complete. □

Exercise 5.5. For $\mathcal{X}$ a Banach space, let a linear operator A: $(\mathcal{D}(A) \subset \mathcal{X}) \to \mathcal{X}$ be the infinitesimal generator of a linear dynamical system on $\mathcal{X}$. Let the injection $\hat{I}$: $(\mathcal{D}(A), \|\cdot\|_g) \to \mathcal{X}$ be compact, where $\|\cdot\|_g$ is a graph norm. Show that J_λ: $\mathcal{X} \to \mathcal{X}$ is compact for $\lambda \in (0, \lambda_0)$, where $J_\lambda \equiv (\hat{I} - \lambda A)^{-1}$.

Example 5.2. Recall the linear heat conduction problem of Example III.3.1, which was considered further in Example 1.1. The state space $\mathcal{X}$ and infinitesimal generator were defined as

$$\mathcal{X} \equiv \{x \in \mathcal{C}[0, 1] | x(0) = 0 = x(1)\}, \qquad \|x\|_{\mathcal{X}} \equiv \max_{\eta} |x(\eta)|,$$

$$Ax \equiv \partial^2 x, \qquad x \in \mathcal{D}(A) \equiv \{x \in \mathcal{C}^2[0, 1] | x(\eta) = 0 = \partial^2 x(\eta) \text{ at } \eta = 0, 1\}.$$

In Example III.3.1 we found that $\|S(t)x\|_{\mathcal{X}} \leq \|x\|_{\mathcal{X}}$ for all $t \in \mathcal{R}^+$, $x \in \mathcal{X}$; hence, $x_e = 0$ is a stable equilibrium, all positive orbits are bounded, and all motions are stable (see Exercise 3.2). In Example 1.1 we also saw that $\|S(t)x\|_{\mathcal{L}_2} \leq \exp\{-\pi^2 t\}\|x\|_{\mathcal{L}_2}$ for all $t \in \mathcal{R}^+$, $x \in \mathcal{X}$; however, we notice that this does not imply exponential stability, or even asymptotic stability, since $\|\cdot\|_{\mathcal{L}_2}$ is not equivalent to $\|\cdot\|_{\mathcal{X}}$. Despite the fact that the eigenvectors of A are $\mathcal{L}_2$-orthogonal and span $\mathcal{X}$, and the eigenvalues are $-n^2\pi^2$, $n = 1, 2, \ldots$, it seems very unlikely that x_e is exponentially stable (in the $\|\cdot\|_{\mathcal{C}}$-topology of $\mathcal{X}$). However, we now can use the Invariance Principle to show that $x_e = 0$ is asymptotically stable.

Employing the continuous Liapunov function $\tilde{V}$ of Example 1.1, we have

$$\tilde{V}(x) \equiv \|x\|^2_{\mathcal{L}_2} \equiv \int_0^1 |x(\eta)|^2 \, d\eta, \qquad \dot{\tilde{V}}(x) \leq -2\pi^2 \|x\|^2_{\mathcal{L}_2} \equiv -W(x),$$

for all $x \in \mathcal{X}$, Hence, the largest positive invariant set in $\mathcal{M}_2 \equiv \{x \in \mathcal{X} | W(x) = 0\} = \{0\}$ is $\{0\} \subset \mathcal{X}$, since $x_e = 0$ is an equilibrium, and the Invariance Principle (Theorem 4.2) shows that $\Omega(x) \subset \{0\}$ for every $x \in \mathcal{X}$. Noting that the injection $\hat{I}$: $(\mathcal{D}(A), \|\cdot\|_g) \to \mathcal{X}$ is compact, Exercise 5.5 shows that J_λ: $\mathcal{X} \to \mathcal{X}$ is compact for all $\lambda \in (0, \lambda_0)$; hence, (b) of Theorem 5.2 shows that bounded positive orbits are precompact. As we already know that all positive orbits are bounded, all must be precompact; hence, by Theorem 4.2, $S(t)x \to^{\mathcal{X}} \Omega(x) = \{0\}$ as $t \to \infty$, for every $x \in \mathcal{X}$. As $x_e = 0$ is already known to be stable, we now see that it is globally asymptotically stable (in the $\|\cdot\|_{\mathcal{C}}$-topology of $\mathcal{X}$). By Exercise 3.2 it also follows that all motions are globally asymptotically stable.

Referring to Examples 5.2 and III.3.2, we see that we have obtained somewhat different stability conclusions for the same physical problem; in Example III.3.2, with $\mathcal{X} \equiv \mathcal{L}_2(0, 1)$, we found global exponential stability of all motions, with exponent $-\pi^2 t$. This demonstrates that, when the same physical system leads to dynamical systems on two or more state spaces having different topologies, such properties as boundedness and stability do not automatically carry over from one state space to another. This leads to a number of interesting questions, a few of which have been studied in Refs. 23 and 29. Here we only point out that, as a result, the term "stability" has no intrinsic meaning for a given physical system (without reference to a particular topology), unless the physical system is finite dimensional.

Example 5.3. Recall the linear delay equation of Example III.3.3, which was considered further in Example 3.3. The state space and infinitesimal generator were defined by

$$\mathfrak{X} = \mathfrak{R} \times \mathcal{L}_2(0,\tau), \qquad \|x\|^2 \equiv y^2 + |\beta| \int_0^\tau |u(\eta)|^2 \, d\eta, \qquad x = (y,u) \in \mathfrak{X},$$

$$Ax \equiv (\alpha y + \beta u(\tau), -\partial u),$$

$$x = (y,u) \in \mathfrak{D}(A) \equiv \{(y,u) \in \mathfrak{R} \times \mathfrak{W}_2^1(0,\tau) \,|\, y = u(0)\},$$

where $\tau > 0$ and $\alpha, \beta \in \mathfrak{R}$ were given.

For $\alpha \leqslant -|\beta|$ we found in Example III.3.3 that $\|S(t)x\| \leqslant \|x\|$ for all $t \in \mathfrak{R}^+$, $x \in \mathfrak{X}$; hence, $V(x) \equiv \|x\|^2$ is a Liapunov function everywhere, $x_e = 0$ is a stable equilibrium, all positive orbits are bounded, and all motions are stable. By constructing another Liapunov function $\tilde{V}$ in Example 3.3, we were able to show that $x_e = 0$ is a globally exponentially stable equilibrium for all $\tau > 0$, provided that $\alpha < -|\beta|$. We now claim that, without constructing the second Liapunov function $\tilde{V}$, we can at least show global asymptotic stability for all $\tau > 0$, $\alpha < -|\beta|$, by using only the original Liapunov function $V(x) \equiv \|x\|^2$.

Assuming $\alpha < -|\beta|$, we see that

$$V(x) \equiv \|x\|^2, \qquad x \in \mathfrak{X},$$

$$\dot{V}(x) = V_x' Ax = -|\beta|[u(\tau) - y]^2 + 2(\alpha + |\beta|)y^2, \qquad x = (y,u) \in \mathfrak{D}(A),$$

and consequently $\dot{V}(x) \leqslant 2(\alpha + |\beta|)y^2 \equiv -W(x) \leqslant 0$ for all $x = (y,u) \in \mathfrak{X}$, using Theorem 2.1. We now wish to locate the largest positive invariant set $\mathfrak{M}^+$ in $\mathfrak{M}_2 \equiv \{x \in \mathfrak{X} \,|\, W(x) = 0\} = \{(y,u) \in \mathfrak{X} \,|\, y = 0\}$, since $\alpha + |\beta| < 0$. As we do not have $\mathfrak{M}_2 \subset \mathfrak{D}(A)$, the evolution equation $\dot{x}(t) = Ax(t)$ does not apply to some of the motions orginating in $\mathfrak{M}_2$. However, for any strong solution $(y(t), u(t,\cdot))$ we have

$$\dot{y}(t) = \alpha y(t) + \beta u(t,\tau), \qquad t \geqslant 0,$$

$$u(t,\tau) = u(0, \tau - t), \qquad \tau \geqslant t \geqslant 0,$$

and integration leads to

$$y(t_2) = y(t_1) + \int_{t_1}^{t_2} [\alpha y(s) + \beta u(0, \tau - s)] \, ds, \qquad \tau \geqslant t_2 \geqslant t_1 \geqslant 0.$$

Continuity of $S(t)$: $\mathfrak{X} \to \mathfrak{X}$ now implies that this equation holds along every motion; hence, if $S(t)x_0 = (y(t), u(t, \cdot)) \in \mathfrak{M}_2$ for all $t \in \mathfrak{R}^+$, then $y(t) \equiv 0$ and this equation implies that

$$\beta \int_{t_1}^{t_2} u(0, \tau - s)\, ds = 0$$

for all $t_2 \geqslant t_1$ such that $\tau \geqslant t_2$, $t_1 \geqslant 0$. Assuming $\beta \neq 0$, this means that $u_0(\eta) = u(0, \eta) = 0$ a.e. $\eta \in [0, \tau]$, and we find that $\mathfrak{M}^+ \subset \{0\}$; in fact, $\mathfrak{M}^+ = \{0\}$ since $x_e = 0$ is an equilibrium. Applying the Invariance Principle (Theorem 4.2) we find $\Omega(x) \subset \{0\}$ for every $x \in \mathfrak{X}$.

Noting that the injection $\hat{I}$: $(\mathfrak{D}(A), \|\cdot\|_g) \to \mathfrak{X}$ is compact, we see that J_λ: $\mathfrak{X} \to \mathfrak{X}$ is compact for all $\lambda \in (0, \lambda_0)$; as all positive orbits are bounded, (b) of Theorem 5.2 now shows that all positive orbits are precompact. Applying Theorem 4.2, we see that $S(t)x \to \{0\}$ as $t \to \infty$, for every $x \in \mathfrak{X}$. As we already know $x_e = 0$ to be stable, it follows that $x_e = 0$ is globally asymptotically stable for $\alpha < -|\beta| \neq 0$, $\tau > 0$.

Exercise 5.6. Consider a linear heat conduction problem for a completely insulated rod, formally described by

$$\frac{\partial}{\partial t}\theta(t, \eta) = \frac{\partial^2}{\partial \eta^2}\theta(t, \eta), \qquad 0 \leqslant \eta \leqslant 1, \quad t \geqslant 0,$$

$$\frac{\partial}{\partial \eta}\theta(t, \eta) = 0 \qquad \text{at } \eta = 0, 1, \quad t \geqslant 0.$$

Show that this problem leads to a linear dynamical system on $\mathfrak{X} \equiv \mathfrak{L}_2(0, 1)$, that all positive orbits are bounded, and that all motions are stable. Determine the set $\mathfrak{Q}_e$ of all equilibria, and show that for each $x \in \mathfrak{X}$ there exists exactly one $y(x) \in \mathfrak{Q}_e$ such that $S(t)x \to y(x)$ as $t \to \infty$. Does this imply asymptotic stability of any motion? Notice that we do *not* have

$$\int_0^1 |\partial x(\eta)|^2\, d\eta \geqslant c \int_0^1 |x(\eta)|^2\, d\eta$$

for all $x \in \mathfrak{D}(A)$, and $c > 0$. Also, defining

$$\tilde{V}(x) = \int_0^1 x(\eta)\, d\eta, \qquad x \in \mathfrak{X},$$

show that both $\tilde{V}$ and $-\tilde{V}$ are Liapunov functions on $\mathfrak{X}$, and find $y(x)$.

For nonlinear dynamical systems, existing concrete results on orbital precompactness are much less general than those of Theorem 5.2 for linear dynamical systems. The following result was obtained in Ref. 22, using a slightly different proof.

Theorem 5.3. For $\mathcal{B}$ a Banach space, let a multivalued operator $A: (\mathcal{D}(A) \subset \mathcal{B}) \to \mathcal{B}$ generate a dynamical system $\{S(t)\}_{t \geqslant 0}$ on $\mathcal{X} \equiv \mathrm{Cl}\,\mathcal{D}(A)$, in the sense of Theorem III.5.1. If $0 \in \mathcal{R}(A)$, $-A$ is accretive, and $J_\lambda \equiv (I - \lambda A)^{-1}$ is compact for all sufficiently small $\lambda > 0$, then all motions are stable, there exists at least one equilibrium $x_e \in \mathcal{D}(A)$, and all positive orbits are precompact.

Proof. As $-A$ is accretive, J_λ is nonexpansive for all $\lambda > 0$; moreover, a result of Crandall (Ref. 30) shows that, for all $\lambda > \mu > 0$,

$$\|x - J_\lambda x\| \leqslant (\lambda/\mu)\|x - J_\mu x\|$$

for all $x \in \mathcal{X}$. Consequently, for $x \in \mathcal{D}(A)$, $n = 1, 2, \ldots,$ and $0 \leqslant t < n\lambda$,

$$\begin{aligned}\|(I - J_\lambda)J^n_{t/n}x\| &\leqslant (n\lambda/t)\|J^n_{t/n}x - J^{n+1}_{t/n}x\|\\ &= (n\lambda/t)\|J^{n+1}_{t/n}(I - (t/n)A)x - J^{n+1}_{t/n}x\| \leqslant \lambda|Ax|\end{aligned}$$

and we see that $\|S(t)x - J_\lambda S(t)x\| \leqslant \lambda|Ax|$ for all $x \in \mathcal{D}(A)$, $t \in \mathcal{R}^+$. Noting that $P_n \equiv J_{1/n}$ is compact for all sufficiently large n, and applying (b) of Theorem 5.1, we see that $\gamma(x)$ is precompact if $x \in \mathcal{D}(A)$ and $\gamma(x)$ is bounded.

Since $0 \in \mathcal{R}(A)$, there exists $x_e \in \mathcal{D}(A)$ such that $Ax_e = 0$, and $x(t) \equiv x_e$ is a strong solution of (6) in Chapter III. By (i) of Theorem III.5.2, $S(t)x_e \equiv x_e$ is a motion of $\{S(t)\}_{t \geqslant 0}$ and, hence, an equilibrium. As $-A$ is accretive, all motions are stable; by the existence of x_e, it also follows that all positive orbits are bounded. Consequently, Lemma 5.1 shows that $\gamma(x)$ is precompact for all $x \in \mathrm{Cl}\,\mathcal{D}(A) = \mathcal{X}$, and the proof is complete. □

Example 5.4. In Exercise III.5.7 we considered a simply supported beam, surrounded by a viscous fluid of negligible density, which was formally described by

$$\rho\frac{\partial^2}{\partial t^2}u(\eta, t) + f\left(\frac{\partial}{\partial t}u(\eta, t)\right) + \alpha\frac{\partial^4}{\partial \eta^4}u(\eta, t) = 0, \qquad 0 \leqslant \eta \leqslant 1, \quad t \geqslant 0,$$

$$u(\eta, t) = 0 = \frac{\partial^2}{\partial \eta^2}u(\eta, t) \quad \text{at } \eta = 0, 1, \quad t \geqslant 0,$$

where $\alpha > 0$, $\rho > 0$, and the continuous function $f: \mathcal{R} \to \mathcal{R}$ were given; moreover, we assumed that

$$(\xi - \hat{\xi})[f(\xi) - f(\hat{\xi})] \geqslant 0, \qquad |f(\xi)| \leqslant \beta|\xi|,$$

for all $\xi, \hat{\xi} \in \mathcal{R}$, some $\beta \in \mathcal{R}^+$.

If we define

$$\hat{\mathfrak{W}}_2^{2n}(0, 1) \equiv \{u \in \mathfrak{W}_2^{2n}(0, 1) | \partial^{2m}u(0) = \partial^{2m}u(0), m = 0, 1, \ldots, n-1\}$$

and

$$\mathfrak{X} \equiv \hat{\mathfrak{W}}_2^2(0, 1) \times \mathfrak{L}_2(0, 1), \qquad \|x\|^2 \equiv \int_0^1 [\alpha|\partial^2 u(\eta)|^2 + \rho|v(\eta)|^2] \, d\eta,$$

$$x = (u, v) \in \mathfrak{X},$$

$$Ax \equiv (v, -\rho^{-1}\partial^4 u), \qquad x = (u, v) \in \mathfrak{D}(A) \equiv \hat{\mathfrak{W}}_2^4(0, 1) \times \hat{\mathfrak{W}}_2^2(0, 1),$$

$$Fx \equiv (0, -\rho^{-1}f(v)), \qquad x = (u, v) \in \mathfrak{X},$$

Proposition III.5.2 can be used to show that $(A + F)$: $(\mathfrak{D}(A) \subset \mathfrak{X}) \to \mathfrak{X}$ generates a dynamical system $\{S(t)\}_{t \geq 0}$ on $\mathfrak{X}$, in the sense of Theorem III.5.1; moreover, $-(A + F)$ is accretive and $\|S(t)x - S(t)\hat{x}\| \leq \|x - \hat{x}\|$ for all $x, \hat{x} \in \mathfrak{X}$, $t \in \mathfrak{R}^+$. We see that all motions are stable, $x_e = 0$ is an equilibrium, and all positive orbits are bounded. Note that the injection I: $\mathfrak{W}_2^4 \times \mathfrak{W}_2^2 \to \mathfrak{W}_2^2 \times \mathfrak{L}_2$ is compact, $\|\cdot\|_{\mathfrak{W}_2^4 \times \mathfrak{W}_2^2}$ is equivalent to $\|\cdot\| + \|A\cdot\|$ on $\mathfrak{D}(A) = \mathfrak{R}(J_\lambda)$, and

$$\begin{aligned} \lambda\|J_\lambda x\| + \lambda\|AJ_\lambda x\| &\leq \lambda\|J_\lambda x\| + \|(I - \lambda A - \lambda F)J_\lambda x\| + \|J_\lambda x\| + \lambda\|FJ_\lambda\| \\ &\leq (\lambda + 1 + \beta\lambda)\|J_\lambda x\| + \|x\| \\ &\leq (\lambda + 2 + \beta\lambda)\|x\| \end{aligned}$$

for all $x \in \mathfrak{X}$, where $J_\lambda \equiv (I - \lambda A - \lambda F)^{-1}$ for $\lambda > 0$. Then it follows from Proposition 5.1 that J_λ is compact for all $\lambda > 0$. Hence, Theorem 5.3 shows that every positive orbit is precompact, since $(A + F)0 = 0$.

Making the additional assumption that $f(\xi) = 0$ only for $\xi = 0$, and defining $V(x) \equiv \|x\|^2$, $x \in \mathfrak{X}$, we see by Theorem 2.1 that

$$\dot{V}(x) = V_x'(A + F)x = -2\int_0^1 v(\eta)f(v(\eta)) \, d\eta, \qquad x = (u, v) \in \mathfrak{D}(A),$$

$$\dot{V}(x) \leq -2\int_0^1 v(\eta)f(v(\eta)) \, d\eta \equiv -W(x), \qquad x = (u, v,) \in \mathfrak{X}.$$

Hence, $\mathfrak{M}^2 \equiv \{x \in \mathfrak{X} | W(x) = 0\} = \{(u, v) \in \mathfrak{X} | v = 0\}$, and we can use the evolution equation $\dot{x}(t) = Ax(t)$, a.e. $t \in \mathfrak{R}^+$, to show that the largest positive invariant set in $\mathfrak{M}_2 \cap \mathfrak{D}(A)$ is $\{0\}$. Unfortunately, the Invariance Principle (Theorem 4.2) requires the largest positive invariant set $\mathfrak{M}^+$ in $\mathfrak{M}_2$; however, by employing the theory of distributions (Ref. 15), one can obtain a distributional evolution equation satisfied by all motions of $\{S(t)\}_{t \geq 0}$, and from this equation it can be concluded that, in fact, $\mathfrak{M}^+ = \{0\}$. Hence, by the Invariance Principle, we find that $S(t)x \to \Omega(x) = \{0\}$ as $t \to \infty$, for every $x \in \mathfrak{X}$, and the equilibrium $x_e = 0$ is globally asymptotically stable.

For linear dynamical systems, Theorem 5.3 is much more restrictive than (b) of Theorem 5.2. Theorem 5.3 requires $\{S(t)\}_{t \geq 0}$ to be nonexpansive ($-A$ accretive), and this is a very restrictive assumption, particularly for nonlinear dynamical systems.

Very recently two other precompactness results have been obtained for certain classes of nonlinear systems (Refs. 13, 24); both results are based on a "variation-of-constants formula," such as (6) of the following lemma (also see Section I.6).

Lemma 5.2. Let a linear operator $A: (\mathfrak{D}(A) \subset \mathfrak{B}) \to \mathfrak{B}$, $\mathfrak{B}$ a Banach space, be the infinitesimal generator of a linear dynamical system $\{S_A(t)\}_{t \geq 0}$ on $\mathfrak{B}$, and let $F: (\mathfrak{D}(F) \subset \mathfrak{B}) \to \mathfrak{B}$ be any function.

(a) If $u(\cdot): \mathfrak{R}^+ \to \mathfrak{X}$ is a strong solution of

$$\dot{x}(t) = (A + F)x(t), \qquad \text{a.e. } t \in \mathfrak{R}^+,$$
$$x(0) = x_0 \in \text{Cl}\, \mathfrak{D}(A + F), \tag{5}$$

then

$$u(t) = S_A(t)x_0 + \int_0^t S_A(t - \tau)Fu(\tau)d\tau \tag{6}$$

for all $t \in \mathfrak{R}^+$ (the Bochner integral exists).

(b) If $A + F$ generates a dynamical system $\{S(t)\}_{t \geq 0}$ on the set $\mathfrak{X} \equiv \text{Cl}\mathfrak{D}(A + F)$, in the sense of Theorem III.5.1, then

$$S(t)x = S_A(t) + \lim_{n \to \infty} \left(\frac{t}{n} \sum_{m=0}^{n-1} H_{t/n}^{n-m} F J_{t/n}^m x \right) \tag{7}$$

for all $t \in \mathfrak{R}^+$, $x \in \mathfrak{X}$, where $H_\lambda \equiv (I - \lambda A)^{-1}$ and $J_\lambda \equiv (I - \lambda A - \lambda F)^{-1}$ for $\lambda \in (0, \lambda_0)$ (the limit exists).

(c) If F is continuous, $\mathfrak{D}(F) \supset \text{Cl}\mathfrak{D}(A + F) \equiv \mathfrak{X}$, and $A + F$ generates $\{S(t)\}_{t \geq 0}$ as in (b), then

$$S(t)x = S_A(t)x + \int_0^t S_A(t - \tau)FS(\tau)x d\tau \tag{8}$$

for all $t \in \mathfrak{R}^+$, $x \in \mathfrak{X}$ (the Riemann integral exists).

Proof. If $u(\cdot): \mathfrak{R}^+ \to \mathfrak{X}$ is a strong solution of (5), then $u(\cdot) \in \mathcal{C}(\mathfrak{R}^+; \mathfrak{X}) \cap \mathfrak{W}_{\text{loc}}^{1,1}(\mathfrak{R}^+; \mathfrak{X})$ with $u(0) = x_0$; moreover, $u(t) \subset \mathfrak{D}(A + F)$ a.e. $t \in \mathfrak{R}^+$ and $u(\cdot)$ is a.e. differentiable. Fixing $t > 0$ and considering any $s \in (0, t)$ such that $u(s) \in \mathfrak{D}(A)$ and $\dot{u}(s)$ exists, we see that

$$\begin{aligned} S_A(t - s)Fu(s) &= S_A(t - s)[Au(s) + Fu(s)] - S_A(t - s)Au(s) \\ &= S_A(t - s)\dot{u}(s) - AS_A(t - s)u(s) \\ &= \frac{d}{ds}[S_A(t - s)u(s)]. \end{aligned}$$

It follows that $v(s) \equiv S_A(t-s)u(s)$ is a.e. differentiable on $(0, t)$, and it is not difficult to use the properties of $\{S_A(t)\}_{t \geq 0}$ and the Lipschitz continuity of $u(\cdot)$: $\mathcal{R}^+ \to \mathcal{X}$ to show that $v(\cdot)$ is Lipschitz continuous on $[0, t]$; hence, $v(\cdot) \in \mathcal{C}([0, t]; \mathcal{X}) \cap \mathcal{W}^{1,1}_{\text{loc}}([0, t]; \mathcal{X})$, and Bochner integration leads to

$$u(t) - S_A(t)x_0 = \int_0^t \dot{v}(s)\,ds = \int_0^t S_A(t-s)Fu(s)\,ds$$

for all $t \in \mathcal{R}^+$. Hence, (a) has been shown.

Now assuming that $A + F$ generates $\{S(t)\}_{t \geq 0}$, in the sense of Theorem III.5.1, we recall that $S(t)x \equiv \lim_{n\to\infty} J^n_{t/n}x$, $S_A(t)x \equiv \lim_{n\to\infty} H^n_{t/n}x$ for all $x \in \mathcal{X}$, $t \in \mathcal{R}^+$, where $J_\lambda \equiv (I - \lambda A - \lambda F)^{-1}$, $H_\lambda \equiv (I - \lambda A)^{-1}$ for $\lambda \in (0, \lambda_0)$. By the linearity of A and H_λ,

$$\begin{aligned} J^n_\lambda - H^n_\lambda &= \sum_{m=0}^{n-1} \left(H^{n-m-1}_\lambda J^{m+1}_\lambda - H^{n-m}_\lambda J^m_\lambda\right) \\ &= \sum_{m=0}^{n-1} \left[H^{n-m}_\lambda (I - \lambda A) J^{m+1}_\lambda - H^{n-m}_\lambda (I - \lambda A - \lambda F) J^m_\lambda\right] \\ &= \lambda \sum_{m=0}^{n-1} H^{n-m}_\lambda F J^m_\lambda, \end{aligned}$$

and it follows that

$$S(t)x - S_A(t)x = \lim_{n\to\infty} \left(\frac{t}{n} \sum_{m=0}^{n-1} H^{n-m}_{t/n} F J^m_{t/n} x \right), \tag{9}$$

where the limit must exist for each $t \in \mathcal{R}^+$, $x \in \mathcal{X}$. Fixing $t > 0$ and defining a family $\{w_n\}$ of piecewise-constant functions $w_n(\cdot)$: $[0, t + t/n) \to \mathcal{X}$ as

$$w_n(\tau) \equiv H^{n-m}_{t/n} F J^{m+1}_{t/n} x, \qquad \tau \in \left[mt/n, (m+1)t/n\right], m = 0, 1, \ldots, n,$$

we see that

$$S(t)x - S_A(t)x = \lim_{n\to\infty} \frac{t}{n} \sum_{m=0}^{n-1} w_n(mt/n) = \lim_{n\to\infty} \int_0^t w_n(\tau)\,d\tau.$$

It can be shown (Ref. 31) that $\lim_{h\searrow 0} H^{[s/\lambda]}_\lambda x = \lim_{n\to\infty} \mathrm{H}^n_{s/n} x = S_A(s)x$, where $[s/\lambda]$ denotes the largest positive integer $\leq s/\lambda$, and we see that for each $\tau \in [0, t]$, $x \in \mathfrak{X}$, we have $mt \leq n\tau < (m+1)t$ and

$$\begin{aligned}\|S_A(t-\tau)FS(\tau)x - w_n(\tau)\| &\leq \|S_A(t-\tau)FS(\tau)x - H^n_{(t-\tau)/n}FS(\tau)x\| \\ &\quad + \|(H^n_{(t-\tau)/n} - H^{n-m}_{t/n})FS(\tau)x\| \\ &\quad + \|H^{n-m}_{t/n}(FS(\tau)x - FJ^{n-m}_{t/n}x)\| \\ &\to 0 \qquad \text{as } n \to \infty,\end{aligned}$$

assuming that F: $\mathfrak{X} \to \mathfrak{B}$ is continuous. As $[0, t]$ is compact, $w_n(\tau) \to S_A(t-\tau)FS(\tau)x$ uniformly for $\tau \in [0, t]$; hence, it follows that

$$S(t)x - S_A(t)x = \int_0^t S_A(t-\tau)FS(\tau)x\,d\tau \tag{10}$$

for all $t \in \mathfrak{R}^+$, $x \in \mathfrak{X}$, when F is continuous. The proof is complete. □

Using the variation-of constants formula, results (a) and (b) of the following theorem were recently obtained in Refs. 24 and 13, respectively.

Theorem 5.4. Let a linear operator A: $(\mathfrak{D}(A) \subset \mathfrak{B}) \to \mathfrak{B}$, for $\mathfrak{B}$ a Banach space, be the infinitesimal generator of a linear dynamical system $\{S_A(t)\}_{t \geq 0}$ on $\mathfrak{B}$, with $\|S_A(t)x\| \leq Me^{\mu t}\|x\|$ for all $t \in \mathfrak{R}^+$, $x \in \mathfrak{B}$. Let F: $(\mathfrak{D}(F) \subset \mathfrak{B}) \to \mathfrak{B}$, $\mathfrak{D}(F) \supset \mathrm{Cl}\,\mathfrak{D}(A+F)$, be any bounded and continuous function such that $A + F$ generates a dynamical system $\{S(t)\}_{t \geq 0}$ on $\mathfrak{X} \equiv \mathrm{Cl}\,\mathfrak{D}(A+F)$ in the sense of Theorem III.5.1. Let $\gamma(x)$, $\mathfrak{Q}_b$, and $\mathfrak{Q}_c$ refer to $\{S(t)\}_{t \geq 0}$.

(a) If $S_A(t)$ is compact for every $t > 0$, then $\mathfrak{Q}_c = \mathfrak{Q}_b$.

(b) If $\mu < 0$ and F is a compact operator, then $\mathfrak{Q}_c = \mathfrak{Q}_b$.

Proof. We choose $x_0 \in \mathfrak{Q}_b$. As F is continuous, it follows from (c) of Lemma 5.2 that for every $n = 1, 2, \ldots,$ $t > 0$,

$$S(t+n)x_0 = S_A(n)S(t)x_0 + \int_0^n S_A(\tau)FS(t+n-\tau)x_0\,d\tau.$$

We also note that

$$\gamma(x_0) = \Big(\bigcup_{0 \leq t \leq n} S(t)x_0\Big) \cup \Big(\bigcup_{t > 0} S(t+n)x_0\Big),$$

and the first union is a compact set because $[0, n]$ is compact and $S(\cdot)x_0$: $\mathfrak{R}^+ \to \mathfrak{X}$ is continuous. As $x_0 \in \mathfrak{Q}_b$ and F is a bounded operator, there exist $N_1, N_2 \in \mathfrak{R}^+$ such that $\|S(t)x_0\| < N_1$ and $\|FS(t)x_0\| < N_2$ for all $t \in \mathfrak{R}^+$.

In (a) we assume that $S_A(t)\colon \mathcal{B} \to \mathcal{B}$ is compact for every $t \in \mathcal{R}^+$. We see that

$$S(t+n)x_0 = S_A(n)S(t)x_0 + \int_0^{1/n} S_A(\tau)FS(t+n-\tau)x_0 d\tau$$

$$+ S_A\left(\frac{1}{n}\right)\int_{1/n}^{n} S_A\left(\tau - \frac{1}{n}\right)FS(t+n-\tau)x_0 d\tau,$$

$$\|S(t)x_0\| < N_1,$$

$$\left\|\int_0^{1/n} S_A(\tau)FS(t+n-\tau)x_0 d\tau\right\| < (MN_2/n)\max\{1, e^{\mu/n}\},$$

$$\left\|\int_{1/n}^{n} S_A\left(\tau - \frac{1}{n}\right)FS(t+n-\tau)x_0 d\tau\right\| < MN_2\left(n - \frac{1}{n}\right)\max\{e^{\mu n - \mu/n}, 1\},$$

for all $t > 0$, $n = 1, 2, \ldots$; hence, defining

$$\mathcal{P}_n \equiv \left(\bigcup_{0<t<n} S(t)x_0\right) \cup \left(S_A(n)\gamma(x_0) + S_A\left(\frac{1}{n}\right)\right.$$

$$\left.\times \bigcup_{t>0} \int_{1/n}^{n} S_A\left(\tau - \frac{1}{n}\right)FS(t+n-\tau)x_0 d\tau\right),$$

$$\varepsilon_n \equiv (MN_2/n)\max\{1, e^{\mu/n}\}, \qquad n = 1, 2, \ldots,$$

we see that $\mathcal{P}_n$ is precompact and $\varepsilon_n \searrow 0$ as $n \to \infty$. As $\gamma(x_0) \subset \mathcal{P}_n + \mathcal{S}_{\varepsilon_n}(0) = \hat{\mathcal{S}}_{\varepsilon_n}(\mathcal{P}_n)$, $n = 1, 2, \ldots$, (c) of Theorem 5.1 implies that $\gamma(x_0)$ is precompact.

In (b) we assume instead that $\mu > 0$ and F is a compact operator, and we see that

$$S(t+n)x_0 = S_A(n)S(t)x_0 + \int_0^{n} S_A(\tau)FS(t+n-\tau)x_0 d\tau,$$

$$\|S_A(n)S(t)x_0\| < MN_1 e^{\mu n},$$

for all $t > 0$, $n = 1, 2, \ldots$. For any $\mathcal{S} \subset \mathcal{B}$, let $\overline{\text{co}}\,\mathcal{S}$ denote the smallest closed convex set containing $\mathcal{S}$; then by the definition of the Riemann integral,

$$\int_0^{n} S_A(\tau)FS(t+n-\tau)x_0 d\tau \in \overline{\text{co}} \bigcup_{0<\tau<n} nS_A(\tau)FS(t+n-\tau)x_0,$$

which implies that

$$\bigcup_{t>0}\int_0^n S_A(\tau)FS(t+n-\tau)x_0d\tau \subset \bigcup_{t>0}\left\{\overline{\text{co}}\bigcup_{0<\tau<n} nS_A(\tau)FS(t+n-\tau)x_0\right\}$$

$$\subset \overline{\text{co}}\bigcup_{0<\tau<n} nSA(\tau)\left\{\bigcup_{s>0} FS(s)x_0\right\} = \overline{\text{co}}\bigcup_{0<\tau<n} nS_A(\tau)F\gamma(x_0).$$

As $F\gamma(x_0)$ is precompact in $\mathfrak{B}$, and $[0, n]$ is compact in $\mathfrak{R}$, continuity of $S_A(\cdot)$: $\mathfrak{R}^+ \times \mathfrak{B} \to \mathfrak{B}$ can be used to show that $\bigcup_{0<\tau<n} nS_A(\tau)F\gamma(x_0)$ is precompact; hence, Masur's theorem (Ref. 32) shows that the set $\overline{\text{co}}\bigcup_{0<\tau<n} nS_A(\tau)F\gamma(x_0)$ is compact. Consequently, defining

$$\mathfrak{P}_n \equiv \left(\bigcup_{0<t<n} S(t)x_0\right) \cup \left(\bigcup_{t>0}\int_0^n S_A(\tau)FS(t+n-\tau)x_0d\tau\right),$$

$$\varepsilon_n \equiv MN_1e^{\mu n}, \qquad n = 1, 2, \ldots,$$

result (b) follows by the argument used in the proof of (a). The proof is complete. □

Result (a) of Theorem 5.4 requires $S_A(t)$ to be compact for all $t > 0$; this condition is closely related to the theory of holomorphic semigroups (Ref. 15). Here, we only remark that a linear partial differential equation of parabolic type normally leads to a dynamical system $\{S_A(t)\}_{t\geq 0}$ satisfying this condition. The following example utilizes result (a) of Theorem 5.4.

Example 5.5. Again consider the nonlinear heat conduction problem of Example III.5.2,

$$\frac{\partial}{\partial t}\theta(\eta, t) = \frac{\partial^2}{\partial\eta^2}\theta(\eta, t) + f(\theta(\eta, t)), \qquad 0 \leq \eta \leq 1, \quad t \geq 0,$$

$$\theta(0, t) = 0 = \theta(1, t), \qquad t \geq 0,$$

under the (different) assumptions that $f(0) = 0$ and $f \in \mathcal{C}^2(\mathfrak{R})$, with $|f''(\xi)| \leq \beta$ for all $\xi \in \mathfrak{R}$, some $\mathfrak{B} \in \mathfrak{R}^+$; we also assume f to be such that $A + F$ generates a dynamical system $\{S(t)\}_{t\geq 0}$ on $\mathfrak{X}$, in the sense of Theorem III.5.1, where

$$\mathfrak{X} \equiv \{x \in \mathfrak{W}_2^1(0, 1) | x(0) = 0 = x(1)\}, \qquad \|x\|^2 \equiv \int_0^1 |\partial x(\eta)|^2 d\eta,$$

$$Ax \equiv \partial^2 x, \qquad x \in \mathfrak{D}(A) \equiv \{x \in \mathfrak{X} | \partial^2 x \in \mathfrak{X}\},$$

$$Fx(\eta) \equiv f(x(\eta)) \quad \text{a.e. } \eta \in [0, 1], \quad x \in \mathfrak{X}.$$

Exercise III.3.3 shows that $-\pi^2 I - A$ is accretive, $\mathfrak{R}(I - \lambda A) = \mathfrak{X}$ for all $\lambda > 0$, and A is the infinitesimal generator of a linear dynamical system $\{S_A(t)\}_{t\geq 0}$ on $\mathfrak{X}$. It is known that $S_A(t)$ is a compact operator for every $t > 0$. Also noting that $F(0) = 0$ and

$$\max_{\eta} |x(\eta)| \leqslant \frac{1}{2}\int_0^1 |\partial x(\eta)|d\eta \leqslant \frac{1}{2}\|x\|,$$

$x \in \mathfrak{X}$, we see that

$$\|Fx - F\hat{x}\|^2 \equiv \int_0^1 |f'(x(\eta))\cdot \partial x(\eta) - f'(\hat{x}(\eta))\cdot \partial\hat{x}(\eta)|^2 d\eta$$

$$\leqslant 2|f'(0) + \beta\|x\|\,|^2\int_0^1|\partial x(\eta) - \partial\hat{x}(\eta)|^2 d\eta + 2\beta^2\|x - \hat{x}\|^2\int_0^1|\partial\hat{x}(\eta)|^2 d\eta$$

$$\leqslant 2(|f'(0) + \beta\|x\|\,|^2 + 2\beta^2\|\hat{x}\|^2)\|x - \hat{x}\|^2$$

for all $x, \hat{x} \in \mathfrak{X}$; hence, $F: \mathfrak{X} \to \mathfrak{X}$ is continuous and bounded, and (a) of Theorem 5.4 shows that bounded positive orbits of $\{S(t)\}_{t\geqslant 0}$ are precompact ($\mathfrak{Q}_b = \mathfrak{Q}_c$).

As $A + F$ is a closed operator, (iii) of Theorem III.5.2 shows that $\mathfrak{Q}_e \subset \mathfrak{D}(A)$, where $\mathfrak{Q}_e$ is the set of all equilibria, hence,

$$\mathfrak{Q}_e = \{x \in \mathfrak{D}(A) | \partial^2 x(\eta) + f(x(\eta)) = 0 \text{ a.e. } \eta \in [0, 1]\}.$$

Let us now define a continuous function $V: \mathfrak{X} \to \mathfrak{R}$,

$$V(x) \equiv \|x\|^2 - 2\int_0^1\left[\int_0^{x(\eta)} f(\xi)d\xi\right]d\eta, \qquad x \in \mathfrak{X},$$

and employ Theorem 2.1 to obtain

$$\dot{V}(x) = -2\int_0^1|\partial^2 x(\eta) + f(x(\eta))|^2 d\eta, \qquad x \in \mathfrak{D}(A),$$

$$\leqslant -W(x), \qquad x \in \mathfrak{X},$$

where the lower semicontinuous function $W: \mathfrak{X} \to \overline{\mathfrak{R}}^+$ is defined by

$$W(x) \equiv 2\int_0^1|\partial^2 x(\eta) + f(x(\eta))|^2 d\eta, \qquad x \in \mathfrak{X} \cap \mathfrak{W}_2^2(0, 1),$$

$$W(x) \equiv \infty, \qquad x \in \mathfrak{X},\ x \notin \mathfrak{W}_2^2(0, 1).$$

We see that $\mathfrak{M}_2 \equiv \{x \in \mathfrak{X} | W(x) = 0\} = \mathfrak{Q}_e$, and $\mathfrak{Q}_e$ is not only positive invariant but also invariant. The Invariance Principle (Theorem 4.2) now shows that $\Omega(x) \subset \mathfrak{Q}_e$ for every $x \in \mathfrak{Q}_b = \mathfrak{Q}_c$.

If there exist $\alpha, \delta \in \mathfrak{R}$ (depending on f) such that

$$\overline{\mathfrak{G}}_\alpha \equiv \{x \in \mathfrak{X} | V(x) \leqslant \alpha\} \subset \mathfrak{S}_\delta(0) \equiv \{x \in \mathfrak{X} |\ \|x\| < \delta\},$$

then the positive invariance of $\overline{\mathfrak{G}}_\alpha$ implies that $\overline{\mathfrak{G}}_\alpha \subset \mathfrak{Q}_b = \mathfrak{Q}_c$, and Theorem 4.2 implies that $S(t)x \to \mathfrak{Q}_e$ as $t \to \infty$ for every $x \in \overline{\mathfrak{G}}_\alpha$; if $\mathfrak{Q}_e$ is a countable set, then to every $x \in \overline{\mathfrak{G}}_\alpha$ there corresponds $x_e \in \mathfrak{Q}_e \cap \overline{\mathfrak{G}}_\alpha$ such that $S(t)x \to x_e$ as $t \to \infty$.

Notice that our conclusions here are qualitatively different from those of Example 4.3 for a similar problem; in Example 4.3 the approach to equilibrium was in the $\mathcal{C}$-topology whereas here it is in the (stronger) $\mathcal{W}_2^1$-topology. Also notice that here V is continuous and may be used with Liapunov's Direct Method to investigate stability in the $\mathcal{W}_2^1$-topology; in Example 4.3, V was not continuous and stability was defined in the $\mathcal{C}$-topology.

Exercise 5.7. Referring to Example 5.5, assume that

$$\int_0^\varepsilon f(\xi)d\xi \leqslant \alpha\varepsilon^2 \qquad \text{for all } \varepsilon \in \mathfrak{R}, \text{ some } \alpha \in \left[0, \pi^2/2\right).$$

Show that all positive orbits are bounded ($\mathcal{Q}_b = \mathfrak{X}$), $x_e = 0$ is stable, and $S(t)x \to \mathcal{Q}_e$ as $t \to \infty$ for every $x \in \mathfrak{X}$. *Hint*: Show that $V(x) \geqslant (1 - 2\alpha/\pi^2)\|x\|^2$ for all $x \in \mathfrak{X}$.

Exercise 5.8. Referring to Example 5.5, show that $x_e = 0$ is stable and all positive orbits are bounded if $f'(\xi) \leqslant \alpha$ for all $\xi \in \mathfrak{R}$, some $\alpha \leqslant \pi^2$. Show that $x_e = 0$ is globally exponentially stable, with exponent $(-\pi^2 + \alpha)t$, if $\alpha < \pi^2$. *Hint*: Define $\tilde{V}(x) \equiv \|x\|^2$, $x \in \mathfrak{X}$.

The advantage of result (b) of Theorem 5.4 is that $S_A(t)$: $\mathfrak{B} \to \mathfrak{B}$ need not be compact, $t > 0$. This is fortunate if $\{S_A(t)\}_{t \geqslant 0}$ arises from a linear partial differential equation of hyperbolic type, as in the following example.

Example 5.6. Consider the sine-Gordon equation (Ref. 11),

$$\frac{\partial^2}{\partial t^2}u(\eta, t) + 2\alpha\frac{\partial}{\partial t}u(\eta, t) - \frac{\partial^2}{\partial \eta^2}u(\eta, t) - \beta\sin(u(\eta, t)) = 0, \quad 0 \leqslant \eta \leqslant 1, \quad t \geqslant 0,$$

$$u(0, t) = 0 = u(1, t), \qquad t \geqslant 0,$$

where $\alpha > 0$, $\beta \in \mathfrak{R}$. We define

$$\hat{\mathcal{W}}_2^n(0, 1) \equiv \{x \in \mathcal{W}_2^n(0, 1) | \partial^m x(0) = 0 = \partial^m x(1);\ m = 0, 2, 4, \ldots, \leqslant n - 1\},$$

and

$$\mathfrak{X} \equiv \hat{\mathcal{W}}_2^1(0, 1) \times \mathfrak{L}_2(0, 1), \qquad \|x\|_{\mathfrak{X}}^2 \equiv \int_0^1 [|\partial u(\eta)|^2 + |v(\eta)|^2]d\eta,$$

$$Ax \equiv (v, -2\alpha v + \partial^2 u), \qquad x \in \mathfrak{D}(A) \equiv \hat{\mathcal{W}}_2^2(0, 1) \times \hat{\mathcal{W}}_2^1(0, 1),$$

$$Fx(\eta) \equiv (0, \beta\sin u(\eta)) \quad \text{a.e. } \eta \in [0, 1], \quad x = (u, v) \in \mathfrak{X}.$$

By Example 4.4 we know that $-A$ is accretive and A: $(\mathfrak{D}(A) \subset \mathfrak{X}) \to \mathfrak{X}$ is the infinitesimal generator of a linear dynamical system $\{S_A(t)\}_{t \geqslant 0}$ on $\mathfrak{X}$, such that $\|S_A(t)x\|_{\mathfrak{X}} \leqslant Me^{\mu t}\|x\|_{\mathfrak{X}}$ for all $t \geqslant 0$, $x \in \mathfrak{X}$, some $M > 0$, $\mu < 0$.

We see as in Example 2.1 that $A + F$ generates a dynamical system $\{S(t)\}_{t\geq 0}$ on $\mathfrak{X}$, and $\|S(t)x - S(t)\hat{x}\|_{\mathfrak{X}} \leq e^{|\beta|t/\pi}\|x - \hat{x}\|_{\mathfrak{X}}$ for all $x, \hat{x} \in \mathfrak{X}$, $t \in \mathfrak{R}^+$. Also noting that F: $\mathfrak{X} \to \mathfrak{X}$ is continuous and

$$\|Fx\|^2_{\mathfrak{W}_2^2 \times \mathfrak{W}_2^1} = \beta^2 \int_0^1 [|\sin u(\eta)|^2 + |\cos u(\eta)|^2] d\eta$$

$$\leq 2\beta^2 \int_0^1 |u(\eta)|^2 d\eta \leq (2\beta^2/\pi^2)\|x\|^2_{\mathfrak{X}},$$

it follows from the compactness of the injection $\hat{I}$: $\mathfrak{W}_2^2 \times \mathfrak{W}_2^1 \to \mathfrak{W}_2^1 \times \mathfrak{L}_2$ that F: $\mathfrak{X} \to \mathfrak{X}$ is compact.

Also defining V: $\mathfrak{X} \to \mathfrak{R}$ as in Example 2.1, we have

$$V(x) \equiv \|x\|^2_{\mathfrak{X}} + 2\beta \int_0^1 \cos(u(\eta)) d\eta,$$

$$\dot{V}(x) \leq -4\alpha \int_0^1 |v(\eta)|^2 d\eta \equiv -W(x),$$

for all $x = (u, v) \in \mathfrak{X}$, and this shows that all positive orbits are bounded.

Now applying (b) of Theorem 5.4, it follows that all positive orbits are precompact. We are interested in the largest positive invariant set $\mathfrak{M}^+$ in $\mathfrak{M}_2 \equiv \{x \in \mathfrak{X} | W(x) = 0\} = \{(u, v) \in \mathfrak{X} | v = 0\}$. Using the evolution equation $\dot{x}(t) = (A + F)x(t)$ a.e. $t \in \mathfrak{R}^+$, $x(0) \in \mathfrak{D}(A)$, it is easy to see that

$$\mathfrak{M}^+ \cap \mathfrak{D}(A) = \{(u, v) \in \mathfrak{D}(A) | v(\eta) = 0 = \partial^2 u(\eta) + \beta \sin u(\eta) \text{ a.e. } \eta \in (0, 1)\},$$

which is the set $\mathfrak{Q}_e$ of all equilibria. Using a distributional equation (Ref. 15) satisfied by every motion, it can be shown that $\mathfrak{M}^+ = \mathfrak{Q}_e$. The set $\mathfrak{Q}_e$ is finite for each $\beta \in \mathfrak{R}$, and $\mathfrak{Q}_e = \{0\}$ if $|\beta| < \pi^2$. Applying the Invariance Principle (Theorem 4.2), we conclude that to each $x \in \mathfrak{X}$ there corresponds $x_e \in \mathfrak{Q}_e$ such that $S(t)x \to x_e$ as $t \to \infty$; if $|\beta| < \pi^2$, then $x_e = 0$ is globally asymptotically stable.

6. Comments and Extensions

In this chapter we have studied the behavior of motions of dynamical systems, and our basic approach has been that of Liapunov. Useful Liapunov functions are not always easy to construct, but we have seen that a Liapunov function, once found, can be very informative. For linear dynamical systems, it is worth noticing that once a single Liapunov function has been found, it may be easy to produce a large family of Liapunov functions. The following theorem is related to Exercise I.9.4.

Theorem 6.1. Let A: $(\mathfrak{D}(A) \subset \mathfrak{X}) \to \mathfrak{X}$, $\mathfrak{X}$ a Banach space, be the infinitesimal generator of a linear dynamical system $\{S(t)\}_{t\geq 0}$ on $\mathfrak{X}$, and let V: $\mathfrak{X} \to \overline{\mathfrak{R}}$ be a Liapunov function for $\{S(t)\}_{t\geq 0}$ on $\mathcal{G} \subset \mathfrak{X}$. For some $p = 0, 1, \ldots,$ let $\hat{\mathfrak{X}}$ be the Banach space $(\mathfrak{D}(A^p), \|\cdot\|_p)$ with either

$\|x\|_p \equiv \Sigma_{m=0}^{p}\|A^m x\|$ or $\|x\|_p^2 \equiv \Sigma_{m=0}^{p}\|A^m x\|^2$. Consider any bounded linear operator $G\colon \hat{\mathfrak{X}} \to \mathfrak{X}$ such that $Gx \in \mathfrak{D}(A)$ and $GAx = AGx$ for all $x \in \mathfrak{D}(A^{p+1})$. Then $\mathfrak{R}(G)$ and $\mathrm{Cl}\,\mathfrak{R}(G)$ are positive invariant; moreover, if there exists a lower semicontinuous function $\tilde{V}\colon \mathfrak{X} \to \overline{\mathfrak{R}}$ such that $\tilde{V}(x) = V(Gx)$ for all $x \in \mathfrak{D}(A^p)$, then $\tilde{V}$ is a Liapunov function on $\tilde{\mathcal{G}} \equiv \{x \in \mathfrak{D}(A^p) | Gx \in \mathcal{G}\}$ with $\dot{\tilde{V}}(x) = \dot{V}(Gx) \leqslant 0$ for all $x \in \tilde{\mathcal{G}}$.

Proof. With $J_\lambda \equiv (I - \lambda A)^{-1}$, $0 \in (0, \lambda_0)$, we see that our assumptions on G imply that $GJ_\lambda x = J_\lambda Gx$ for all $x \in \mathfrak{D}(A^p)$, $\lambda \in (0, \lambda_0)$; hence, for $x \in \mathfrak{D}(A^p)$, $t \geqslant 0$,

$$S(t)Gx = \lim_{n\to\infty} J_{t/n}^n Gx = \lim_{n\to\infty} GJ_{t/n}^n x.$$

As $G\colon \hat{\mathfrak{X}} \to \mathfrak{X}$ is continuous, Theorem III.4.1 implies that the last limit is $GS(t)x$; hence, G and $S(t)$ commute on $\mathfrak{D}(A^p)$ for $t \in \mathfrak{R}^+$. It follows that if $y \in \mathfrak{R}(G)$, there exists $x \in \mathfrak{D}(A^p)$ such that $S(t)y = GS(t)x \in \mathfrak{R}(G)$ for all $t > 0$; hence, $\mathfrak{R}(G)$ is positive invariant. As a positive invariant set has positive invariant closure, $\mathrm{Cl}\,\mathfrak{R}(G)$ is positive invariant as well.

If the lower semicontinuous function $\tilde{V}\colon \mathfrak{X} \to \overline{\mathfrak{R}}$ is such that $\tilde{V}(x) = V(Gx)$, $x \in \mathfrak{D}(A^p)$, we see that for all $x \in \mathfrak{D}(A^p)$ such that $|V(Gx)| \neq \infty$ we have

$$\dot{\tilde{V}}(x) \equiv \liminf_{t\searrow 0} \frac{1}{t}\left[\tilde{V}(S(t)x) - \tilde{V}(x)\right]$$

$$= \liminf_{t\searrow 0} \frac{1}{t}\left[V(S(t)Gx) - V(Gx)\right] = \dot{V}(Gx).$$

By the definition of a Liapunov function (Definition 1.1), the proof is complete. □

In Theorem III.4.1 we saw that when a linear dynamical system $\{S(t)\}_{t\geqslant 0}$ has an unbounded infinitesimal generator $A\colon (\mathfrak{D}(A) \subset \mathfrak{X}) \to \mathfrak{X}$, $\mathfrak{X}$ a Banach space, then various restrictions of $\{S(t)\}_{t\geqslant 0}$ are dynamical systems on a family of Banach spaces $\hat{\mathfrak{X}}$ which are algebraic and topological subspaces of $\mathfrak{X}$; moreover, the original exponential estimate carries over from $\mathfrak{X}$ to $\hat{\mathfrak{X}}$. It is interesting that all Liapunov functions carry over in a similar manner on this family.

Theorem 6.2. Let $A\colon (\mathfrak{D}(A) \subset \mathfrak{X}) \to \mathfrak{X}$, $\mathfrak{X}$ a Banach space, be the infinitesimal generator of a linear dynamical system $\{S(t)\}_{t\geqslant 0}$ on $\mathfrak{X}$, and let $V\colon \mathfrak{X} \to \mathfrak{R}$ be a Liapunov function for $\{S(t)\}_{t\geqslant 0}$ on $\mathcal{G} \subset \mathfrak{X}$. For some $p = 1, 2, \ldots,$ let $\hat{S}(t)$ denote the restriction of $S(t)$ to $\hat{\mathfrak{X}}$, where $\hat{\mathfrak{X}}$ is the Banach space $(\mathfrak{D}(A^p), \|\cdot\|_p)$ with either $\|x\|_p \equiv \Sigma_{m=0}^{p}\|A^m x\|$ or $\|x\|_p^2 \equiv$

$\sum_{m=0}^{p}\|A^m x\|^2$. Consider any bounded linear operator $G: \hat{\mathfrak{X}} \to \mathfrak{X}$ such that $GAx = AGx$ for all $x \in \mathfrak{D}(\hat{A}) \equiv \mathfrak{D}(A^{p+1}) \subset \hat{\mathfrak{X}}$. Defining $\hat{V}(x) \equiv V(Gx)$, $x \in \hat{\mathfrak{X}}$, the function $\hat{V}: \hat{\mathfrak{X}} \to \overline{\mathfrak{R}}$ is a Liapunov function on $\hat{\mathcal{G}} \equiv \{x \in \hat{\mathfrak{X}} | Gx \in \mathcal{G}\}$ for the linear dynamical system $\{\hat{S}(t)\}_{t\geq 0}$ on $\hat{\mathfrak{X}}$ generated by the restriction of $\hat{A}$ of A to $\mathfrak{D}(\hat{A}) \equiv \mathfrak{D}(A^{p+1})$. Moreover, $\dot{\hat{V}}(x) = \dot{V}(Gx)$ for every $x \in \hat{\mathcal{G}}$.

Proof. As $V: \mathfrak{X} \to \overline{\mathfrak{R}}$ is lower semicontinuous, and $G: \hat{\mathfrak{X}} \to \mathfrak{X}$ is continuous, it follows that $\hat{V}: \hat{\mathfrak{X}} \to \overline{\mathfrak{R}}$ is lower semicontinuous. All conclusions now follow from Theorems 6.1 and III.4.1. □

In Sections 4 and 5 we saw how useful the Invariance Principle can be in studying the asymptotic behavior of a motion $S(\cdot)x: \mathfrak{R}^+ \to \mathfrak{X}$, provided that the positive orbit $\gamma(x)$ is known to be precompact. It is worth asking if any similar result applies when $\gamma(x)$ is not, or has not been shown to be, precompact. Results in this direction are rather weak.

Theorem 6.3. Let $\{S(t)\}_{t\geq 0}$ be a dynamical system on a metric space $\mathfrak{X}$, and let $V: \mathfrak{X} \to \overline{\mathfrak{R}}$ be a Liapunov function on a set $\mathcal{G} \subset \mathfrak{X}$. Let $\mathfrak{Y}$ be a complete metric space such that $\mathfrak{X} \subset \mathfrak{Y}$ and the injection $\hat{I}: \mathfrak{X} \to \mathfrak{Y}$ is compact and continuous. Let there exist lower semicontinuous functions $\hat{V}: \mathfrak{Y} \to \overline{\mathfrak{R}}$, $\hat{W}: (\mathrm{Cl}_{\mathfrak{Y}}\mathcal{G} \subset \mathfrak{Y}) \to \overline{\mathfrak{R}}^+$, such that

$$V(x) = \hat{V}(x), \qquad x \in \mathcal{G} \subset \mathfrak{Y},$$

$$\dot{V}(x) \leq -\hat{W}(x), \qquad x \in \mathcal{G} \subset \mathfrak{Y},$$

$$\hat{V}(y) > -\infty, \qquad y \in \mathrm{Cl}_{\mathfrak{Y}}\mathcal{G}.$$

If $\gamma(x) \subset \mathcal{G}$ and $\gamma(x)$ is $\mathfrak{X}$ – bounded, then $\hat{d}_{\mathfrak{Y}}(S(t)x, \mathfrak{M}_3) \to 0$ as $t \to \infty$, where

$$\mathfrak{M}_3 \equiv \{y \in \mathrm{Cl}_{\mathfrak{Y}}\mathcal{G} \,|\, \hat{\mathfrak{W}}(y) = 0\} \subset \mathfrak{Y}.$$

If, in addition, $\hat{V}: \mathfrak{Y} \to \overline{\mathfrak{R}}$ is $\mathfrak{R}$-valued and continuous, then $\hat{d}_{\mathfrak{Y}}(S(t)x, \mathfrak{M}_3 \cap \hat{V}^{-1}(\beta)) \to 0$ as $t \to \infty$, for some $\beta \in \mathfrak{R}$, $\beta \leq V(x)$.

Proof. Given $x \in \mathcal{G}$ such that $\gamma(x) \subset \mathcal{G}$ and $\gamma(x)$ is $\mathfrak{X}$-bounded, define $\Omega_{\mathfrak{Y}}(x) \equiv \cap_{\tau>0}\mathrm{Cl}_{\mathfrak{Y}}\gamma(S(\tau)x) \subset \mathfrak{Y}$ and note that $\hat{I}S(\cdot)x: \mathfrak{R}^+ \to \mathfrak{Y}$ is continuous, $\gamma(x)$ is $\mathfrak{Y}$-precompact. Working in $\mathfrak{Y}$ and reasoning as in the proof of Theorem 4.4, it follows that $\Omega_{\mathfrak{Y}}(x)$ is $\mathfrak{Y}$-compact, nonempty, $\mathfrak{Y}$-connected, and $d_{\mathfrak{Y}}(S(t)x, \Omega_{\mathfrak{Y}}(x)) \to 0$ as $t \to \infty$. Replacing V and W by $\hat{V}$ and $\hat{W}$, respectively, all conclusions now follow by essentially the same arguments used in the proof of Theorem 4.2; hence, the proof is complete. □

Although Theorem 6.3 requires only $\mathcal{X}$-boundedness of $\gamma(x)$, rather than $\mathcal{X}$-precompactness, it suffers from two severe disadvantages relative to the Invariance Principle. First, convergence of $S(t)x$ to $\mathcal{M}_3$ is in the $\mathcal{Y}$-topology as $t \to \infty$, rather than in the original (and generally stronger) $\mathcal{X}$-topology. Second, $\mathcal{M}_3$ may be quite large; it is not even certain that $\Omega_{\mathcal{Y}}(x) \subset \mathcal{X}$, and invariance (or even positive invariance) ideas do not apply to $\Omega_{\mathcal{Y}}(x)$. However, if $\gamma(x)$ is $\mathcal{X}$-bounded but not $\mathcal{X}$-precompact, then Theorem 6.3 says something about asymptotic behavior, whereas the Invariance Principle does not.

We have said very little about the determination of periodic motions of a dynamical system $\{S(t)\}_{t \geq 0}$ on a metric space $\mathcal{X}$. Considering this question, suppose that by using the Invariance Principle (Theorem 4.2) we are able to determine sets $\mathcal{G}$ and $\mathcal{M}$, with $\mathcal{M} \subset \mathcal{G} \subset \mathcal{X}$ and $\mathcal{M}$ closed, such that $x \in \mathcal{G}$ implies $S(t)x \to \mathcal{M}$ as $t \to \infty$. Then it is apparent that if a motion $S(\cdot)y$ is periodic for some $y \in \mathcal{G}$, we must have $y \in \mathcal{M}$; it follows that we need only study $\mathcal{M}$ in order to determine which (if any) initial $y \in \mathcal{G}$ lead to periodic motions. It is also apparent that similar reasoning can be applied in conjunction with Theorem 6.3. Hence, we do possess effective means of studying periodic motions.

In Section III.6 we mentioned that many of the ideas of Chapter III admit extensions to nonautonomous evolution equations; here we similarly mention that most of the ideas of the present chapter can be extended to processes. The study of processes involves several additional difficulties, currently the subject of very active investigation (Refs. 1, 2, 19, 24, 33–35). Liapunov functions (of slightly extended form) continue to provide the basic instruments for obtaining hard information in applications involving processes.

References

1 Dafermos, C. M., Uniform processes and semicontinuous Liapunov functionals, *Journal of Differential Equations*, Vol. 11, pp. 401–415, 1972.

2. Dafermos, C. M., Applications of the invariance principle for compact processes. II: Asymptotic behavior of solutions of a hyperbolic conservation law, *Journal of Differential Equations*, Vol. 11, pp. 416–424, 1972.

3. Walker, J. A., Some results on Liapunov functions and generated dynamical systems, *Journal of Differential Equations*, Vol. 30, pp. 424–440, 1978.

4 McShane, E. J., *Integration*, Princeton University Press, Princeton, New Jersey, 1947.

5. Datko, R., Extending a theorem of A. M. Liapunov to Hilbert space, *Journal of Mathematical Analysis and Applications*, Vol. 32, pp. 610–616, 1970.

6. Pazy, A., On the applicability of Liapunov's theorem in Hilbert space, *SIAM Journal on Mathematical Analysis*, Vol. 3, pp. 291–294, 1972.

7. Pao, C. V., Semigroups and asymptotic stability of nonlinear differential equations, *SIAM Journal on Mathematical Analysis*, Vol. 3, pp. 371–379, 1972.

8. WONG, J. W., A remark on a theorem of Liapunov, *Canadian Mathematical Bulletin*, Vol. 13, pp. 141–143, 1970.
9. WALKER, J. A., On the application of Liapunov's Direct Method to linear dynamical systems, *Journal of Mathematical Analysis and Applications*, Vol. 53, pp. 187–220, 1976.
10. YOSHIZAWA, T., *Stability by Liapunov's Direct Method*, Mathematical Society of Japan, Tokyo, 1966.
11. CALLEGARI, A. J., and REISS, E. L., Nonlinear stability problems for the sine-Gordon equation, *Journal of Mathematical Physics*, Vol. 14, pp. 267–276, 1973.
12 DICKEY, R. W., Stability theory for the damped sine-Gordon equation, *SIAM Journal on Applied Mathematics*, Vol. 30, pp. 248–262, 1976.
13. WEBB, G. F., A bifurcation problem for a nonlinear hyperbolic partial differential equation, *SIAM Journal on Mathematical Analysis*, Vol 10, pp. 922–932, 1979.
14. CHAFEE, N., and INFANTE, E. F., A bifurcation problem for a nonlinear parabolic equation, *Applicable Analysis*, Vol. 4, pp. 17–37, 1974.
15. YOSIDA, K., *Functional Analysis*, Springer-Verlag, New York, 1971.
16. SLEMROD, M., Asymptotic behavior of C_0-semigroups as determined by the spectrum of the generator, *Indiana University Mathematics Journal*, Vol. 25, pp. 783–792, 1976.
17. INFANTE, E. F., and WALKER, J. A., A Liapunov functional for a scalar differential difference equation, *Proceedings of the Royal Society of Edinburgh*, Vol. 79A, pp. 307–316, 1977.
18. BIRKHOFF, G. D., *Dynamical Systems*, American Mathematical Society Colloquium Publications, Vol. 9, Providence, Rhode Island, 1927; revised edition, 1966.
19. LASALLE, J. P., *The Stability of Dynamical Systems*, CBMS Regional Conference Series in Applied Mathematics, Society for Industrial and Applied Mathematics, Philadelphia, 1976.
20. LASALLE, J. P., and LEFSCHETZ, S., *Stability by Liapunov's Direct Method*, Academic Press, New York, 1961.
21. HALE, J. K., Dynamical systems and stability, *Journal of Mathematical Analysis and Applications*, Vol. 26, pp. 39–59, 1969.
22. DAFERMOS, C. M., and SLEMROD, M., Asymptotic behavior of nonlinear contraction semigroups, *Journal of Functional Analysis*, Vol. 13, pp. 97–106, 1973.
23. WALKER, J. A., and INFANTE, E. F., Some results on the precompactness of orbits of dynamical systems, *Journal of Mathematical Analysis and Applications*, Vol. 51, pp. 56–67, 1975.
24. BALL, J., On the asymptotic behavior of generalized processes, with applications to nonlinear evolution equations, *Journal of Differential Equations*, Vol. 27, pp. 224–265, 1978.
25. SLEMROD, M., Asymptotic behavior of a class of abstract dynamical systems, *Journal of Differential Equations*, Vol. 7, pp. 584–600, 1970.
26. KURATOWSKI, K., *Topology I*, Academic Press, New York, 1966.
27. DARBO, G., Punti uniti in transformazioni con dominio non compatto, *Rendiconti del Seminario Matematico dell'Universita di Padova*, Vol. 24, pp. 84–92, 1955.
28. CARROLL, R. W., *Abstract Methods in Partial Differential Equations*, Harper and Row, New York, 1969.
29. WALKER, J. A., On state transformation and stability analysis of distributed-parameter systems, *Quarterly of Applied Mathematics*, Vol. 32, pp. 333–336, 1974.
30. CRANDELL, M. G., A generalized domain for semigroup generators, *Proceedings of the American Mathematical Society*, Vol. 37, pp. 434–440, 1973.
31. CRANDALL, M. G., and LIGGETT, T. M., Generation of nonlinear semigroups of nonlinear transformations on general Banach spaces, *American Journal of Mathematics*, Vol. 93, pp. 265–298, 1971.

32. Dunford, N., and Schwartz, J. T., *Linear Operators*, Vol. 1, Wiley-Interscience, New York, 1958.
33. Artstein, Z., The limiting equations of nonautonomous ordinary differential equations, *Journal of Differential Equations*, Vol. 25, pp. 184–202, 1977.
34. Miller, R. K., and Sell, G. R., *Volterra Integral Equations and Topological Dynamics*, Memoirs of the American Mathematical Society, Vol. 102, Providence, Rhode Island, 1970.
35. Miller, R. K., and Sell, G. R., Topological dynamics and its relations to integral equations and nonautonomous systems, *Dynamical Systems: An International Symposium*, Vol. I. Edited by L. Cesari, J. K. Hale, and J. P. LaSalle, Academic Press, New York, pp. 223–249, 1976.

V

Applications and Special Topics

We have consistently implied that most physical systems can be associated with mathematical objects which we have called dynamical systems (autonomous case) or processes (nonautonomous case). It follows that the techniques described in Chapters III and IV can be very useful in analyzing the time behavior of autonomous physical systems; in Sections III.6 and IV.6 we also mentioned that these techniques admit nonautonomous extensions. With increasing frequency, these methods and ideas are being applied to concrete problems in physics, economics, and engineering. We shall describe a small sample of such applications in the first five sections of this chapter.

We have concentrated on the autonomous case because extensions to the nonautonomous case involve additional difficulties, some of which are quite complex and are currently under intensive investigation. Our techniques can also be "extended" in another direction; in Section 6 we discuss discrete dynamical systems, showing them to be much simpler. Discrete dynamical systems are "discrete" in time, meaning that we consider only nonnegative integer values for time t, rather than all $t \in \mathfrak{R}^+$ as for a dynamical system. Discrete dynamical systems are often interesting as approximations of dynamical systems, and they have other uses as well.

Another approximation question arises when one is considering a dynamical system on a (linear) metric space having infinite dimension; in applications one often attempts to approximate such a dynamical system by another dynamical system on a finite-dimensional state space. We shall discuss one approach to this problem in Section 7.

1. A Feedback Control Problem

For those unfamiliar with control theory, we shall sketch the general problem of controller design (see Figure 7). We are given a physical device called the *plant*, denoting its state at $t \in \mathcal{R}^+$ by $x(t) \in \mathcal{X}_p$. The plant state depends on $x(0)$ and certain *control forces* $u(\cdot)$: $\mathcal{R}^+ \to \mathcal{U}$, whose values we are free to select. At each $t \in \mathcal{R}^+$, there is a *plant output* $y(t) \in \mathcal{Y}$, depending only on the present value of $x(t)$, and we wish $y(\cdot)$: $\mathcal{R}^+ \to \mathcal{Y}$ to "follow" a *reference signal* $r(\cdot)$: $\mathcal{R}^+ \to \mathcal{Y}$, in some given sense, where $r(\cdot)$ is any member of some given class of reference signals; this is the *control objective*. In many cases we are able to monitor some *feedback signal* $v(t) \in \mathcal{V}$, with $v(t)$ depending only on the plant state $x(t)$; $v(t)$ may or may not be $y(t)$. The *controller design* problem is to construct another physical device (the *controller*), having state $z(t) \in \mathcal{Z}$ at $t \in \mathcal{R}^+$, which accepts $r(t)$ and $v(t)$ as inputs and produces $u(t)$ as an output, with $u(t)$ depending only on $r(t)$, $v(t)$, and $z(t)$, in such a manner that the control objective is automatically achieved for each reference signal $r(\cdot)$ in the given class. Connection of the controller with the plant produces a *control system* with state $(x(t), z(t)) \in \mathcal{X}_p \times \mathcal{Z}$ at $t \in \mathcal{R}^+$, having input $r(t)$ and output $y(t)$. If the controller design is successful, $y(\cdot)$: $\mathcal{R}^+ \to \mathcal{Y}$ will resemble $r(\cdot)$: $\mathcal{R}^+ \to \mathcal{Y}$ in the desired manner, for every $r(\cdot)$ in the given class of reference signals. If this class consists only of constant signals, $r(t) \equiv r_0$ for all $t \in \mathcal{R}^+$, then the controller is called a *regulator*.

As the controller is a physical device that we must design and construct, a desire for simplicity and economy usually leads to a controller's having a finite-dimensional state space $\mathcal{Z} = \mathcal{R}^p$; in fact, we generally want the dimension p to be as low as possible. However, the plant is given and unchangeable (e.g., an engine), and $\mathcal{X}_p$ might be a general metric space. Most present design theory is based on the assumption that $\mathcal{X}_p = \mathcal{R}^n$ (Refs. 1, 2); however, a great deal of recent work has been directed to the case in which $\mathcal{X}_p$ is a general Banach space (Refs. 3–9). Here we shall give an example of this type of control problem, described earlier in Refs. 10 and 11.

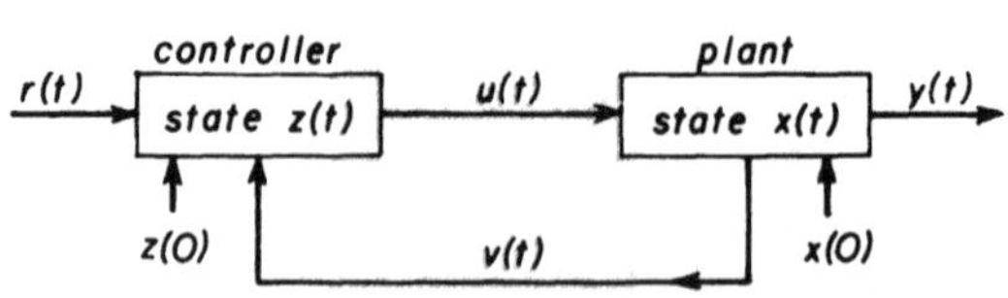

Fig. 7. A control system.

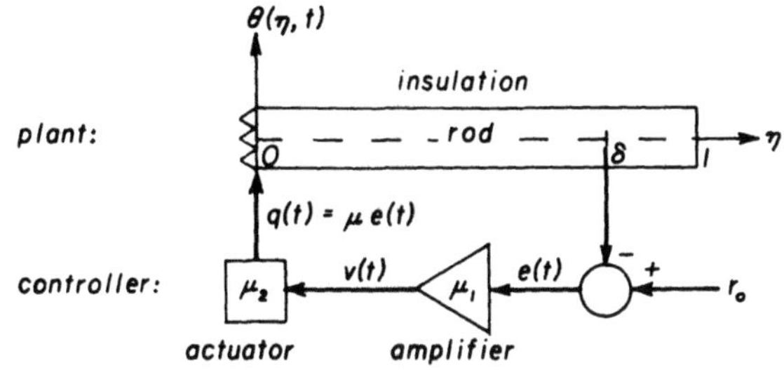

Fig. 8. Temperature control of a rod.

We consider the problem of regulating the temperature distribution of an insulated rod to a desired uniform temperature distribution r_0, regardless of the initial temperature distribution $\theta(\eta, 0)$, $0 \leqslant \eta \leqslant 1$. Referring to Figure 8, we assume that the temperature can be continuously monitored at some given location δ along the rod $(0 < \delta \leqslant 1)$. The temperature $\theta(\delta, t)$ is to be compared with the desired temperature r_0, both in the form of voltages, and the local error $e(t) \equiv r_0 - \theta(\delta, t)$ is to be fed into an amplifier having output voltage $v(t) = \mu_1 e(t)$, where μ_1 is a parameter that is to be selected appropriately. We assume that an *actuator* can be constructed in such a way that heat can be transferred at rate $q(t) = \mu_2 v(t)$, for some fixed parameter μ_2, to the left end of the rod ($q(t)$ may be negative). We are attempting to construct a zero-dimensional controller and, defining $\mu = \mu_1 \mu_2$, our design problem has been reduced to the selection of a sutiable value of μ.

Defining $\psi(\eta, t) \equiv \theta(\eta, t) - r_0$, the control system is formally described by

$$\frac{\partial}{\partial t}\psi(\eta, t) = \alpha \frac{\partial}{\partial \eta^2}\psi(\eta, t), \qquad 0 \leqslant \eta \leqslant 1, \quad t \geqslant 0,$$

$$\frac{\partial}{\partial \eta}\psi(0, t) - \beta\mu\psi(\delta, t) = 0 = \frac{\partial}{\partial \eta}\psi(1, t), \qquad t \geqslant 0,$$

where $\alpha > 0$, $\beta > 0$, and $\delta \in (0, 1]$ are fixed parameters. Our design objective is to choose $\mu \in \mathfrak{R}$ such that $\psi(\cdot, t)$ decays to zero, in some sense, as $t \to \infty$. To this end, we define

$$\mathfrak{X} \equiv \mathfrak{L}_2(0, 1), \qquad \|x\|^2 \equiv \int_0^1 |x(\eta)|^2 \, d\eta,$$

$$Ax \equiv \alpha\partial^2 x, \qquad x \in \mathfrak{D}(A) \equiv \{x \in \mathfrak{W}_2^2(0, 1) | \partial x(0) - \beta\mu x(\delta) = 0 = \partial x(1)\}.$$

We see that $A : (\mathfrak{D}(A) \subset \mathfrak{X}) \to \mathfrak{X}$ is linear with $\mathfrak{D}(A)$ dense. It is not too difficult to show that $\mathfrak{R}(I - \lambda A) = \mathfrak{X}$ for all sufficiently small $\lambda > 0$; moreover, for $x \in \mathfrak{D}(A)$,

$$\langle x, Ax \rangle = \alpha \int_0^1 x(\eta) \cdot \partial^2 x(\eta) = -\alpha x(0) \cdot \partial x(0) - \alpha \int_0^1 |\partial x(\eta)|^2 \, d\eta,$$

$$|x(\delta) - x(0)|^2 = \left(\int_0^\delta \partial x(\eta) \, d\eta \right)^2 \leqslant \delta \int_0^\delta |\partial x(\eta)|^2 \, d\eta,$$

where we have employed Schwarz's inequality. Hence, for $x \in \mathfrak{D}(A)$,

$$\begin{aligned}\langle x, Ax \rangle &\leqslant -\alpha\beta\mu x(0) \cdot x(\delta) - (\alpha/\delta)|x(\delta) - x(0)|^2 - \alpha \int_\delta^1 |\partial x(\eta)|^2 \, d\eta \\ &= -(\alpha/\delta)[x(\delta) - (1 - \beta\mu\delta/2)x(0)]^2 \\ &\quad - \alpha\beta\mu(1 - \beta\mu\delta/4)|x(0)|^2 - \alpha \int_\delta^1 |\partial x(\eta)|^2 \, d\eta.\end{aligned}$$

and we see that $-A$ is accretive if $0 \leqslant \beta\mu\delta \leqslant 4$. Assuming this condition on μ, A is the infinitesimal generator of a linear dynamical system on $\mathfrak{X}$, such that $\|S(t)x\| \leqslant \|x\|$ for all $t \in \mathfrak{R}^+$; hence, the equilibrium $x_e = 0$ is stable and all positive orbits are bounded. These are essentially the conclusions reached in Refs. 10 and 11, where this problem was first studied.

Unfortunately, we still do not know if the controller will operate as desired, even if $\mu \in [0, 4/\beta\delta]$; we want $x_e = 0$ to be asymptotically stable rather than merely stable. We have simple stability for $\mu = 0$, in which case the controller does nothing at all; in fact, for $\mu = 0$ the set of all equilibria is unbounded and the system certainly does not operate as desired (see Exercise IV.5.6).

In order to resolve this question, for some $\mu \in (0, 4/\beta\delta]$, note that the injection $\hat{I} : (\mathfrak{D}(A), \|\cdot\|_g) \to \mathfrak{X}$ is compact; hence, J_λ is compact and Theorem IV.5.2 shows that all positive orbits are precompact. Wishing to employ the Invariance Principle, we define $V(x) \equiv \|x\|^2$, $x \in \mathfrak{X}$, and notice that for $\mu \in (0, 4/\beta\delta)$ our earlier computation implies the existence of $\varepsilon_1 > 0$, $\varepsilon_2 > 0$, such that

$$\dot{V}(x) = 2\langle x, Ax \rangle \leqslant -\varepsilon_1 |x(0)|^2 - \varepsilon_2 \int_0^1 |\partial x(\eta)|^2 \, d\eta, \qquad x \in \mathfrak{D}(A).$$

Consequently, by Theorem IV.1.3, $\dot{V}(x) \leqslant -W(x)$ for all $x \in \mathfrak{X}$, where $W : \mathfrak{X} \to \overline{\mathfrak{R}}^+$ is lower semicontinuous and defined by

$$W(x) \equiv \varepsilon_1 |x(0)|^2 + \varepsilon_2 \int_0^1 |\partial x(\eta)|^2 \, d\eta, \qquad x \in \mathcal{W}_2^1(0, 1) \subset \mathfrak{X},$$

$$W(x) = \infty, \qquad x \in \mathfrak{X}, x \notin \mathcal{W}_2^1(0, 1).$$

We wish to find the largest positive invariant set $\mathfrak{M}^+$ in

$$\mathfrak{M}_2 \equiv \{x \in \mathfrak{X} \,|\, W(x) = 0\}$$
$$= \{x \in \mathcal{W}_2^1(0, 1) | x(0) = 0, \partial x(\eta) = 0 \text{ a.e. } \eta \in (0, 1)\} = \{0\};$$

hence, $\mathfrak{M}^+ = \{0\}$ and the Invariance Principle (Theorem IV.4.2) now implies that $S(t)x \to 0$ as $t \to \infty$, for every $x \in \mathfrak{X}$. Therefore, for any choice of $\mu \in (0, 4/\beta\delta)$, $x_e = 0$ is a globally asymptotically stable equilibrium, and the controller will operate as we originally intended. Notice that our conclusions are independent of the value of the desired uniform temperature r_0; this is a property of linear regulator design for linear plants.

2. The Thermoelastic Stability Problem

There is a general belief, based on experience, that the mechanical energy of any elastic system will eventually be dissipated in the form of thermal energy (ignoring any kinetic energy of rigid motion, and assuming no external work is done). Accounting for such dissipation represents a significant problem in the mathematical modeling of elastic systems; "pure" elasticity theory leads to conservation of mechanical energy. If the general belief is correct, a good model should confirm it. One approach, long conjectured to be successful in this respect, was provided by the theory of thermoelasticity (Refs. 12–15). We now wish to see if this model does indeed assure "global asymptotic stability" (Refs. 16, 17); our treatment is based on the results of Ref. 18, which confirm those of Ref. 19. We shall see that the usual linear thermoelastic model does not quite fill the bill.

We consider an elastic body occupying an open region $\boldsymbol{R} \subset \mathfrak{R}^3$ while in some reference configuration of zero stress and uniform temperature $T_0 > 0$. For a material point having position $\eta \in \boldsymbol{R}$ when in the reference configuration, let $u(\eta, t) \in \mathfrak{R}^3$ and $\theta(\eta, t) \in \mathfrak{R}$ denote the displacement and temperature deviation, respectively, from the reference position and temperature. By $\rho(\eta)$ and $c(\eta)$ we mean the mass density and specific heat at η in the reference configuration. We denote the closure of $\boldsymbol{R}$ by $\bar{\boldsymbol{R}}$, and the boundary by $\bar{\boldsymbol{R}} \setminus \boldsymbol{R}$.

The basic constitutive equations of linear thermoelasticity now lead to

$$\rho(\eta)\frac{\partial^2}{\partial t^2} u_i(\eta, t) = \frac{\partial}{\partial \eta_j}\left[\alpha_{ijkl}(\eta)\frac{\partial}{\partial \eta_l} u_k(\eta, t) - m_{ij}(\eta)\theta(\eta, t)\right],$$

$$\rho(\eta)c(\eta)\frac{\partial}{\partial t}\theta(\eta, t) = -T_0 m_{ij}(\eta)\frac{\partial^2}{\partial \eta_j \partial t} u_i(\eta, t) + \frac{\partial}{\partial \eta_i}\left[k_{ij}(\eta)\frac{\partial}{\partial \eta_j}\theta(\eta, t)\right],$$

for all $\eta = (\eta_1, \eta_2, \eta_3) \in \boldsymbol{R}$, $t \geqslant 0$, using Cartesian coordinates. All coefficients are smooth functions of η on $\bar{\boldsymbol{R}}$ and

$$\alpha_{ijkl}(\eta) = \alpha_{klij}(\eta), \qquad m_{ij}(\eta) = m_{ji}(\eta), \qquad k_{ij}(\eta) = k_{ji}(\eta),$$

for all $\eta \in \bar{\boldsymbol{R}}$. A repeated index in a single term indicates summation on that index. We also assume that

$$\min_{\eta \in \bar{\boldsymbol{R}}} \rho(\eta) = \rho_0 > 0, \qquad \min_{\eta \in \bar{\boldsymbol{R}}} c(\eta) = c_0 > 0.$$

Slightly strengthening the Clausius–Duhem inequality (Ref. 20) and using a general property of elastic moduli (Ref. 21), we assume the existence of $k_0 > 0$, $\alpha_0 > 0$, such that

$$k_{ij}(\eta)\xi_i\xi_j \geqslant k_o\xi_i\xi_i \qquad \text{for all } \eta \in \boldsymbol{R}, \quad (\xi_1, \xi_2, \xi_3) \in \mathcal{R}^3,$$

and

$$\int_{\boldsymbol{R}} \alpha_{ijkl}(\eta)\left(\frac{\partial}{\partial \eta_j} u_i(\eta)\right)\left(\frac{\partial}{\partial \eta_l} u_k(\eta)\right) d\boldsymbol{R} \geqslant \alpha_0 \int_{\boldsymbol{R}}\left(\frac{\partial}{\partial \eta_j} u_i(\eta)\right)\left(\frac{\partial}{\partial \eta_j} u_i(\eta)\right) d\boldsymbol{R}$$

for all $u_i \in \mathcal{C}^1(\bar{\boldsymbol{R}})$, $i = 1, 2, 3$. This problem may be studied under various assumptions on the region $\boldsymbol{R}$ and various boundary conditions on (u_1, u_2, u_3, θ). Here we shall assume $\boldsymbol{R}$ to be open, bounded, connected, and properly regular (Ref. 22) with boundary conditions

$$u_i(\eta) = 0 = \theta(\eta), \qquad \eta \in \bar{\boldsymbol{R}} \setminus \boldsymbol{R}, \quad i = 1, 2, 3.$$

To obtain an abstract description, we define

$$(\mathcal{L}_2(\boldsymbol{R}))^3 \equiv \{u = (u_1, u_2, u_3) | u_i \in \mathcal{L}_2(\boldsymbol{R})\},$$

$$(\mathcal{W}_2^1(\boldsymbol{R}))^3 \equiv \{u \in (\mathcal{L}_2(\boldsymbol{R}))^3 | \partial_j u_i \in \mathcal{L}_2(\boldsymbol{R})\},$$

$$\mathcal{X} \equiv \Big\{(u, v, \theta) \in (\mathcal{W}_2^1(\boldsymbol{R}))^3 \times (\mathcal{L}_2(\boldsymbol{R}))^3 \times \mathcal{L}_2(\boldsymbol{R}) \Big| u_i(\eta) = 0 \quad \text{for } \eta \in \bar{\boldsymbol{R}} \setminus \boldsymbol{R}\Big\},$$

$$\|x\|^2 \equiv \int_{\boldsymbol{R}} \big[\alpha_{ijkl} \cdot (\partial_j u_i)(\partial_l u_k) + \rho v_i v_i + (\rho c / T_0)|\theta|^2\big]\, d\boldsymbol{R},$$

$$\mathfrak{D}(A) \equiv \Big\{(u, v, \theta) \in \mathcal{X} \Big| \partial_j \partial_k u_i, \partial_j v_i, \partial_j \theta \in \mathcal{L}_2(\boldsymbol{R}); \quad 0 = \theta(\eta) = v_i(\eta) = \partial_j(\alpha_{ijkl}\partial_l u_k) \text{ for } \eta \in \bar{\boldsymbol{R}} \setminus \boldsymbol{R}\Big\}.$$

The space $\mathcal{X}$ is a Hilbert space and, making the obvious definition of $A : (\mathfrak{D}(A) \subset \mathcal{X}) \to \mathcal{X}$, we see that our problem is modeled by the linear evolution equation

$$\dot{x}(t) = Ax(t), \qquad t \in \mathcal{R}^+,$$
$$x(0) \in \mathfrak{D}(A) \subset \mathcal{X}.$$

It can be verified that $-A$ is accretive and that $\mathcal{R}(I - \lambda A) = \mathcal{X}$ for all $\lambda > 0$; hence, A generates a nonexpansive linear dynamical system $\{S(t)\}_{t \geq 0}$ on $\mathcal{X}$.

Defining $V(x) \equiv |x|^2$ and applying Theorem IV.2.1, we have

$$\dot{V}(x) = V_x' Ax = -(2/T_0) \int_{\boldsymbol{R}} k_{ij}(\partial_i \theta)(\partial_j \theta)\, d\boldsymbol{R},$$

$$\leq -(2k_0/T_0) \int_{\boldsymbol{R}} \partial_i \theta \cdot \partial_i \theta\, d\boldsymbol{R}, \qquad x \in \mathfrak{D}(A).$$

Therefore, defining a lower semicontinuous function $W : \mathcal{X} \to \bar{\mathcal{R}}$ as

$$W(x) \equiv (2k_0/T_0) \int_{\boldsymbol{R}} \partial_i \theta \cdot \partial_i \theta\, d\boldsymbol{R}, \qquad x = (u, v, \theta) \in \mathcal{X}, \theta \in \mathcal{W}_2^1(\boldsymbol{R}),$$

$$W(x) \equiv \infty, \qquad x = (u, v, \theta) \in \mathcal{X}, \theta \notin \mathcal{W}_2^1(\boldsymbol{R}),$$

we see that

$$\mathcal{M}_2 \equiv \{x \in \mathcal{X} \mid W(x) = 0\}$$
$$= \left\{(u, v, \theta) \in \mathcal{X} \mid \theta \in \mathcal{W}_2^1(\boldsymbol{R}),\ \partial_i\theta(\eta) = 0 \text{ a.e. } \eta \in \boldsymbol{R}\right\}.$$

By the compactness of the injection $\hat{I} : (\mathcal{D}(A), \|\cdot\|_g) \to \mathcal{X}$, $J_\lambda \equiv (I - \lambda A)^{-1}$ is compact for all $\lambda > 0$; hence, by Theorem IV.5.2, all positive orbits are precompact. Applying the Invariance Principle, we see that $S(t)x \to \mathcal{M}^+$ as $t \to \infty$, for every $x \in \mathcal{X}$, where $\mathcal{M}^+$ is the largest positive invariant set in $\mathcal{M}_2$.

To find $\mathcal{M}^+$, we first notice that

$$\mathcal{M}^+ \cap \mathcal{D}(A) \subset \left\{x \in \mathcal{D}(A) \mid \theta(\eta) = 0 = m_{ij}(\eta)\partial_j v_i(\eta) \text{ a.e. } \eta \in \boldsymbol{R}\right\},$$

since a motion $S(\cdot)x : \mathcal{R}^+ \to \mathcal{X}$ is a strong solution if $x \in \mathcal{D}(A)$. Using a distributional evolution equation, $\mathcal{M}^+$ can be precisely defined: $\mathcal{M}^+$ is the union of the positive orbits of all motions $S(\cdot)\bar{x}$ which are generalized solutions of

$$\frac{\partial}{\partial t}\bar{u}_i(\eta, t) = \bar{v}_i(\eta, t),$$

$$\rho(\eta)\frac{\partial}{\partial t}\bar{v}_i(\eta, t) = \frac{\partial}{\partial \eta_j}\left[\alpha_{ijkl}(\eta)\frac{\partial}{\partial \eta_l}\bar{u}_k(\eta, t)\right],$$

$$m_{ij}(\eta)\frac{\partial}{\partial \eta_j}\bar{v}_i(\eta, t) \equiv 0 \equiv \bar{\theta}(\eta, t).$$

Such motions $S(\cdot)\bar{x}$ are isothermal oscillations having constant mechanical energy (first two terms in $\|x\|^2$). However, assuming that some $m_{ij} \neq 0$, these equations are formally overdetermined. Under the usual thermoelastic assumptions that $m_{ij} = m_0\delta_{ij}$, $a_{ijkl} = \alpha\delta_{ij}\delta_{kl} + \mu(\delta_{ik}\delta_{jl} + \delta_{jk}\delta_{il})$, for some positive m_0, σ, μ, one usually finds that the only solution is the trivial one, in which case $\mathcal{M}^+ = \{0\}$ and the equilibrium $x_e = 0$ is globally asymptotically stable. However, there are some notable exceptions to this rule, occurring whenever $\boldsymbol{R}$ has a geometry that permits pure shear waves (Refs. 18, 19).

Our general conclusion is that every motion tends to the set of isothermal shear waves, and $x_e = 0$ is (globally) asymptotically stable if and only if this set is $\{0\}$, which is true in many but not all specific cases. This means that the usual linear thermoelastic model does not fully confirm the general belief described earlier. It is possible that this belief is wrong, but it is more probable that the usual linear thermoelastic model is simply inadequate.

3. The Viscoelastic Stability Problem

We reconsider the question posed in Section 2, concerning the dissipation of the "vibrational energy" of an elastic body. There we found that the usual linear thermoelastic model fails to fully confirm the general belief that this energy must be eventually dissipated if no external work is done; this failure, if it is a failure, may be due to an overly simple thermoelastic model. However, another displeasing feature of the thermoelastic theory is the absence of any physical description of the dissipation mechanism. An interesting and plausible description of a possible dissipation mechanism is provided by the viscoelastic theory (Refs. 23–25), and here we shall see that one implication of linear viscoelasticity is that the general belief is probably correct. Our treatment here is related to those of Refs. 26–29.

We continue to use much of the notation introduced in Section 2. The basic constitutive equations of linear viscoelasticity lead to the description

$$\rho(\eta)\frac{\partial^2}{\partial t^2}u_i(\eta, t) = \frac{\partial}{\partial \eta_j}\left[\alpha_{ijkl}(\eta)\frac{\partial}{\partial \eta_l}u_k(\eta, t) - \int_0^\gamma g_{ijkl}(\tau, \eta)\frac{\partial}{\partial \eta_l}u_k(\eta, t-\tau)\, d\tau\right]$$

for all $\eta = (\eta_1, \eta_2, \eta_3) \in \boldsymbol{R}$, $t \geqslant 0$, where the memory length $\gamma > 0$ might be infinite. All coefficients are smooth functions of $\eta \in \bar{\boldsymbol{R}}$, $\tau \in \mathfrak{R}^+$, and

$$\alpha_{ijkl}(\eta) = \alpha_{klij}(\eta), \qquad g_{ijkl}(\tau, \eta) = g_{klij}(\tau, \eta),$$

for all $\eta \in \bar{\boldsymbol{R}}$, $\tau \in \mathfrak{R}^+$. We assume the existence of positive numbers ρ_0, ε_0, and a continuous function $g_0 : \mathfrak{R}^+ \to \mathfrak{R}^+$, such that

$$\min_{\eta \in \bar{\boldsymbol{R}}} \rho(\eta) = \rho_0,$$

$$g_0(\tau) > 0 \quad \text{for } 0 \leqslant \rho < \gamma, \text{ where } 0 < \gamma \leqslant \infty,$$

$$g_0(\tau) = 0 \quad \text{for } \tau \geqslant \gamma \text{ if } \gamma < \infty,$$

$$\int_{\boldsymbol{R}} g_{ijkl}(\tau, \eta)\left(\frac{\partial}{\partial \eta_j}u_j(\eta)\right)\left(\frac{\partial}{\partial \eta_l}u_k(\eta)\right) d\boldsymbol{R}$$

$$\geqslant g_o(\tau)\int_{\boldsymbol{R}}\left(\frac{\partial}{\partial \eta_j}u_i(\eta)\right)\left(\frac{\partial}{\partial \eta_j}u_i(\eta)\right) d\boldsymbol{R} \qquad \text{for all } \tau \in [0, \gamma),$$

$$\int_{\boldsymbol{R}} \varepsilon_{ijkl}(\eta)\left(\frac{\partial}{\partial \eta_j}u_i(\eta)\right)\left(\frac{\partial}{\partial \eta_l}u_k(\eta)\right) d\boldsymbol{R} \geqslant \varepsilon_0 \int \left(\frac{\partial}{\partial \eta_j}u_i(\eta)\right)\left(\frac{\partial}{\partial \eta_j}u_i(\eta)\right) d\boldsymbol{R},$$

where

$$\varepsilon_{ijkl}(\eta) \equiv \alpha_{ijkl} - \int_0^\gamma g_{ijkl}(\tau, \eta)\, d\tau.$$

The existence of $\varepsilon_0 > 0$ follows from a general property of elastic moduli (Ref. 21). We also make a "decaying memory" assumption (Ref. 25),

$$\int_{\boldsymbol{R}} \left(\frac{\partial}{\partial \tau} g_{ijkl}(\tau, \eta) \right) \left(\frac{\partial}{\partial \eta_j} u_i(\eta) \right) \left(\frac{\partial}{\partial \eta_l} u_k(\eta) \right) d\boldsymbol{R}$$

$$\leqslant -\mu g_0(\tau) \int_{\boldsymbol{R}} \left(\frac{\partial}{\partial \eta_j} u_i(\eta) \right) \left(\frac{\partial}{\partial \eta_j} u_i(\eta) \right) d\boldsymbol{R}$$

for all $\tau \in [0, \gamma)$, some $\mu > 0$. We assume $\boldsymbol{R}$ to be open, bounded, connected, and properly regular (Ref. 22), with boundary conditions

$$u_i(\eta) = 0, \qquad \eta \in \bar{\boldsymbol{R}} \setminus \boldsymbol{R}, \qquad i = 1, 2, 3.$$

To pose this problem abstractly, we define $(\mathcal{L}_2(\boldsymbol{R}))^3$ and $(\mathcal{W}_2^1(\boldsymbol{R}))^3$ as in Section 2. Also considering the set of $w \in \mathcal{C}([0, \gamma); (\mathcal{W}_2^1(\boldsymbol{R}))^3)$ such that $w(0, \cdot) = 0 \in (\mathcal{W}_2^1(\boldsymbol{R}))^3$, $w(\tau, \eta) = 0$ for all $\eta \in \bar{\boldsymbol{R}} \setminus \boldsymbol{R}$, $\tau \in [0, \gamma)$, and

$$\|w\|_{\mathcal{K}}^2 \equiv \int_0^\gamma \int_{\boldsymbol{R}} g_{ijkl}(\eta, \tau) \left(\frac{\partial}{\partial \eta_j} w_i(\tau, \eta) \right) \left(\frac{\partial}{\partial \eta_l} w_k(\tau, \eta) \right) d\boldsymbol{R}\, d\tau < \infty,$$

we define $\mathcal{K}$ to be the $\|\cdot\|_{\mathcal{K}}$-completion of this set. Identifying $v(\eta, t)$ and $w(\tau, \eta, t)$ with $(\partial/\partial t)u(\eta, t)$ and $u(\eta, t) - u(\eta, t - \tau)$, respectively, we now define

$$\mathcal{X} \equiv \left\{ (u, v, w) \in (\mathcal{W}_2^1(\boldsymbol{R}))^3 \times (\mathcal{L}_2(\boldsymbol{R}))^3 \times \mathcal{K} \,|\, u_i(\eta) = 0 \text{ for } \eta \in \bar{\boldsymbol{R}} \setminus \boldsymbol{R} \right\}.$$

$$\|x\|^2 = \int_{\boldsymbol{R}} \left[\varepsilon_{ijkl}(\partial_j u_i)(\partial_l u_k) + \rho v_i v_i \right] d\boldsymbol{R} + \|w\|_{\mathcal{K}}^2$$

and

$$Ax \equiv \left\{ \left(v_i, \frac{1}{\rho} \partial_j \left[\varepsilon_{ijkl} \partial_l u_k + \int_0^\gamma g_{ijkl} \partial_l w_k(\tau, \cdot)\, d\tau \right], v_i - \partial_\tau w_i \right) \right\}_{i=1,2,3}$$

for all $x = (u, v, w) \in \mathcal{X}$ such that this quantity lies in $\mathcal{X}$.

Although it is not easy, it can be shown that $\mathcal{D}(A)$ is dense in $\mathcal{X}$ and $\mathcal{R}(I - \lambda A) = \mathcal{X}$ for all $\lambda > 0$. Clearly $\mathcal{X}$ is a Hilbert space and, using the natural inner product, we find that

$$\begin{aligned}
2\langle x, Ax\rangle &= -2\int_0^\gamma \int_R g_{ijkl} \cdot (\partial_l w_k)(\partial_\tau \partial_j w_i)\, dR\, d\tau \\
&= -\int_0^\gamma \partial_\tau \left(\int_R g_{ijkl} (\partial_l w_k)(\partial_j w_i)\, dR \right) d\tau \\
&\quad + \int_0^\gamma \int_R (\partial_\tau g_{ijkl})(\partial_l w_k)(\partial_j w_i)\, dR\, d\tau \\
&\leqslant \int_0^\gamma \int_R (\partial_\tau g_{ijkl})(\partial_l w_k)(\partial_j w_i)\, dR\, d\tau \\
&\leqslant -\mu \int_0^\gamma g_0(\tau) \int_R (\partial_j w_i)(\partial_j w_i)\, dR\, d\tau \leqslant 0
\end{aligned}$$

for every $x \in \mathfrak{D}(A)$, and it follows that $-A$ is monotone (accretive). Hence, A generates a linear dynamical system $\{S(t)\}_{t\geqslant 0}$ on $\mathfrak{X}$, $x_e = 0$ is a stable equilibrium, and all positive orbits are bounded.

Defining $V(x) = \|x\|^2$, $x \in \mathfrak{X}$, and noting that $\dot{V}(x) \leqslant -W(x)$ for all $x \in \mathfrak{X}$, where

$$W(x) \equiv \mu \int_0^\gamma g_0(\tau) \int_R (\partial_i w_i)(\partial_j w_i)\, dR\, d\tau \geqslant 0, \qquad x = (u, v, w) \in \mathfrak{X},$$

we see by Theorem IV.4.2 that $\Omega(x) \subset \mathfrak{M}^+$ for every $x \in \mathfrak{X}$, where $\mathfrak{M}^+$ is the largest positive invariant set in

$$\mathfrak{M}_2 \equiv \{x \in \mathfrak{X} \,|\, W(x) = 0\} = \left\{(u, v, w) \in \mathfrak{X} \,|\, \partial_j w_i = 0\right\}.$$

For all $x_0 \in \mathfrak{D}(A)$, $S(\cdot)x_0$ is a strong solution of $\dot{x}(t) = Ax(t)$, $t \in \mathfrak{R}^+$, and inspection of this equation shows that $\mathfrak{M}^+ \cap \mathfrak{D}(A) = \{0\}$. By using a distributional evolution equation, valid for all motions, we can show that $\mathfrak{M}^+ = \{0\}$; hence, $\Omega(x) \subset \{0\}$ for every $x \in \mathfrak{X}$.

Orbital precompactness is difficult to prove, since $J_\lambda \equiv (I - \lambda A)^{-1}$ is not a compact operator. To investigate this question without exerting undue effort, we shall assume that $\gamma < \infty$ and also that the material is homogeneous, implying that ρ, α_{ijkl}, g_{ijkl}, and ε_{ijkl} are independent of $\eta \in \bar{R}$. Defining $K : (\mathfrak{D}(K) \subset (\mathfrak{L}_2(R))^3) \to (\mathfrak{L}_2(R))^3$ as

$$Ku \equiv \{-\partial_j \partial_j u_k\}_{k=1,2,3},$$

$$u \in \mathfrak{D}(K) \equiv \left\{u \in \left(\mathfrak{W}_2^2(R)\right)^3 | u_i(\eta) = 0 \text{ for } \eta \in \bar{R} \setminus R\right\},$$

we find that K is invertible with compact K^{-1} defined on $(\mathcal{L}_2(\boldsymbol{R}))^3$. Now defining $\boldsymbol{P} : \mathcal{X} \to \mathcal{X}$ by

$$Px = (K^{-1}u, K^{-1}v, K^{-1}w), \qquad x = (u, v, w) \in \mathcal{X},$$

we can show that PJ_λ is a compact operator and that $PJ_\lambda A = APJ_\lambda$ on $\mathfrak{D}(A)$. Consequently, Theorem IV.5.2 implies that all positive orbits are precompact, and we conclude that $S(t)x \to 0$ as $t \to \infty$, for every $x \in \mathcal{X}$.

We see that $x_e = 0$ is globally asymptotically stable, agreeing with the general belief about dissipation.

4. A Fission Reactor Stability Problem

A nuclear fission reactor usually acts as the high-temperature heat reservoir for a heat engine; that is, the reactor produces heat that is transmitted by a coolant to some type of heat engine. The coolant enters the near vicinity of the reactor at a fixed low temperature T_L, absorbs heat, and is pumped to the engine at a fixed high temperature T_H; the engine does work, thereby reducing the coolant temperature. The coolant is then further cooled to T_L (e.g., by a cooling tower) and returns to the vicinity of the reactor. If $p(t)$ denotes the total power being produced by fission within the reactor, we see that $p(t)$ serves to heat the reactor, and the reactor heats the coolant by conduction; hence, the temperature field $T(\cdot, t)$ within the reactor is described by a heat conduction equation driven by $p(t)$. Also, the fission reaction rate varies with the temperature field of the reactor, and this leads to another evolution equation for $\dot{p}(t)$. This equation may be further complicated by the effect of "delayed neutrons," which are ejected some time after fission occurs (Ref. 30).

If anything goes wrong during operation of the reactor, the reaction can be automatically or manually shut down by significantly increasing the coolant flow rate and/or by inserting control rods. However, it is highly desirable that the normal operating condition $(T(\cdot, t) \equiv T_s(\cdot), p(t) \equiv p_0)$ be asymptotically stable, in some sense, so that such measures are not required during normal operation. Various studies suggest that this desirable situation exists for many (and possibly all) reactor configurations (Refs. 30–40). Here we will follow the analysis of Refs. 34 and 39.

To demonstrate the assumptions usually employed, we first consider a very simple model. To this end, let the entire reactor (fuel elements, moderators, reflectors) be considered a rod, heat being transmitted to the coolant only at the ends of the rod (see Figure 9).

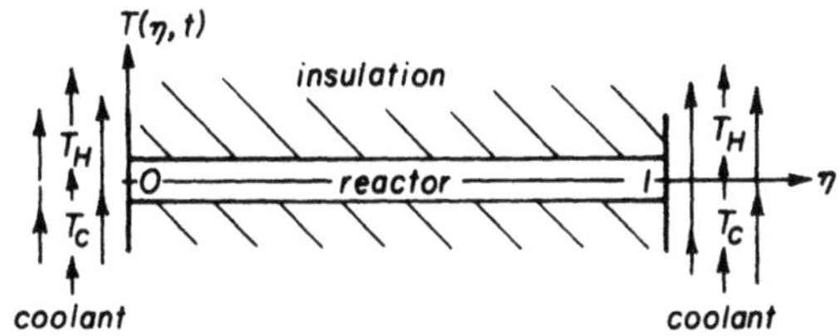

Fig. 9. Reactor modeled as a rod.

We assume that the high temperature T_H is held constant, with $T(0, t) = T_H = T(1, t)$ for all $t \geqslant 0$. We consider $T(\eta, t)$ to be described by

$$c_0 \frac{\partial}{\partial t} T(\eta, t) = k_0 \frac{\partial^2}{\partial \eta^2} T(\eta, t) + g(\eta) p(t), \qquad 0 < \eta < 1, t \geqslant 0,$$

where $c_0 > 0$ and $k_0 > 0$ are the heat capacity and conductivity, respectively, and $g(\eta)p(t)\,\Delta\eta$ is the fraction of fission power $p(t)$ produced in an element of length $\Delta\eta$ at location η; hence, the continuous function $g : ([0, 1] \subset \mathcal{R}) \to \mathcal{R}^+$ is such that $\int_0^1 g(\eta)\, d\eta = 1$. During steady-state operation, we see that $T_s(\eta)$ and p_0 must satisfy $p_0 g(\eta) = -k_0(\partial^2/\partial\eta^2)T_s(\eta)$, $0 < \eta < 1$, with $T_s(0) = T_H = T_s(1)$.

An equation describing $p(t)$ is considerably less obvious; ignoring the effect of delayed neutrons, it is usually assumed that

$$\frac{d}{dt} \ln p(t) = -\int_0^1 f(\eta) T(\eta, t)\, d\eta + c_1, \qquad t \geqslant 0,$$

for some $c_1 > 0$ and some continuous $f : ([0, 1] \subset \mathcal{R}) \to \mathcal{R}^+$; $-f(\eta)$ is called the temperature coefficient of reactivity at η, and we see that $c_1 = \int_0^1 f(\eta) T_s(\eta)\, d\eta$.

Defining $u(t) \equiv \ln(p(t)/p_0)$, $\theta(\eta, t) \equiv T(\eta, t) - T_s(\eta)$, our formal equations become

$$\frac{d}{dt} u(t) = -\int_0^1 f(\eta) \theta(\eta, t)\, d\eta,$$

$$c_0 \frac{\partial}{\partial t} \theta(\eta, t) = k_0 \frac{\partial^2}{\partial \eta^2} \theta(\eta, t) + p_0(e^{u(t)} - 1) g(\eta), \qquad 0 < \eta < 1,$$

with $\theta(0, t) = 0 = \theta(1, t)$, for all $t \geqslant 0$. Now defining $\hat{g}(\eta) \equiv (p_0/c_0) g(\eta)$,

$0 < \eta < 1$, and

$$\langle y, \hat{y} \rangle_{\mathcal{H}} \equiv \int_0^1 y(\eta) \cdot \hat{y}(\eta)\, d\eta, \qquad y, \hat{y} \in \mathcal{L}_2(0, 1),$$

$$Fy \equiv (k_0/c_0)\partial^2_\eta y, \qquad y \in \mathfrak{D}(F) \equiv \{ y \in \mathcal{W}_2^2(0, 1) | y(0) = 0 = y(1) \},$$

we see that $F : (\mathfrak{D}(F) \subset \mathcal{L}_2(0, 1)) \to \mathcal{L}_2(0, 1)$ and we are interested in

$$\dot{u}(t) = -\langle y(t), f \rangle_{\mathcal{H}},$$

$$\dot{y}(t) = Fy(t) + (e^{u(t)} - 1)\hat{g}, \qquad \text{a.e. } t \in \mathcal{R}^+,$$

with $u(0) \in \mathcal{R}$, $y(0) \in \mathrm{Cl}_{\mathcal{H}}\mathfrak{D}(F) = \mathcal{L}_2(0, 1)$. We can easily verify that the range $\mathcal{R}(I - \lambda F) = \mathcal{L}_2(0, 1)$ for all $\lambda > 0$, and $(-\pi^2 k_0/c_0)I - F$ is accretive.

We wish to embed our example into a much more general class of reactor configurations; hence, we shall generalize our definitions of f, $\hat{g}$, and F. To this end, let $\mathcal{H}$ be any real Hilbert space, and let $F : (\mathfrak{D}(F) \subset \mathcal{H}) \to \mathcal{H}$ be any (possibly nonlinear) operator such that $\mathrm{Cl}_{\mathcal{H}}\mathfrak{D}(F) = \mathcal{H}$, $F(0) = 0$, $\mu I - F$ is accretive for some $\mu \in \mathcal{R}$, and $\mathcal{R}(I - \lambda F) = \mathcal{H}$ for all $\lambda \in (0, \lambda_0)$; let f, $\hat{g}$ be given elements of $\mathcal{H}$. We now define

$$\mathfrak{X} \equiv \mathcal{R} \times \mathcal{H}, \qquad \langle x, \hat{x} \rangle \equiv u\hat{u} + \langle y, \hat{y} \rangle_{\mathcal{H}},$$

$$Ax \equiv (-\langle y, f \rangle_{\mathcal{H}}, Fy + (e^u - 1)\hat{g}), \qquad (u, y) = x \in \mathfrak{D}(A) \equiv \mathcal{R} \times \mathfrak{D}(F),$$

and we are interested in the evolution equation

$$\dot{x}(t) = Ax(t), \qquad \text{a.e. } t \in \mathcal{R}^+,$$

$$x(0) = x_0 \in \mathrm{Cl}\, \mathfrak{D}(A) = \mathfrak{X}.$$

We see that $\mathfrak{X}$ is a Hilbert space and $\mathfrak{D}(A)$ is dense. Unfortunately, $\omega I - A$ is not accretive for any $\omega \in \mathcal{R}$, and even the range condition of Theorem III.5.1 poses problems. Seeking to employ Theorem III.6.1 to show that A generates a dynamical system, we make an assumption that will be discussed later.

Assumption 4.1. Let there exist a symmetric bounded linear operator $G : \mathcal{H} \to \mathcal{H}$ such that

(i) $G\hat{g} = f$.
(ii) $\langle y, Gy \rangle_{\mathcal{H}} \geq \varepsilon_1 \|y\|^2_{\mathcal{H}}$ for all $y \in \mathcal{H}$, some $\varepsilon_1 > 0$.
(iii) $-\langle y, GFy \rangle_{\mathcal{H}} \geq \varepsilon_2 \|y\|^2_{\mathcal{H}}$ for all $y \in \mathcal{H}$, some $\varepsilon_2 \geq 0$.

Under Assumption 4.1, consider the continuous function $V : \mathfrak{X} \to$

$\mathcal{R}^+$ defined by

$$V(x) \equiv 2(e^u - 1 - u) + \langle y, Gy \rangle_{\mathcal{K}}, \qquad x = (u, y) \in \mathcal{X},$$

noting that $V(0) = 0$, $V(x) > 0$ if $x \neq 0$, and $V(x) \to \infty$ as $\|x\| \to \infty$. Since

$$V(x - \lambda Ax) \geqslant V(x) + 2e^u(e^{\lambda\langle y, f\rangle_{\mathcal{K}}} - 1) - 2\lambda\langle y, f\rangle_{\mathcal{K}} - 2\lambda\langle y, GFy\rangle_{\mathcal{K}}$$

$$-2\lambda(e^u - 1)\langle y, G\hat{g}\rangle_{\mathcal{K}} + \varepsilon_1\lambda^2\|Fy + (e^u - 1)\hat{g}\|_{\mathcal{K}}^2$$

$$\geqslant V(x) + 2e^u\lambda\langle y, f\rangle_{\mathcal{K}} - 2\lambda\langle y, f\rangle_{\mathcal{K}} + 2\lambda\varepsilon_2\|y\|_{\mathcal{K}}^2 - 2\lambda(e^u - 1)\langle y, G\hat{g}\rangle_{\mathcal{K}}$$

$$= V(x) + 2\lambda\varepsilon_2\|y\|_{\mathcal{K}}^2, \qquad x = (u, y) \in \mathcal{D}(A), \lambda > 0,$$

condition (i) of Theorem III.6.1 is satisfied.

Now consider $\bar{\mathcal{G}}_\alpha \equiv \{x \in \mathcal{X} \mid V(x) \leqslant \alpha\}$ for some $\alpha \in \mathcal{R}$. If $x = (u, y) \in \bar{\mathcal{G}}_\alpha$, we see that there exists $\delta(\alpha)$ such that $|u| \leqslant \|x\| \leqslant \delta$. We now ask if, given $\hat{x} = (\hat{u}, \hat{y}) \in \bar{\mathcal{G}}_\alpha$, there exists $x = (u, y) \in \mathcal{D}(A)$ such that $(I - \lambda A)x = \hat{x}$ for sufficiently small $\lambda > 0$; equivalently,

$$u + \lambda\langle y, f\rangle_{\mathcal{K}} = \hat{u}, \qquad y - \lambda Fy - \lambda(e^u - 1)\hat{g} = \hat{y}.$$

A solution does exist if $h_\lambda : \mathcal{R} \to \mathcal{R}$ has range $\mathcal{R}(h_\lambda) \supset (-\delta - 1, \delta + 1)$ for all sufficiently small $\lambda > 0$, where

$$h_\lambda v \equiv v + \lambda\langle (I - \lambda F)^{-1}\lambda(e^v - 1)\hat{g}, f\rangle_{\mathcal{K}}, \qquad v \in \mathcal{R}.$$

As $F(0) = 0$ and $\mu I - F$ is accretive,

$$|\lambda \langle (I - \lambda F)^{-1}\lambda(e^v - 1)\hat{g}, f\rangle_{\mathcal{K}}| \leqslant \frac{\lambda^2}{1 - \lambda\mu}\|\hat{g}\|_{\mathcal{K}}\|f\|_{\mathcal{K}}(e^v - 1)$$

for $\lambda \in (0, \lambda_0)$; hence, there exists $\lambda_1(\delta) < \lambda_0$ such that $\mathcal{R}(h_\lambda) \supset (-\delta - 1, \delta + 1)$ for all $\lambda \in (0, \lambda_1)$, and it follows that $\mathcal{R}(I - \lambda A) \supset \bar{\mathcal{G}}_\alpha$ for all $\lambda \in (0, \lambda_1)$. Hence, condition (ii) of Theorem III.6.1 is satisfied.

For $x, \hat{x} \in \bar{\mathcal{G}}_\alpha \cap \mathcal{D}(A)$, we see that

$$\langle x - \hat{x}, Ax - A\hat{x}\rangle_{\mathcal{K}} = -(u - \hat{u})\langle y - \hat{y}, f\rangle_{\mathcal{K}} + \langle y - \hat{y}, Fy - F\hat{y}\rangle_{\mathcal{K}}$$

$$+ \langle y - \hat{y}, (e^u - e^{\hat{u}})\hat{g}\rangle_{\mathcal{K}}$$

$$\leqslant \|f\|_{\mathcal{K}}|u - \hat{u}|\|y - \hat{y}\|_{\mathcal{K}} + \mu\|y - \hat{y}\|_{\mathcal{K}}^2$$

$$+ \|g\|_{\mathcal{K}}\|y - \hat{y}\||e^u - e^{u}|$$

$$\leqslant \omega\|x - \hat{x}\|^2$$

for some $\omega(\alpha)$. Now applying Theorem III.6.1, and noting that $\cup_{\alpha>0}\bar{\mathcal{G}}_\alpha = \mathfrak{X}$, we conclude that A generates a dynamical system $\{S(t)\}_{t\geq 0}$ on $\mathfrak{X}$, via the product formula $S(t)x \equiv \lim_{n\to\infty} J^n_{t/n}x$, $t \in \mathfrak{R}^+$, $x \in \mathfrak{X}$. Moreover, $S(\cdot)x$ is a strong solution of $\dot{x}(t) = Ax(t)$ a.e. $t \in \mathfrak{R}^+$ for every $x \in \mathfrak{D}(A)$, and $\|S(t)x - S(t)\hat{x}\| \leq e^{\omega(\alpha)t}\|x - \hat{x}\|$ for all $x, \hat{x} \in \bar{\mathcal{G}}_\alpha$, $t \in \mathfrak{R}^+$.

Recalling that $\varepsilon_2 \geq 0$, it is also apparent that V is a Liapunov function on $\mathfrak{X}$, with $\dot{V}(x) \leq -2\varepsilon_2\|y\|^2_{\mathcal{H}}$ for all $x = (u, y) \in \mathfrak{X}$. Hence, every $\bar{\mathcal{G}}_\alpha$ is positive invariant, the equilibrium $x_e = 0$ is stable, and the boundedness of $\bar{\mathcal{G}}_\alpha$ implies that all positive orbits are bounded, since $\cup_{\alpha>0}\bar{\mathcal{G}}_\alpha = \mathfrak{X}$.

If $\varepsilon_2 > 0$ and $\hat{g} \neq 0 \in \mathcal{H}$, the largest positive invariant set $\mathfrak{M}^+$ in $\mathfrak{M}_1 \equiv \{x \in \mathfrak{X} \mid \dot{V}(x) = 0\} \subset \{(u, y) \in \mathfrak{X} \mid y = 0\}$ can be shown to be $\mathfrak{M}^+ = \{0\}$. Hence, the Invariance Principle implies that the positive limit set $\Omega(x) \subset \{0\}$ for every $x \in \mathfrak{X}$. If $\varepsilon_2 > 0$ and F is linear, then (b) of Theorem IV.5.4 can be used to show that all positive orbits are precompact; hence $S(t)x \to 0$ as $t \to \infty$ for every $x \in \mathfrak{X}$, and $x_e = 0$ is globally asymptotically stable. For $\varepsilon_2 > 0$ and nonlinear F, exponential stability of $x_e = 0$ is shown in Refs. 34 and 39 by constructing a "better" Liapunov function $\tilde{V}$; the exponent approaches zero as $\alpha \to \infty$, $x \in \bar{\mathcal{G}}_\alpha$.

We see that under Assumption 4.1, with $\varepsilon_2 > 0$, the reactor has the desired property; the normal operating condition ($x_e = 0$) is globally asymptotically stable in the $\|\cdot\|_{\mathfrak{X}}$-sense. The validity of Assumption 4.1 clearly depends on the choice of $\mathcal{H}$ and $F : (\mathfrak{D}(F) \subset \mathcal{H}) \to \mathcal{H}$; Assumption 4.1 has been verified in certain special cases.

If $\mathcal{H} = \mathfrak{R}^n$ and F is linear (a real $n \times n$ matrix), Assumption 4.1 is related to the Lur'e control problem (Refs. 41, 42). If all (complex) eigenvalues λ_n of F have negative real parts, $\hat{g} \neq 0 \in \mathfrak{R}^n$, and

$$\operatorname{Re}\langle f, (\xi I - F)^{-1}\hat{g}\rangle_{\mathcal{C}^n} \geq 0$$

for all $\xi \in \hat{\mathcal{C}}$ such that $\operatorname{Re}\xi \geq 0$, then it is known that Assumption 4.1 holds (see Lemma 2 of Ref. 43). For extensions of this result, see Refs. 41–44.

If $\mathcal{H}$ is any real Hilbert space and $F : (\mathfrak{D}(F) \subset \mathcal{H}) \to \mathcal{H}$ is linear, suppose that all $\operatorname{Re}\lambda_n \leq -\varepsilon_2$ for every eigenvalue λ_n and that the set $\{e_n\}$ of eigenvectors of F is orthonormal and spans $\mathcal{H}$ (Ref. 34). Let f and $\hat{g}$ be represented as $f = \sum_n \alpha_n e_n$, $\hat{g} = \sum_n \beta_n e_n$ for some $\{\alpha_n\}$, $\{\beta_n\} \in \mathfrak{R}^\infty$. Let us also suppose that $\beta_n = 0$ if and only if $\alpha_n = 0$, and $\mu_1 \leq \alpha_n/\beta_n \leq \mu_2$ for all n, some $\mu_2 \geq \mu_1 > 0$, if $\beta_n \neq 0$. Then defining

$$Gy \equiv \sum_n \delta_n \langle y, e_n\rangle_{\mathcal{H}} e_n, \qquad y \in \mathcal{H},$$

with

$$\delta_n \equiv \alpha_n/\beta_n \quad \text{if } \alpha_n\beta_n > 0,$$
$$\delta_n \equiv 1 \qquad \text{if } \alpha_n\beta_n = 0,$$

we see that

$$G\hat{g} = \sum_n \delta_n \beta_n e_n = \sum_n \alpha_n e_n = f,$$

$$\left(\inf_n \delta_n\right)\|y\|_{\mathcal{H}}^2 \leqslant \langle y, Gy\rangle_{\mathcal{H}} \leqslant \left(\sup_n \delta_n\right)\|y\|_{\mathcal{H}}^2,$$

$$\langle Gy, \hat{y}\rangle_{\mathcal{H}} = \sum_n \delta_n \langle y, e_n\rangle_{\mathcal{H}}\langle e_n, \hat{y}\rangle_{\mathcal{H}} = \sum_n \delta_n \langle \hat{y}, e_n\rangle_{\mathcal{H}}\langle e_n, y\rangle_{\mathcal{H}}$$
$$= \langle G\hat{y}, y\rangle_{\mathcal{H}} = \langle y, G\hat{y}\rangle_{\mathcal{H}}$$

$$\langle y, GFy\rangle_{\mathcal{H}} = \sum_n \lambda_n \delta_n \langle e_n, e_n\rangle \leqslant -\varepsilon_2\left(\inf_n \delta_n\right)\|y\|^2,$$

for all $y, \hat{y} \in \mathcal{H}$. We notice that F of our original example meets these requirements; then $\lambda_n = k_0 n^2\pi^2/c_0$ and $e_n(\eta) = \sin n\pi\eta$ for $n = 1, 2, \ldots$. Hence, Assumption 4.1 is fulfilled by our original example for many $f, \hat{g} \in \mathcal{H} = \mathcal{L}_2(0, 1)$.

For the general case, it is often physically plausible to assume that $f = \nu\hat{g}$ for some real number $\nu > 0$ (Ref. 35); then Assumption 4.1 is fulfilled by $G \equiv \nu I$ if $-\langle y, Fy\rangle_{\mathcal{H}} \geqslant \varepsilon_2\|y\|_{\mathcal{H}}^2$ for all $y \in \mathcal{H}$, some $\varepsilon_2 \geqslant 0$, even when F is nonlinear.

To extend the foregoing idea, suppose that the reactor occupies an open region $\boldsymbol{R} \subset \mathfrak{R}^m$, $m = 1$, 2, or 3, with $\mathcal{H} \equiv \mathcal{L}_2(\boldsymbol{R})$. Recalling that $\hat{g} : (\boldsymbol{R} \subset \mathfrak{R}^m) \to \mathfrak{R}$ and $f : (\boldsymbol{R} \subset \mathfrak{R}^m) \to \mathfrak{R}$ are continuous and $\mathfrak{R}^+$-valued, suppose that $h(\eta) \equiv f(\eta)/\hat{g}(\eta)$ is measurable and essentially bounded on $\boldsymbol{R}$, with $\inf_{\eta\in\boldsymbol{R}} h(\eta) \geqslant \varepsilon_1$ for some $\varepsilon_1 > 0$. Then, defining $G : \mathcal{H} \to \mathcal{H}$ by $Gy(\eta) \equiv h(\eta)y(\eta)$, $\eta \in \boldsymbol{R}$, $y \in \mathcal{H} \equiv \mathcal{L}_2(\boldsymbol{R})$, we see that Assumption 4.1 holds if

$$-\int_{\boldsymbol{R}} y(\eta)\cdot h(\eta)\cdot Fy(\eta)\, d\boldsymbol{R} \geqslant \varepsilon_2 \int_{\boldsymbol{R}} |y(\eta)|^2\, d\boldsymbol{R}$$

for all $y \in \mathfrak{D}(F) \subset \mathcal{H} \equiv \mathcal{L}_2(\boldsymbol{R})$, some $\varepsilon_2 \geqslant 0$ (Ref. 34).

We see that condition (i) of Assumption 4.1 may create difficulties; using a "better" Liapunov function $\tilde{V}$ developed in Refs. 34 and 39, we can weaken this condition. If $\|G\hat{g} - f\|_{\mathcal{H}} \leqslant \varepsilon_3$ for sufficiently small $\varepsilon_3 \geqslant 0$, and if conditions (ii) and (iii) hold for sufficiently large $\varepsilon_2 > 0$, all our conclusions can be shown to remain true by using $\tilde{V}$ in place of V. Essentially, one uses $\tilde{V}$ to prove exponential stability under Assumption 4.1 and then considers a "perturbed problem" for which condition (i) is not quite satisfied, with ε_3 limiting the magnitude of the perturbation. The resulting perturbation analysis can be based on the ideas of Theorem I.10.1, Proposition III.5.1, and Theorem IV.3.3.

Finally, let us mention the effect of delayed neutrons. If we model this effect as in Refs. 31, 37, 39, considering N "groups" of delayed neutrons having concentrations proportional to $e^{q_i(t)}$, $i = 1, 2, \ldots, N$, and if we define F and $\mathcal{H}$ as before, we now are interested in another evolution equation $\dot{x}(t) = Ax(t)$ a.e. $t \in \mathcal{R}^+$, where

$$\mathcal{X} \equiv \mathcal{R} \times \mathcal{R}^N \times \mathcal{H},$$

$$\|x\|^2 \equiv |u|^2 + \sum_{i=1}^{N} |q_i|^2 + \langle y, y \rangle_{\mathcal{H}}, \qquad x = (u, q, y) \in \mathcal{X},$$

$$Ax \equiv \left(-\langle y, f \rangle_{\mathcal{H}} + \sum_{i=1}^{N} (\beta_i / l^*) \sigma(u - q_i), \lambda_j \sigma(u - q_j), \right.$$

$$\left. Fy + (e^u - 1)\hat{g} \right)_{j=1,\ldots,N}, \qquad x \in \mathcal{D}(A) \equiv \mathcal{R} \times \mathcal{R}^N \times \mathcal{D}(F),$$

The numbers l^*, β_i, λ_i are all positive and $\sigma(z) \equiv e^z - 1$, $z \in \mathcal{R}$ (Ref. 39). If Assumption 4.1 holds and if we define

$$V(x) \equiv 2\int_0^u \sigma(z)\, dz + 2\sum_{i=1}^{N} (\beta_i / \lambda_i l^*) \int_0^{q_i} \sigma(z)\, dz + \langle y, Gy \rangle_{\mathcal{H}},$$

we can show, just as before, that all our earlier conclusions remain valid (Ref. 39). Hence, under Assumption 4.1, the effect of delayed neutrons is not destabilizing.

5. A Supersonic Panel-Flutter Problem

An interesting class of problems is encountered when considering an elastic body that interacts with the environment in such a way that the interaction forces depend only on the motion of the body; then the equations of motion are autonomous, describing a problem in "nonconservative elastic stability" (Ref. 45) when the interaction forces are not derivable from a "potential energy function." Such problems arise, for example, in the vibration of helicopter rotor blades and in the motions of pipes conveying fluids. In many cases, an adjacent fluid is capable of doing an unlimited amount of positive work on the elastic body and, hence, the mechanical energy of the elastic body may grow without bound. This phenomenon is related to that of "elastic buckling" and includes it as a special case.

As a practical example, consider a thin rectangular flat panel exposed on one side to a supersonic airstream parallel to one edge of the panel, as shown in Figure 10. To be specific, the panel might be part of the skin of a supersonic aircraft. If the panel remains undeformed, we assume that there is essentially no aerodynamic force on the panel; however, if the panel deflects, there is an aerodynamic pressure that is influenced by this deflection $w(\eta, \zeta, t)$, as well as by its time rate of change. We notice that the kinetic energy of the airstream is unlimited, and the panel may absorb energy from the airstream. To avoid destruction of the panel, it is essential that the undeformed configuration be stable in some sense; to avoid vibration and metal fatigue, we strongly prefer some sort of asymptotic stability.

For simplicity, we shall assume that only cylindrical deformations occur ($w(\eta, \zeta, t) = y(\eta, t)$, independent of ζ), and we shall model the panel as a uniform pin-ended beam of thickness h, under constant tension T, with $l_2 \equiv 1$.

$$\rho h \frac{\partial^2}{\partial t^2} y(\eta, t) + (Eh^3/12)\frac{\partial^4}{\partial \eta^4} y(\eta, t) - T\frac{\partial^2}{\partial \eta^2} y(\eta, t) = -f(\eta, t),$$

$$0 \leqslant \eta \leqslant l_1, \quad t \geqslant 0,$$

$$y(\eta, t) = 0 = \frac{\partial^2}{\partial \eta^2} y(\eta, t) \qquad \text{at } \eta = 0, l_1, \quad t \geqslant 0.$$

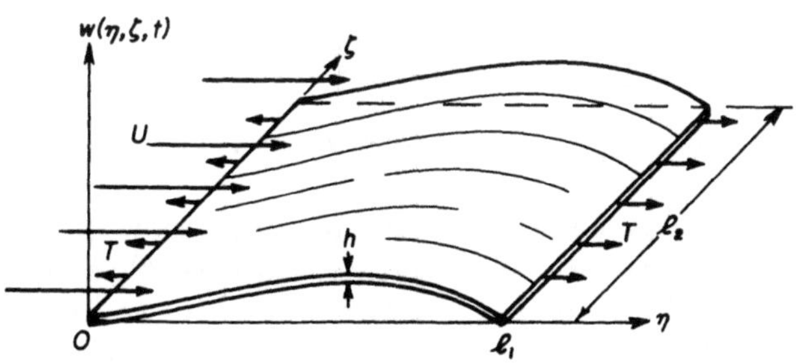

Fig. 10. Supersonic panel.

The mass density ρ and the elastic modulus E are positive, as is h, and $f(\eta, t)$ denotes the aerodynamic pressure at $\eta \in (0, l_1)$. Both $f(\eta, t)$ and T may be negative.

We model the aerodynamic pressure $f(\eta, t)$ by using a linearized "law of plane sections" (Refs. 45, 46), assuming that the airstream has speed $U > c_0$, where c_0 is the speed of sound; hence, we assume that

$$f(\eta, t) = \rho_0 c_0 \left(\frac{\partial}{\partial t} + U \frac{\partial}{\partial \eta} \right) y(\eta, t),$$

where ρ_0 is the mass density of the air. Now defining $\hat{\eta} \equiv \eta / l_1$ and

$$\alpha_1 \equiv Eh^2/12\rho l_1^4 > 0, \qquad \alpha_2 \equiv T/l_1^2 \rho h,$$

$$\alpha_3 \equiv \rho_0 c_0 U/\rho h l_1 > 0, \qquad \alpha_4 \equiv \rho_0 c_0/\rho h > 0,$$

our formal equation becomes

$$\frac{\partial^2}{\partial t^2} \hat{y}(\hat{\eta}, t) + \alpha_4 \frac{\partial}{\partial t} \hat{y}(\hat{\eta}, t) + \alpha_1 \frac{\partial^4}{\partial \hat{\eta}^4} \hat{y}(\hat{\eta}, t) - \alpha_2 \frac{\partial^2}{\partial \hat{\eta}^2} \hat{y}(\hat{\eta}, t)$$

$$+ \alpha_3 \frac{\partial}{\partial \hat{\eta}} \hat{y}(\hat{\eta}, t) = 0, \qquad 0 \leqslant \hat{\eta} \leqslant 1, \quad t \geqslant 0,$$

$$\hat{y}(\hat{\eta}, t) = 0 = \frac{\partial^2}{\partial \hat{\eta}^2} \hat{y}(\hat{\eta}, t) = 0 \qquad \text{at } \hat{\eta} = 0, 1, \quad t \geqslant 0.$$

Defining a linear operator $A : (\mathfrak{D}(A) \subset \mathfrak{X}) \to \mathfrak{X}$,

$$\mathfrak{X} \equiv \hat{\mathfrak{W}}_2^2(0, 1) \times \mathfrak{L}_2(0, 1),$$

$$\|x\|^2 \equiv \int_0^1 \left[\alpha_1 |\partial^2 u(\hat{\eta})|^2 + |v(\hat{\eta})|^2 \right] d\hat{\eta}, \qquad x = (u, v) \in \mathfrak{X},$$

$$\hat{\mathfrak{W}}_2^{2n}(0, 1) \equiv \left\{ u \in \mathfrak{W}_2^{2n}(0, 1) | \partial^{2m} x(0) = 0 = \partial^{2m} x(1);\ m = 0, 1, \ldots, n \right\},$$

$$Ax \equiv \left(v, -\alpha_4 v - \alpha_1 \partial^4 u + \alpha_2 \partial^2 u - \alpha_3 \partial u \right),$$

$$x = (u, v) \in \mathfrak{D}(A) \equiv \hat{\mathfrak{W}}_2^4(0, 1) \times \hat{\mathfrak{W}}_2^2(0, 1),$$

we see that $\mathfrak{D}(A)$ is dense and we can easily show (e.g., by using Proposition III.5.1) that $\omega I - A$ is accretive, some $\omega \in \mathfrak{R}$, and $\mathfrak{R}(I - \lambda A) = \mathfrak{X}$ for all $\lambda \in (0, \lambda_0)$, some $\lambda_0 > 0$. Unfortunately, we find that $\omega > 0$ unless $\alpha_2 = \alpha_3 = 0$, in which case $\omega = 0$. In any case, A generates a linear dynamical system $\{S(t)\}_{t \geqslant 0}$ on $\mathfrak{X}$.

To investigate stability of $x_e = 0$, we define a continuous $V : \mathcal{X} \to \mathcal{R}$ as in Ref. 47,

$$V(x) \equiv \|x\|^2 + \int_0^1 \alpha_2\big[|\partial u(\hat{\eta})|^2 + \alpha_4 v(\hat{\eta})\cdot u(\hat{\eta}) + (\alpha_4^2/2)|u(\hat{\eta})|^2\big]\, d\hat{\eta},$$

$$x = (u, v) \in \mathcal{X}.$$

Hence, for $x \in \mathcal{D}(A)$,

$$\dot{V}(x) = V_x' Ax = -\alpha_4 \int_0^1 \big[\alpha_1|\partial^2 u(\hat{\eta})|^2 + \alpha_2|\partial u(\hat{\eta})|^2$$

$$+ (\alpha_3/\alpha_4)v(\hat{\eta})\cdot \partial u(\hat{\eta}) + |v(\hat{\eta})|^2\big]\, d\hat{\eta}$$

$$\leqslant -\alpha_4 \int_0^1 \big[(\pi^2\alpha_1 + \alpha_2)|\partial u(\hat{\eta})|^2 + (\alpha_3/\alpha_4)v(\hat{\eta})\cdot \partial u(\hat{\eta}) + |v(\hat{\eta})|^2\big]\, d\hat{\eta},$$

and it follows that V is a Liapunov function on all of $\mathcal{D}(A)$ (hence, on all of $\mathcal{X}$) if

$$4(\pi^2\alpha_1 + \alpha_2) \geqslant (\alpha_3/\alpha_4)^2. \tag{1}$$

Similarly, we see that there exists $\varepsilon_1 > 0$ such that $\dot{V}(x) \leqslant -\varepsilon_1\|x\|^2$ for all $x \in \mathcal{X}$ if

$$4(\pi^2\alpha_1 + \alpha_2) > (\alpha_3/\alpha_4)^2. \tag{2}$$

By similar means we see that there exists $\varepsilon_2 > 0$ such that $V(x) \geqslant \varepsilon_2\|x\|^2$ for all $x \in \mathcal{X}$ if $4\pi^2(\pi^2\alpha_1 + \alpha_2) + \alpha_4^2 > 0$, which is implied by (1) since $\alpha_4 \neq 0$. Hence, applying Liapunov's Direct Method, condition (1) implies that $x_e = 0$ is stable and all positive orbits are bounded, while condition (2) implies that $x_e = 0$ is globally exponentially stable with exponent $-\varepsilon_1 t/2\varepsilon_2$.

In terms of the original parameters, we notice that our sufficient condition (2) for exponential stability is

$$4\big[(\pi^2 EH^2/12\rho l_1^2) + T/\rho h\big] > U^2,$$

which is a restriction on the airspeed $U > c_0$. Despite our crude model for the panel and our very questionable model for the aerodynamic pressure, experiments seem to confirm this sufficient condition (Ref. 48). Moreover, these experiments also suggest that this condition is not too far from being necessary; when (1) is "strongly violated," the panel tends to oscillate with increasing amplitude, a highly undesirable phenomenon known as *flutter*.

The quantity $D \equiv Eh^3/12 > 0$ is called the *stiffness* of the beam, and we see that (1) can be written

$$4\left[\pi^2 D + Tl_1^2\right] \geqslant U^2 \rho h l_1^2,$$

a condition that becomes more restrictive as the stiffness D decreases. Again, this trend agrees with experiment, where a panel seems more likely to flutter if it is made less stiff. On the basis of these experimental results, it was once thought that sufficient conditions for panel stability could be obtained by modeling a thin flat panel as a flat membrane, corresponding to the case of zero stiffness. It was later found that this is not true; using the same description of the aerodynamic pressure, it can be shown that the membrane always has a globally exponentially stable equilibrium (Refs. 49, 50). This conclusion is called the membrane flutter paradox (Refs. 45, 49, 50).

To give a simple example related to the supersonic membrane problem, we need only replace our beam model with that of a string, assuming of necessity that $T > 0$. We can describe the problem formally by

$$\frac{\partial^2}{\partial t^2}\hat{y}(\hat{\eta}, t) + \alpha_4 \frac{\partial}{\partial t}\hat{y}(\hat{\eta}, t) - \alpha_2 \frac{\partial^2}{\partial \hat{\eta}}\hat{y}(\hat{\eta}, t) + \alpha_3 \frac{\partial}{\partial \hat{\eta}}(\hat{\eta}, t) = 0,$$

$$0 \leqslant \hat{\eta} \leqslant 1, \quad t \geqslant 0,$$

$$\hat{y}(0, t) = 0 = \hat{y}(1, t), \qquad t \geqslant 0,$$

for some $\alpha_4 > 0$, $\alpha_2 > 0$, $\alpha_3 > 0$. Now defining

$$\mathcal{X} \equiv \hat{\mathcal{W}}_2^1(0, 1) \times \mathcal{L}_2(0, 1),$$

$$\|x\|^2 \equiv \int_0^1 \left[\alpha_2 |\partial u(\hat{\eta})|^2 + |v(\hat{\eta})|^2\right] d\hat{\eta}, \qquad x = (u, v) \in \mathcal{X},$$

$$\hat{\mathcal{W}}_2^n(0, 1) \equiv \left\{u \in \mathcal{W}_2^n(0, 1) | \partial^{2m} u(0) = 0 = \partial^{2m} u(1);\ m = 0, 1, \ldots, \leqslant n/2\right\},$$

$$Ax \equiv \left(v, -\alpha_4 v + \alpha_2 \partial^2 u - \alpha_3 \partial u\right),$$

$$x = (u, v) \in \mathcal{D}(A) \equiv \hat{\mathcal{W}}_2^2(0, 1) \times \hat{\mathcal{W}}_2^1(0, 1),$$

it is easy to show that $A : (\mathcal{D}(A) \subset \mathcal{X}) \to \mathcal{X}$ is the infinitesimal generator of a linear dynamical system $\{S(t)\}_{t \geqslant 0}$ on $\mathcal{X}$.

Defining a continuous $V : \mathcal{X} \to \mathcal{R}$ as

$$V(x) = \int_0^1 \exp\{-\alpha_3 \hat{\eta}/\alpha_2\}\left[\alpha_2 |\partial u(\hat{\eta})|^2 + (\alpha_4^2/2)|u(\hat{\eta})|^2 + \alpha_4 v(\hat{\eta}) \cdot u(\hat{\eta}) + |v(\hat{\eta})|^2\right] d\hat{\eta}, \qquad x = (u, v) \in \mathcal{X},$$

we find that, for $x \in \mathfrak{D}(A)$,

$$\dot{V}(x) = V_x' A x = -\alpha_4 \int_0^1 \exp\{-\alpha_3 \hat{\eta}/\alpha_2\}\left[\alpha_2 |\partial u(\hat{\eta})|^2 + |v(\hat{\eta})|^2\right] d\hat{\eta}$$

$$\leqslant -\alpha_4 \exp\{-\alpha_3/\alpha_2\} \|x\|^2.$$

Theorem IV.1.3 shows that this estimate for $\dot{V}(x)$ holds for all $x \in \mathfrak{X}$. Also, noting that

$$V(x) \geqslant \exp\{-\alpha_3/\alpha_2\} \int_0^1 \left[\alpha_2 |\partial u(\hat{\eta})|^2 + \tfrac{1}{2}|\alpha_4 u(\hat{\eta}) + v(\hat{\eta})|^2 + \tfrac{1}{2}|v(\hat{\eta})|^2\right] d\hat{\eta}$$

$$\geqslant \exp\{-\alpha_3/\alpha_2\} \|x\|^2/2, \qquad x \in \mathfrak{X},$$

$$V(x) \leqslant \int_0^1 \left[\alpha_2 |\partial u(\hat{\eta})|^2 + \alpha_4^2 |u(\hat{\eta})|^2 + (3/2)|v(\hat{\eta})|^2\right] d\hat{\eta}$$

$$\leqslant \int_0^1 \left[(\alpha_2 + \alpha_4^2/\pi^2)|\partial u(\hat{\eta})|^2 + (3/2)|v(\hat{\eta})|^2\right] d\hat{\eta}$$

$$\leqslant \max(1 + \alpha_4^2/\alpha_2 \pi^2, 3/2) \|x\|^2, \qquad x \in \mathfrak{X},$$

we see by Theorem IV.3.1 that the equilibrium $x_e = 0$ is globally exponentially stable, with exponent $-\mu t/2$, for all $\alpha_2 > 0$, $\alpha_3 > 0$, $\alpha_4 > 0$. The exponent can be obtained by employing the same type of estimates to show that

$$V(x) < \max(1 + \alpha_4^2/\alpha_2 \pi^2, 3/2) \int_0^1 \{\exp - \alpha_3 \hat{\eta}/\alpha_2\}\left[\alpha_2 |\partial u(\hat{\eta})|^2 + |v(\hat{\eta})|^2\right] d\hat{\eta}$$

$$< -\mu^{-1} \dot{V}(x), \qquad x \in \mathfrak{X},$$

where

$$\mu \equiv \alpha_4/\max(1 + \alpha_4^2/\alpha_2 \pi^2, 3/2).$$

We see that a supersonic plane membrane, modeled as a string, cannot flutter; a similar analysis in Ref. 50 shows that the same conclusion is true for the general plane membrane. This illustrates the "membrane flutter paradox" and strongly suggests that even very thin panels should not be modeled as membranes, a suggestion that is echoed by singular perturbation theory (Ref. 49).

6. Discrete Dynamical Systems

Let us recall the conclusions of Theorem III.5.1: Given a multivalued operator $A : (\mathfrak{D}(A) \subset \mathfrak{B}) \to \mathfrak{B}$, $\mathfrak{B}$ a Banach space, such that $\omega I - A$ is accretive for some $\omega \in \mathfrak{R}$, and $\mathfrak{R}(I - \lambda A) \supset \mathfrak{X} \equiv \mathrm{Cl}_{\mathfrak{B}}\mathfrak{D}(A)$ for all $\lambda \in (0, \lambda_0)$, some $\lambda_0 > 0$, a dynamical system $\{S(t)\}_{t \geq 0}$ is generated on $\mathfrak{X}$ by the product formula $S(t)x \equiv \lim_{n\to\infty} J^n_{t/n}x$, $x \in$, $t \in \mathfrak{R}^+$, where $J_0 \equiv I$ and $J_\lambda \equiv (I - \lambda A)^{-1}$, $\lambda \in (0, \lambda_0)$; further, $\|S(t)x - S(t)\hat{x}\| \leq e^{\omega t}\|x - \hat{x}\|$ for all $x, \hat{x} \in \mathfrak{X}$, $t \in \mathfrak{R}^+$. We also recall that this product formula was directly related to the backward-difference equation (7) of Chapter III,

$$\frac{1}{(t/n)}\big[z(mt/n) - z(mt/n - t/n)\big] \in Az(mt/n),$$

$$t \in \mathfrak{R}^+, m = 1, 2, \ldots, n,$$

$$z(0) = x_0 \in \mathrm{Cl}_{\mathfrak{B}}\mathfrak{D}(A) \equiv \mathfrak{X},$$

which is equivalent to the forward-difference equation

$$z(mt/n) = J_{t/n}z(mt/n - t/n), \qquad t \in \big[0, n\lambda_0), m = 1, 2, \ldots, n,$$

$$z(0) = x_0 \in \mathrm{Cl}_{\mathfrak{B}}\mathfrak{D}(A) \equiv \mathfrak{X}.$$

Defining $\varepsilon \equiv t/n$, $y_\varepsilon(m) \equiv z(m\varepsilon - \varepsilon)$, the last equation becomes

$$y_\varepsilon(m+1) = J_\varepsilon y_\varepsilon(m), \qquad \varepsilon = t/n, \quad m = 0, 1, \ldots, n-1,$$

$$y_\varepsilon(0) = x_0 \in \mathfrak{X}.$$

Hence, if we are interested in approximating $S(t)x_0$ by $J^n_{t/n}x_0$ for each $t \in \mathfrak{R}^+$, while holding the time increment $\varepsilon = t/n$ fixed, we see that we are interested in a difference equation of the form

$$y(m+1) = Hy(m), \qquad m \in \mathfrak{N}^+,$$

$$y(0) = x_0 \in \mathfrak{X}, \tag{3}$$

where $\mathfrak{N}^+ \equiv \{0, 1, 2, \ldots\} \subset \mathfrak{R}^+$ and the function $H : \mathfrak{X} \to \mathfrak{X}$ is continuous. It is apparent that a unique solution $y(m) = H^m x_0$ of (3) exists for every $x_0 \in \mathfrak{X}$.

Aside from approximating differential equations by difference equations, equations of the form (3) also arise in other ways, such as the use of successive approximation schemes on many problems in which "time" is not even involved. Hence, there is considerable interest in the properties of solutions of (3) for given continuous $H : \mathcal{X} \to \mathcal{X}$.

Definition 6.1. A *discrete dynamical system* on a metric space $\mathcal{X}$ is a mapping $u : \mathcal{N}^+ \times \mathcal{X} \to \mathcal{X}$ such that

(i) $u(n, \cdot) : \mathcal{X} \to \mathcal{X}$ is continuous,

(ii) $u(0, x) = x$,

(iii) $u(m + n, x) = u(m, u(n, x))$

for all $m, n \in \mathcal{N}^+$, $x \in \mathcal{X}$.

We see that a discrete dynamical system on $\mathcal{X}$ is a direct analogue of a dynamical system on $\mathcal{X}$, with $\mathcal{R}^+$ replaced by $\mathcal{N}^+$ (see Definition III.1.1). However, defining $Hx \equiv u(1, x)$, $x \in \mathcal{X}$, we also see that a discrete dynamical system is much simpler. By (i) the operator $H : \mathcal{X} \to \mathcal{X}$ is continuous, and (ii) and (iii) imply that $u(n, x) = H^n x$ for all $n \in \mathcal{N}^+$, $x \in \mathcal{X}$, with $H^0 \equiv I$; moreover, every motion of u is a solution of (3), and conversely, since solutions of (3) are unique. Also note that if there exists $\beta \in \mathcal{R}$ such that $d(Hx, H\hat{x}) \leqslant \beta d(x, \hat{x})$ for all $x, \hat{x} \in \mathcal{X}$, then the discrete dynamical system u (or $\{H^n\}_{n\in\mathcal{N}^+}$) satisfies the estimate $d(H^n x, H^n \hat{x}) \leqslant \beta^n d(x, \hat{x})$ for all $n \in \mathcal{N}^+$, $x \in \mathcal{X}$. Hence, we have obtained the analogues of essentially all results in Chapter III.

To parallel the results of Chapter IV, we need only define the *motion* $H^{(\cdot)}x\ \mathcal{N}^+ \to \mathcal{X}$, the *positive orbit* $\gamma(x) \equiv \cup_{n\in\mathcal{N}^+} H^n x$, the *positive limit set* $\Omega(x) \equiv \cap_{n\in\mathcal{N}^+} \text{Cl}\ \gamma(H^n x)$, and the idea of a Liapunov function for $\{H^n\}_{n\in\mathcal{N}^+}$ on $\mathcal{G} \subset \mathcal{X}$.

Definition 6.2. Let $\{H^n\}_{n\in\mathcal{N}^+}$ be a discrete dynamical system on a metric space $\mathcal{X}$, and let $V : \mathcal{X} \to \overline{\mathcal{R}}$ be lower semicontinuous. V is a *Liapunov function* for $\{H^n\}_{n\in\mathcal{N}^+}$ on a set $\mathcal{G} \subset \mathcal{X}$ if $\dot{V}(x) \leqslant 0$ for every $x \in \mathcal{G}$, where $\dot{V} : \mathcal{X} \to \overline{\mathcal{R}}$ is the function defined by

$$\dot{V}(x) \equiv V(Hx) - V(x) \quad \text{if } |V(x)| < \infty,$$

$$\dot{V}(x) \equiv 0 \quad \text{if } V(x) = +\infty, \qquad \dot{V}(x) \equiv 1 \quad \text{if } V(x) = -\infty.$$

We see that this definition is a direct analogue of Definition IV.1.1, and by its use we can obtain analogues of essentially all results of Chapter IV.

Proposition 6.1. Let $\{H^n\}_{n\in\mathcal{N}^+}$ be a discrete dynamical system on a metric space $\mathcal{X}$, and let V be a Liapunov function for $\{H^n\}_{n\in\mathcal{N}^+}$ on

$\mathcal{G} \subset \mathfrak{X}$. If $x \in \mathcal{G}$, $V(x) < \infty$, and $H^n x \in \mathcal{G}$ for all $n = 0, 1, \ldots, N$, some $N \in \mathfrak{N}^+$, then $V(H^{(\cdot)}x) : \mathfrak{N}^+ \to \mathfrak{X}$ ix nonincreasing on $[0, N] \subset \mathfrak{N}^+$, with

$$V(H^n x) = V(x) + \sum_{m=0}^{n-1} \dot{V}(H^m x) \qquad \text{for all } n = 1, 2, \ldots, N.$$

If, in addition, $\dot{V}(y) \leqslant -\alpha V(y)$ for all $y \in \mathcal{G}$, some $\alpha \leqslant 1$, then

$$V(H^n x) \leqslant (1 - \alpha)^n V(x) \qquad \text{for all } n = 1, 2, \ldots, N.$$

Exercise 6.1. Prove Proposition 6.1. *Hint:* See Theorem IV.1.1.

Exercise 6.2. In Definition 6.2, suppose that $\mathcal{G}$ contains the set $\bar{\mathcal{G}}_{\mathfrak{B}} \equiv \{x \in \mathfrak{X} \mid V(x) \leqslant \beta\}$ for some $\beta < \infty$. Show that $\bar{\mathcal{G}}_\delta$ is positive invariant for every $\delta \leqslant \beta$, where $\bar{\mathcal{G}}_\delta \equiv \{x \in \mathfrak{X} \mid V(x) \leqslant \delta\}$. *Hint:* See Theorem IV.1.2.

One can also obtain analogous versions of Liapunov's Direct Method and the Invariance Principle; these and other ideas are discussed in some detail in Ref. 51. Here we mention only that the construction of useful Liapunov functions remains very important. If $\mathfrak{X} = \mathfrak{R}^p$, equipped with the Euclidean inner product $\langle \cdot, \cdot \rangle$, and H is a linear operator represented by a real $p \times p$ matrix, then the analogue of Liapunov's equation (see Section I.9) is given by

$$H^T B H - B = -C,$$

where $C = C^T > 0$ and $B = B^T$ are real $p \times p$ matrices. If suitable B and C exist, then $V(x) \equiv \langle x, Bx \rangle$ is a Liapunov function on all of $\mathfrak{R}^p$, with $\dot{V}(x) = -\langle x, Cx \rangle$, $x \in \mathfrak{R}^p$. It can be shown that a unique solution $B = B^T$ exists for *every* given $C = C^T$ if and only if no (complex) eigenvalue of H has modulus one (Ref. 51).

We notice that the idea of a discrete dynamical system furnishes an interesting way of viewing various fixed-point theorems. An *equilibrium* of $\{H^n\}_{n \in \mathfrak{N}^+}$ is a point $x_e \in \mathfrak{X}$ such that $H^n x_e = x_e$ for all $n \in \mathfrak{N}^+$; hence, x_0 is a fixed point of H if and only if x_0 is an equilibrium of $\{H^n\}_{n \in \mathfrak{N}^+}$.

Autonomous physical systems normally lead to dynamical systems rather than to discrete dynamical systems. Consequently, let us return to the approximation problem described earlier, assuming that a multivalued operator $A : (\mathfrak{D}(A) \subset \mathfrak{B}) \to \mathfrak{B}$, $\mathfrak{B}$ a Banach space, generates a dynamical system $\{S(t)\}_{t \geqslant 0}$ on $\mathfrak{X} \equiv \mathrm{Cl}\, \mathfrak{D}(A)$ in the sense of Theorem III.5.1.

Defining $J_0 \equiv I$, $J_\lambda \equiv (I - \lambda A)^{-1}$ for $\lambda \in (0, \lambda_0)$, and choosing $\varepsilon \in (0, \lambda_0)$, we wish to compare $\{S(t)\}_{t \geqslant 0}$ with the discrete dynamical system $\{J_e^m\}_{m \in \mathfrak{N}^+}$ at $t = 0, \varepsilon, 2\varepsilon, \ldots$. To this end, recall that the estimate (2) of Section III.3 applies, with $\|Ax\|$ replaced by $|Ax|$ for all $x \in \mathfrak{D}(A)$. This implies that

$$\|J^n_{m\varepsilon/n}x - J^m_\varepsilon x\| \leq 2m\varepsilon \exp\{4|\omega|m\varepsilon\}\left(\frac{1}{m} - \frac{1}{n}\right)^{1/2}|Ax|, \qquad x \in \mathfrak{D}(A),$$

for $n > m = 0, 1, \ldots,$ assuming that $\omega I - A$ is accretive and $\mathfrak{R}(I - \lambda A) = \mathfrak{X}$ for all $\lambda \in (0, \varepsilon]$; hence, we see that

$$\|S(m\varepsilon)x - J^m_\varepsilon x\| \leq 2m^{1/2}\varepsilon \exp\{4|\omega|m\varepsilon\}|Ax|, \qquad x \in \mathfrak{D}(A), \quad m \in \mathfrak{N}^+.$$

Moreover, $-A$ is accretive if $\omega I - A$ is accretive for some $\omega \leq 0$, and therefore this estimate implies that

$$\|S(m\varepsilon)x - J^m_\varepsilon x\| \leq 2m^{1/2}\varepsilon|Ax|, \qquad x \in \mathfrak{D}(A), \quad m \in \mathfrak{N}^+,$$

if $\omega \leq 0$.

These "error estimates" are poor for large $m \in \mathfrak{N}^+$; if $\omega < 0$, another estimate can be obtained. If $\omega < 0$, we see as in Exercise III.6.5 that there exists x_e such that $J_\lambda x_e = x_e$ for all $\lambda \in (0, \lambda_0)$; hence, x_e is an equilibrium of both $\{J^m_\varepsilon\}_{m\in\mathfrak{N}^+}$ and $\{S(t)\}_{t\geq 0}$, with

$$\|S(t) - x_e\| \leq e^{\omega t}\|x - x_e\|, \qquad x \in \mathfrak{X}, \quad t \in \mathfrak{R}^+,$$

$$\|J^m_\varepsilon x - x_e\| \leq (1 + \varepsilon\omega)^m\|x - x_e\|, \qquad x \in \mathfrak{X}, \quad m \in \mathfrak{N}^+.$$

It follows that if $\omega < 0$,

$$\|S(m\varepsilon)x - J^m_\varepsilon x\| \leq \left[e^{\omega m\varepsilon} + (1 + \varepsilon\omega)^m\right]\|x - x_e\|, \qquad x \in \mathfrak{X}, \quad m \in \mathfrak{N}^+,$$

$$\to 0 \qquad \text{as } m \to \infty.$$

We see that x_e is a globally asymptotically stable equilibrium of $\{J^m_\varepsilon\}_{m\in\mathfrak{N}^+}$, as well as of $\{S(t)\}_{t\in\mathfrak{R}^+}$, and the "large time" error (in approximating $S(\cdot)x : \mathfrak{R}^+ \to \mathfrak{X}$ by $J^{(\cdot)}_\varepsilon x : \mathfrak{N}^+ \to \mathfrak{X}$) decays to zero as $t = m\varepsilon \to \infty$, $m \in \mathfrak{N}^+$, for each $x \in \mathfrak{X}$, provided that $\omega I - A$ is accretive for some $\omega < 0$.

7. Finite-Dimensional Approximation

Recall the evolution equation (6) of Chapter III,

$$\dot{x}(t) \in Ax(t), \qquad \text{a.e. } t \in \mathfrak{R}^+,$$

$$x(0) = x_0 \in \text{Cl}\,\mathfrak{D}(A) \equiv \mathfrak{X} \subset \mathfrak{B}, \tag{4}$$

for some given multivalued operator $A : (\mathfrak{D}(A) \subset \mathfrak{B}) \to \mathfrak{B}$, $\mathfrak{B}$ a Banach space, and suppose that there exists a strong solution $x(\cdot) : \mathfrak{R}^+ \to \mathfrak{X}$ for some given $x_0 \in \text{Cl}\,\mathfrak{D}(A)$. Suppose that we wish to "approximate" $x(\cdot)$ by a strong solution $z(\cdot) : \mathcal{R}^+ \to \tilde{\mathcal{X}}$ of

$$\dot{z}(t) \in \tilde{A}z(t), \qquad \text{a.e. } t \in \mathcal{R}^+,$$
$$z(0) = z_0 \in \text{Cl } \mathcal{D}(\tilde{A}) \equiv \tilde{\mathcal{X}} \subset \tilde{\mathcal{B}}, \tag{5}$$

where $\tilde{\mathbf{B}}$ is some *finite*-dimensional subspace of $\mathcal{B}$, $\dim \hat{\mathcal{B}} < \dim \mathcal{B}$, $\|z\|_{\tilde{\mathcal{B}}} \equiv \|z\|_{\mathcal{B}}$ for $z \in \tilde{\mathcal{B}}$, and $\tilde{A} : (\mathcal{D}(\tilde{A}) \subset \tilde{\mathcal{B}}) \to \tilde{\mathcal{B}}$ is some multivalued operator. We introduce the lower-dimensional subspace $\hat{\mathcal{B}} \subset \mathcal{B}$ in the hope that the lower dimensionality will enable $z(\cdot)$ to be explicitly determined more easily than $x(\cdot)$, but we obviously require some estimate of the error $e(t) \equiv x(t) - z(t)$, $t \in \mathcal{R}^+$. Clearly, this error depends on our choices of $\tilde{\mathcal{B}} \subset \mathcal{B}$, $\tilde{A} : (\mathcal{D}(\tilde{A}) \subset \tilde{\mathcal{B}}) \to \tilde{\mathcal{B}}$, and $z_0 \in \text{Cl } \mathcal{D}(\tilde{A})$. Any program for making these choices is called an *approximation scheme* for (4), and the value of such a program depends on the existence of a suitable error estimate. Many approximation schemes are based on the use of projections.

Definition 7.1. For $\mathcal{Y}$ a linear space, a *projection* is a linear operator $P : \mathcal{Y} \to \mathcal{Y}$ such that $P^2 = P$; i.e., such that $Py = y$ for every $y \in \mathcal{R}(P)$.

If $P : \mathcal{Y} \to \mathcal{Y}$ is a projection, $\mathcal{Y}$ a linear space, we see that $I - P$ is also a projection, and $P(I - P) = 0 = (I - P)P$. If $\mathcal{Y}$ is normed and P is not the null operator, then $\|P\|_{\mathcal{B}(\mathcal{Y}, \mathcal{Y})} \geqslant 1$.

Suppose that $\mathcal{Y}$ is an inner product space, and consider an n-dimensional subspace $\hat{\mathcal{Y}}$ spanned by $a_1, a_2, \ldots, a_n \in \hat{\mathcal{Y}} \subset \mathcal{Y}$. If $P : \mathcal{Y} \to \mathcal{Y}$ is a projection with range $\mathcal{R}(P) = \hat{\mathcal{Y}}$, we see that P must be of the form $Py \equiv \sum_{m=1}^n \langle y, b_m \rangle a_m$, $y \in \mathcal{Y}$, for some $b_1, b_2, \ldots, b_n \in \mathcal{Y}$ such that $\langle a_p, b_m \rangle = \delta_{pm}$. In general, there are many different projections having the same range $\mathcal{R}(P) = \hat{\mathcal{Y}}$, unless $\hat{\mathcal{Y}} = \mathcal{Y}$. Now suppose that $\hat{\mathcal{Y}}$ is spanned by orthonormal $a_1, a_2, \ldots, a_n$, and make the particular choice $b_m \equiv a_m$, $m = 1, 2, \ldots, n$. Then we see that the projection defined by

$$Py \equiv \sum_{m=1}^{n} \langle y, a_m \rangle a_m, \qquad y \in \mathcal{Y},$$

is symmetric with $\mathcal{R}(P) = \hat{\mathcal{Y}}$. Symmetric projections are often called "orthogonal projections" (Ref. 52), although this terminology leads to some ambiguity when considering families of projections. Given the subspace $\hat{\mathcal{Y}} \subset \mathcal{Y}$, it is not difficult to show that there exists exactly one symmetric projection $P : \mathcal{Y} \to \mathcal{Y}$ such that $\mathcal{R}(P) = \hat{\mathcal{Y}}$; however, its representation obviously varies with the choice of the spanning set $\{a_m\}$ for $\hat{\mathcal{Y}}$. We notice that $\|P\|_{\mathcal{B}(\mathcal{Y}, \mathcal{Y})} = 1$ if P is symmetric and not null, for $\mathcal{Y}$ a pre-Hilbert space.

Returning to the approximation problem for (4), notice that if $x(\cdot) : \mathcal{R} \to \mathcal{X}$ is a strong solution of (4), $\mathcal{X} \equiv \text{Cl } \mathcal{D}(A) \subset \mathcal{B}$, $\mathcal{B}$ a Banach space, and if $P : \mathcal{B} \to \mathcal{B}$ is any given projection, then $P\dot{x}(t) \in PAx(t)$ a.e. $t \in \mathcal{R}^+$. This idea leads to the following approximation scheme: Consider (5) with

$$\tilde{\mathcal{B}} \equiv \mathcal{R}(P) \subset \mathcal{B}, \qquad \|\cdot\|_{\tilde{\mathcal{B}}} \equiv \|\cdot\|_{\mathcal{B}},$$

$$\tilde{A}z \equiv PAz, \qquad z \in \mathcal{D}(\tilde{A}) \equiv \mathcal{D}(A) \cap \mathcal{R}(P),$$

$$z_0 \equiv Px_0,$$

assuming that $P\mathcal{X} \subset \tilde{\mathcal{X}} \equiv \text{Cl } \mathcal{D}(\tilde{A}) \subset \tilde{\mathcal{B}}$, $\mathcal{X} \equiv \text{Cl } \mathcal{D}(A) \subset \mathcal{B}$.

This approximation scheme is a slight generalization of Galerkin's method (Ref. 53); in Galerkin's method it is assumed that $\mathcal{B}$ is a Hilbert space and that P is a symmetric projection with finite-dimensional $\mathcal{R}(P) \equiv \tilde{\mathcal{B}}$. Under these additional assumptions, notice that if $\omega I - A$ is accretive for some $\omega \in \mathcal{R}$, then for all $z, \hat{z} \in \mathcal{D}(\tilde{A})$ we have

$$\begin{aligned} \langle z - \hat{z}, \hat{A}z - \tilde{A}\hat{z} \rangle &= \langle z - \hat{z}, PAz - PA\hat{z} \rangle \\ &= \langle Pz - P\hat{z}, Az - A\hat{z} \rangle \\ &= \langle z - \hat{z}, Az - A\hat{z} \rangle \leqslant \omega \|z - \hat{z}\|^2; \end{aligned}$$

hence, $\omega I - \tilde{A} : (\mathcal{D}(\tilde{A}) \subset \tilde{\mathcal{B}}) \to \mathcal{B}$ is also accretive. However, this result does not always hold for our generalization of Galerkin's method.

If we choose to use this approximation scheme, we shall need an error estimate. Such estimates can be very difficult to obtain for nonlinear problems; hence, let us consider the linear case of (4),

$$\begin{aligned} \dot{x}(t) &= Ax(t), \qquad t \in \mathcal{R}^+, \\ x(0) &= x_0 \in \mathcal{D}(A) \subset \mathcal{B}, \end{aligned} \tag{6}$$

assuming A to be the infinitesimal generator of a linear dynamical system $\{S(t)\}_{t \geqslant 0}$ on the Banach space $\mathcal{B}$. It follows that $\mathcal{D}(A)$ is dense, $\mathcal{R}(I - \lambda A) = \mathcal{B}$ for sufficiently small $\lambda > 0$, and $\|S(t)x\| \leqslant Me^{\omega t}\|x\|$ for all $x \in \mathcal{B}$, some $M, \omega \in \mathcal{R}$; moreover, $S(\cdot)x_0 : \mathcal{R}^+ \to \mathcal{B}$ is the (unique) strong solution of (6). Let $P : \mathcal{B} \to \mathcal{B}$ be any projection with finite-dimensional range $\mathcal{R}(P) \subset \mathcal{D}(A)$, and define the Banach space $\tilde{\mathcal{B}} \equiv \mathcal{R}(P)$, $\|\cdot\|_{\tilde{\mathcal{B}}} \equiv \|\cdot\|_{\mathcal{B}}$. Now defining $\tilde{A}z \equiv PAz$ for $z \in \mathcal{D}(\tilde{A}) \equiv \mathcal{D}(A) \cap \mathcal{R}(P) = \mathcal{R}(P)$, we see that $\tilde{A} : \tilde{\mathcal{B}} \to \tilde{\mathcal{B}}$ is a linear operator which must be bounded (see Proposition II.3.1). Hence, $\tilde{A}$ generates a linear dynamical system $\{\tilde{S}(t)\}_{t \geqslant 0}$ on $\tilde{\mathcal{B}}$, satisfying $\|\tilde{S}(t)z\| \leqslant \tilde{M}e^{\tilde{\omega}t}\|z\|$ for all $z \in \tilde{\mathcal{B}}$, some $\tilde{M}, \tilde{\omega} \in \mathcal{R}$ (e.g., $\tilde{M} = 1$, $\tilde{\omega} = \|\tilde{A}\|$); moreover, $\tilde{S}(\cdot)z_0 : \mathcal{R}^+ \to \tilde{\mathcal{B}}$ is the (unique) strong solution of

$$\dot{z}(t) = \tilde{A}z(t), \qquad t \in \mathcal{R}^+,$$

$$z(0) = Px_0 \in \mathcal{D}(\tilde{A}) = \tilde{\mathcal{B}}.$$

Now defining the error $e(t) \equiv S(t)x_0 - \tilde{S}(t)Px_0$, we see that $e(\cdot):\mathcal{R}^+ \to \mathcal{B}$ is everywhere differentiable, $e(0) = (I - P)x_0$, and

$$\begin{aligned}\dot{e}(t) &= AS(t)x_0 - \tilde{A}\tilde{S}(t)Px_0\\ &= Ae(t) + (I - P)A\tilde{S}(t)Px_0\\ &= Ae(t) + (I - P)AP\tilde{S}(t)Px_0, \qquad t \in \mathcal{R}^+.\end{aligned}$$

We notice that $(I - P)AP$ is bounded (even if A is not), and a variation-of-constants formula (see Lemma IV.5.2) provides an error estimate,

$$\|e(t)\| = \|S(t)e(0) + \int_0^t S(t-\tau)(I - P)AP\tilde{S}(\tau)Px_0\, d\tau\|$$

$$\leqslant \|S(t)e(0)\| + \int_0^t Me^{\omega(t-\tau)}\|(I - P)AP\|\tilde{M}e^{\tilde{\omega}\tau}\|Px_0\|\, d\tau$$

$$\leqslant Me^{\omega t}\Big(\|(I - P)x_0\| + \tilde{M}\|(I - P)AP\|\|Px_0\|\int_0^t e^{(\tilde{\omega}-\omega)\tau}\, d\tau\Big), \qquad t \in \mathcal{R}^+.$$

Notice that we can conclude that $\|e(t)\| \to 0$, as $t \to \infty$, if (and generally only if) both $\omega < 0$ and $\hat{\omega} < 0$. An interesting (but not particularly useful) special case occurs when $\mathcal{B}$ is a Hilbert space, P is symmetric, and $\tilde{\mathcal{B}} = \mathcal{R}(P)$ is spanned by a (finite) set of orthogonal eigenvectors of A; then the operator P commutes with A, $(I - P)AP = A(I - P)P = 0$, $\|e(t)\| \leqslant Me^{\omega t}\|(I - P)x_0\|$, $\mathcal{R}(P)$ is positive invariant under $\{S(t)\}_{t \geqslant 0}$, and $\tilde{S}(t)$ is the restriction of $S(t)$ to $\mathcal{R}(P)$, $t \in \mathcal{R}^+$.

Once an error estimate has been obtained, we might wish to investigate the convergence properties of a given approximation scheme. Suppose, for example, that for our linear problem (6) there exists a family $\{P_n\}_{n \in \mathcal{K}^+}$ of projections such that $\mathcal{R}\{P_n\} \subset \mathcal{D}(A)$ and $P_n x \to x$ as $n \to \infty$, for every $x \in \mathcal{B}$. Suppose also that $(I - P_n)AP_n x \to 0$ as $n \to \infty$, for every $x \in \mathcal{B}$; then the preceding estimates show that

$$\max_{t \in [0,\, T]} \|S(t)x_0 - \tilde{S}_n(t)P_n x_0\| \to 0 \qquad \text{as} \qquad n \to \infty,$$

for each finite $T > 0$.

Many interesting questions in approximation theory remain open. The fundamental problem is to obtain a good error estimate.

References

1. CHEN, C. T., *Introduction to Linear System Theory*, Holt, Rinehart and Winston, New York, 1970.
2. EVELEIGH, V. W., *Introduction to Control Systems Design*, McGraw-Hill, New York, 1972.
3. DATKO, R., A linear control problem in abstract Hilbert space, *Journal of Differential Equations*, Vol. 9, pp. 346–359, 1971.
4. DATKO, R., Unconstrained control problems with quadratic cost, *SIAM Journal on Control*, Vol. 11, pp. 32–52, 1973.
5. FATTORINI, H. O., Boundary control systems, *SIAM Journal on Control*, Vol. 3, pp. 349–385, 1968.
6. RUSSELL, D. L., Boundary control of the higher dimensional wave equation, *SIAM Journal on Control*, Vol. 9, pp. 29–42, 1971.
7. SLEMROD, M., The linear stabilization problem in Hilbert space, *Journal of Functional Analysis*, Vol. 11, pp. 334–345, 1972.
8. SLEMROD, M., An application of the theory of maximal dissipative sets in control theory, *Journal of Mathematical Analysis and Applications*, Vol. 46, pp. 364–387, 1974.
9. SLEMROD, M., Stabilization of boundary control systems, *Journal of Differential Equations*, Vol. 22, pp. 402–415, 1976.
10. PARKS, P. C., and PRITCHARD, A. J., Paper 20.5, *Proceedings of the Fourth IFAC Congress*, held in Warsaw, Poland; International Federation of Automatic Control, London, 1969.
11. PARKS, P. C., Some applications of Liapunov functions, *Instability of Continuous Systems*. Edited by H. Leipholz, Springer-Verlag, New York, 1971.
12. CHADWICK, P., Thermal damping of a vibrating elastic body, *Mathematika*, Vol. 9, pp. 38–48, 1962.
13. PÄSLER, M., *Zeitschrift für Physik*, Vol. 122, pp. 357–386, 1944.
14. ZENER, C., Internal friction in solids. I: Theory of internal friction in rods, *Physical Review*, Vol. 52, pp. 230–235, 1937.
15. ZENER, C., Internal friction in solids. II: General theory of thermoelastic internal friction, *Physical Review*, Vol. 53, pp. 90–99, 1938.
16. ERICKSEN, J. L., A thermo-kinetic view of elastic stability theory, *International Journal of Solids and Structures*, Vol. 2, pp. 573–580, 1966.
17. ERICKSEN, J. L., Thermoelastic stability, *Proceedings of the Fifth U.S. National Congress of Applied Mechanics*, American Society of Chemical Engineers, New York, pp. 187–193, 1966.
18. SLEMROD, M., and INFANTE, E. F., An invariance principle for dynamical systems on Banach space: Application to the general problem of thermoelastic stability, *Instability of Continuous Systems*. Edited by H. Leipholz, Springer-Verlag, New York, 1971.
19. DAFERMOS, C., On the existence and asymptotic stability of solutions to the equations of linear thermoelasticity, *Archive for Rational Mechanics and Analysis*, Vol. 29, pp. 241–271, 1968.
20. LANDAU, L. D., and LIFSHITZ, E. M., *Theory of Elasticity*, Addison-Wesley, Reading, Massachusetts, 1959.
21. TRUESDELL, C., and NOLL, W., *The Nonlinear Field Theories of Mechanics*, Springer-Verlag, New York, 1965.
22. FICHERA, G., *Lectures on Elliptic Boundary Differential Systems and Eigenvalue Problems*, Springer-Verlag, New York, 1965.
23. COLEMAN, B. D., and NOLL, W., An approximation theorem for functionals with applications in continuum mechanics, *Archive for Rational Mechanics and Analysis*, Vol. 6, pp. 355–370, 1960.

24. Coleman, B. D., and Noll, W., Foundations of linear viscoelasticity, *Reviews of Modern Physics*, Vol. 33, pp. 239–249, 1961.
25. Wang, C. C., The principle of fading memory, *Archive for Rational Mechanics and Anaylsis*, Vol. 18, pp. 343–366, 1965.
26. Dafermos, C. M., Asymptotic stability in viscoelasticity, *Archive for Rational Mechanics and Analysis*, Vol. 37, pp. 297–308, 1970.
27. Dafermos, C. M., An abstract Volterra equation with applications to linear viscoelasticity, *Journal of Differential Equations*, Vol. 7, pp. 554–569, 1970.
28. Travis, C. C., and Webb, G. F., Partial differential equations with deviating arguments in the time variable, *Journal of Mathematical Analysis and Applications*, Vol. 56, pp. 397–409, 1976.
29. Webb, G. F., Asymptotic stability for abstract nonlinear functional differential equations, *Proceedings of the American Mathematical Society*, Vol. 54, pp. 225–230, 1976.
30. Ergen, W. K., Kinetics of a circulating fuel nuclear reactor, *Journal of Applied Physics*, Vol. 25, pp. 702–711, 1954.
31. Baran, W., and Meyer, K., Effect of delayed neutrons on the stability of a nuclear reactor, *Nuclear Science and Engineering*, Vol. 24, pp. 356–361, 1966.
32. Bronikowski, T. A., Hall, J. E., and Nohel, J. A., Quantitative estimates for a nonlinear system of integrodifferential equations arising in reactor dynamics, *SIAM Journal on Mathematical Analysis*, Vol. 3, pp. 567–588, 1972.
33. Ergen, W. K., and Nohel, J. A., Stability of a continuous-medium reactor, *Journal of Nuclear Engineering (A): Reactor Science*, Vol. 10, pp. 14–18, 1959.
34. Infante, E. F., and Walker, J. A., On the stability properties of an equation arising in reactor dynamics, *Journal of Mathematical Analysis and Applications*, Vol. 55, pp. 112–124, 1976.
35. Levin, J. J., and Nohel, J. A., On a system of integrodifferential equations occurring in reactor dynamics, *Journal of Mathematics and Mechanics*, Vol. 9, pp. 347–368, 1960.
36. Levin, J. J., and Nohel, J. A., A system of nonlinear integrodifferential equations, *Michigan Mathematical Journal*, Vol. 13, pp. 267–270, 1966.
37. Levin, J. J., and Nohel, J. A., The integrodifferential equations of a class of nuclear reactors with delayed neutrons, *Archive for Rational Mechanics and Analysis*, Vol. 31, pp. 151–171, 1968.
38. Suhadolc, A., On a system of integrodifferential equations, *SIAM Journal on Applied Mathematics*, Vol. 21, pp. 195–206, 1971.
39 Walker, J. A., and Infante, E. F., On the stability of an operator equation modeling nuclear reactors with delayed neutrons, *Quarterly of Applied Mathematics*, Vol. 34, pp. 421–427, 1971.
40. Wexler, D., Frequency domain stability for a class of equations arising in reactor dynamics, *SIAM Journal on Mathematical Analysis*, Vol. 10, pp. 118–138, 1979.
41. Kalman, R. E., Liapunov functions for the problem of Lur'e in automatic control, *Proceedings of the U.S. National Academy of Sciences*, Vol. 49, p. 201, 1963.
42. Yacubovich, V. A., The solution of certain matrix inequalities in automatic control theory, *Doklady Akademii Nauk SSSR*, Vol. 143, pp. 1304–1307, 1962.
43. Meyer, K. R., Liapunov functions for the problem of Lur'e, *Proceedings of the U.S. National Academy of Science*, Vol. 53, pp. 501–503, 1965.
44. Brusin, V. A., Equations of Lur'e in Hilbert space and their solvability, *Prikladnaya Matematika i Mekhanika*, Vol. 40, pp. 947–956, 1976.
45. Bolotin, V. V., *Nonconservative Problems of the Theory of Elastic Stability*, Macmillan Company, New York, 1963.
46. Ashley, H., and Zartarian, G., Piston theory: A new aerodynamic tool for the aeroelastician, *Journal of the Aeronautical Sciences*, Vol. 20, pp. 1109–1118, 1956.

47. Parks, P. C., A stability criterion for panel flutter via the Second Method of Liapunov, *AIAA Journal*, Vol. 4, pp. 175–177, 1966.
48. Fung, Y. C. B., *A Summary of the Theories and Experiments on Panel Flutter*, Grad. Aeronautical Laboratories, California Institute of Technology, Air Force Office of Scientific Research GALCIT-AFOSR TN 60-224, 1960.
49. Spriggs, J. H., Messiter, A. F., and Anderson, W. J., Membrane flutter paradox: An explanation by singular perturbation methods, *AIAA Journal*, Vol. 7, pp. 1704–1709, 1969.
50. Walker, J. A., and Dixon, M., Stability of the general plane membrane adjacent to a supersonic airstream, *ASME Journal of Applied Mechanics*, Vol. 40, pp. 395–398, 1973.
51. LaSalle, J. P., *The Stability of Dynamical Systems*, CBMS Regional Conference Series in Applied Mathematics, Society for Industrial and Applied Mathematics, Philadelphia, 1976.
52. Kato, T., *Perturbation Theory for Linear Operators*, Springer-Verlag, New York, 1966.
53. Oden, J. T., and Reddy, J. N., *An Introduction to the Mathematical Theory of Finite Elements*, Wiley-Interscience, New York, 1976.

Index

Accretive operator, 62
  multivalued, 119
Affine function, 22
Asymptotic behavior of motions, 38-42, 164-174
Asymptotic stability, 23, 154, 155
Attraction, domain of, 37
Autonomous evolution equation, 13, 90, 19

Banach space, 54
Beam problems
  linear, 111, 112
  nonlinear, 125
Bounded operator, 61
Bounded set, 52

Canonical restriction of multivalued operator, 119
Cauchy sequence, 52
Chain rule, 78
Closed operator, 63
  multivalued, 119
Closed set, 52
Closure, 52
Compact function, 61
Compact set, 52, 59
Complete space, 52
Completely observable, 40
Compexification, 73
Composite function, 60
Continuation theorem, 6
Continuity, 60
Continuous spectrum, 72
Contraction, 61
Control problems, 36, 37, 200
Convergent sequence, 52
Crandall-Liggett theorem, 121

Dense set, 52
Derivative, 75
  chain rule for, 78
  Fréchet, 76
  generalized, 58
  left, 75
  right, 75
Difference equation
  backward, 102, 120, 122
  forward, 22
Differential-difference equation, 50, 95, 160, 161, 182
Differential inequality, 45
Dimension of a linear space, 53
Direct Method of Liapunov, 25, 154
Discrete dynamical system, 223
Distribution, 70
Domain, 59
Domain of attraction, 37
Dry friction problem, 117, 121, 159, 170, 171
Dual space, 69
Dynamical system, 13, 86
  approximation of, 222, 224-225
  discrete, 223
  generation of, 14, 15, 91, 116, 121
  infinitesimal generator of, 88
  linear, 15-17, 87

Eigenvalue, 72
Eigenvector, 72
Equation of perturbation, 20
Equilibrium, 19, 86, 224
  domain of attraction for, 37
  existence of, 23, 133-135
  stability of, 23, 45, 154
Euclidean space, 2, 55
Evolution equation, 3, 49, 81, 85, 90, 116-119
  autonomous, 13, 90, 19
  finite-dimensional approximation of, 225
  linearization of, 35, 162
  solution of, 4, 90, 119
    existence of, 6, 9, 90, 122
    maximal interval for, 7
    uniqueness of, 7, 9, 90, 97, 122
Exponential stability, 24, 45, 154, 155
Extension of a function, 60

Feedback control problem, 200
Field of a linear space, 53
Finite-dimensional approximation, 225
Finite escape times, 4, 9
Fission reactor problem, 210
Fixed-point theorem, 6, 61
Fréchet derivative, 76
Function, 59
  accretive, 62
  affine, 22
  bounded, 61
  closed, 63
  compact, 61
  composite, 60
  continuous, 60
  contractive, 61
  domain of, 59
  extension of, 60
  graph of, 59
  identity, 60
  inverse of, 60
  Liapunov: *see* Liapunov function
  linear, 60, 70
  Lipschitz continuous, 7, 61
  lower semicontinuous, 138
  monotone, 42, 62
  multivalued: *see* Multivalued operator
  nonexpansive, 63
  one-to-one, 59
  projective, 226

Function (*cont.*)
  range of, 59
  restriction of, 60
  sum of, 60
Functional, 60

Galerkin's method, 227
Generalized derivative, 58
Generated dynamical systems, 14, 15, 91, 116, 121
Generated process, 12, 129
Generator, infinitesimal, 88
Graph of a function, 59
  of a multivalued operator, 119
Group, 91

Heat conduction problems
  linear, 49, 81, 93, 94, 104, 105, 145, 181, 183
  nonlinear, 123, 125, 151, 172, 190
Hilbert space, 55
Hille-Yosida-Phillips theorem, 93

Identity operator, 60
Infinitesimal generator, 88
Inner product, 2-3, 54
Inner product space, 54
Integral
  Bochner, 81
  Lebesgue, 57, 81
  Riemann, 80-81
Integral solution, 128
Interior point, 52
Invariance Principle, 38, 168
Invariant set, 166
Inverse of a function, 60

Left derivative, 75
Liapunov
  direct method of, 25, 154
  function, 138, 223
    continuous, 24, 139
    quadratic, 24, 139
  matrix equation of, 30
Limit point, 52
Limit set, positive, 165, 223
Linear dynamical system, 13, 86
Linear function, 60, 70
Linear independence, 53
Linear manifold, 53

Linear space, 53
dimension of, 53
field of, 53
normed, 53
Linearization, 35, 162
Lipschitz continuity, 7, 61
Locally compact space, 52
Lower semicontinuity, 138

Mapping: *see* Function
Matrix equation of Liapunov, 30
Membrane flutter paradox, 220
Metric, 46, 52
Metric space, 46, 52
complete, 52
locally compact, 52
Monotone function, 42, 62
Monotone multivalued operator, 119
Motion, 19, 86, 223
asymptotic behavior of, 38–42, 164–174
periodic, 165, 196
stability of, 19, 154
Multivalued operator, 118
accretive, 119
canonical restriction of, 119
closed, 119
graph of, 119
monotone, 119

Neighborhood, 52
Noncompactness, α-measure of, 179
Nonexpansive function, 63
Norm, 2, 53
equivalent, 2, 53
Normed linear space, 53

Open ball, 52
Open set, 52
Operator: *see* Function
Operator norm, 62
Orbital stability, 155

Panel-flutter problem, 216
Parallelogram law, 3, 55
Peano's existence theorem, 6
Periodic motion, 165, 196
Perturbation, equation of, 20
Picard–Lindelöf theorem, 8
Point spectrum, 72
Positive invariant set, 165, 223
Positive limit set, 165, 223
Positive orbit, 137, 223
precompactness of, 175–193
Precompact set, 52
Pre-Hilbert space, 55
Process, 12
generation of, 12, 129
simplest, 17, 19
Product space, 58
Projection, 226

Quadratic Liapunov function, 24, 139

Range, 59
Reflexive space, 69
Residual spectrum, 72
Resolvent, 71
Resolvent set, 71
Restriction of a function, 60
Right derivative, 75

Schwarz inequality, 3, 55
Semidynamical system, 13, 86
Semiflow, 13, 86
Semigroup, 18
Seminorm, 113
Sequence
Cauchy, 52
convergent, 52
Sequentially compact, 59
Set
bounded, 52
closed, 52
compact, 52, 59
dense, 52
open, 52
precompact, 52
positive limit, 165, 223
resolvent, 71
stability of, 154–155
Sine-Gordon equation, 125, 147, 192
Sobolev space, 58
Space, 51
Banach, 54
of bounded linear operators, 62
dual, 69
Euclidean, 2, 55
Hilbert, 55
inner product, 54
linear, 53
metric, 46, 52
pre-Hilbert, 55

Space (*cont.*)
  product, 58
  reflexive, 69
  Sobolev, 58
  state: *see* State space
  topological, 51
Spectrum, 72
  continuous, 72
  point, 72
  residual, 72
Stability
  asymptotic, 23, 154, 155
  of an equilibrium, 23, 45, 154
  exponential, 24, 45, 154, 155
  of a motion, 19, 154
  orbital, 155
  of a set, 154-155
  structural, 45
State, 4, 86
State space, 4, 86
  choice of, 103, 126
Strong solution, 90, 119
Structural stability, 45
Sum of functions, 60
Sylvester's lemma, 33

Thermoelastic problem, 203
Topological space, 51
Triangle inequality, 52

Uniform operator topology, 71
Unstable: *see* Stability

Van der Pol's equation, 11
Variation-of-constants formula, 186
Vibration problems
  linear, 28, 39
  nonlinear, 29
Viscoelastic problem, 207

Wave equation, 50, 106, 107, 109, 111
  damped, 174
  nonlinear, 132

MIX
Papier aus verantwortungsvollen Quellen
Paper from responsible sources
FSC® C105338

If you have any concerns about our products,
you can contact us on
ProductSafety@springernature.com

In case Publisher is established outside the EU,
the EU authorized representative is:
Springer Nature Customer Service Center GmbH
Europaplatz 3, 69115 Heidelberg, Germany

Printed by Libri Plureos GmbH
in Hamburg, Germany